D0148102

Strength Of Materials

Strength Of Materials

Fifth Edition

William A. Nash, Ph.D.

Former Professor of Civil Engineering
University of Massachusetts

Merle C. Potter, Ph.D.

Professor Emeritus of Mechanical Engineering
Michigan State University

Schaum's Outline Series

New York Chicago San Francisco Lisbon London
Madrid Mexico City Milan New Delhi San Juan
Seoul Singapore Sydney Toronto

The McGraw·Hill Companies

Copyright © 2011, 1998, 1994, 1972 by The McGraw-Hill Companies, Inc. All rights reserved. Printed in the United States of America. Except as permitted under the Copyright Act of 1976, no part of this publication may be reproduced or distributed in any forms or by any means, or stored in a database or retrieval system, without the prior written permission of the publisher.

1 2 3 4 5 6 7 8 9 0 CUS/CUS 1 9 8 7 6 5 4 3 2 1 0

ISBN 978-0-07-163508-0
MHID 0-07-163508-4

Schaum's Outline of STRENGTH OF MATERIALS

This publication is designed to provide accurate and authoritative information in regard to the subject matter covered. It is sold with the understanding that neither the author nor the publisher is engaged in rendering legal, accounting, securities trading, or other professional services. If legal advice or other expert assistance is required, the services of a competent professional person should be sought.

—From a Declaration of Principles Jointly Adopted by a Committee of the American Bar Association and a Committee of Publishers and Associations

Trademarks: McGraw-Hill, the McGraw-Hill Publishing logo, Schaum's and related trade dress are trademarks or registered trademarks of The McGraw-Hill Companies and/or its affiliates in the United States and other countries and may not be used without written permission. All other trademarks are the property of their respective owners. The McGraw-Hill Companies is not associated with any product or vendor mentioned in this book.

McGraw-Hill books are available at special quantity discounts to use as premiums and sales promotions or for use in corporate training programs. To contact a representative, please e-mail us at bulksales@mcgraw-hill.com.

Preface

This fifth edition of *Schaum's Strength of Materials* book has been substantially modified by the second author to better fit the outline of the introductory Strength of Materials (Solid Mechanics) course, and to better fit the presentation of material in most introductory textbooks on the subject. In addition, the following changes have been made:

1. Problem solutions and Supplementary Problems are presented using the metric SI units only.
2. The computer programs have been omitted. The use of MATLAB or other programs are available to students if more complicated problems are of interest.
3. The more advanced materials and problems that are not found in an introductory course have been omitted for simplicity of presentation. This book is intended to be used in an introductory course only.
4. A short chapter on Fatigue, a subject included on the Fundamentals of Engineering Examination, has been added. It is a modified chapter, based on a section on Fatigue written by my friend and previous colleague, Charlie Muvdi, from "Engineering Mechanics of Materials," by B. B. Muvdi and J. W. McNabb.
5. A section on Combined Loading has been added.
6. The chapter on Centroids and Moments of Inertia has been omitted; it is assumed to have been part of a Statics course that precedes Solid Mechanics.

Strength of Materials, also called The Mechanics of Materials or Solid Mechanics, provides the basis for the design of the components that make up machines and load-bearing structures. In Statics, the forces and moments acting at various points in a structural component or at points of contact with other structures were determined. The forces, stresses, and strains existing within a component were not of interest. In Solid Mechanics, we will consider questions like, "What load will cause this structure to fail?", "What maximum torque can this shaft transmit?", "What material should be selected for this component?", "At what load will this column buckle?" Such questions were not of interest in a Statics course. But, before any of these questions can be answered, we must calculate the forces and moments acting on the components that make up a structure or machine. So, Statics always precedes the study of Strength of Materials. Sometimes Statics is combined with Strength of Materials in one course since they are so closely related.

I would like to thank the estate of the late William Nash for allowing me to create this fifth edition of a book that obviously required much diligent work by Professor Nash. Many thanks are also given to Dr. Charlie Muvdi who provided good advice on the content of this revision. It was a pleasure working with Kimberly Eaton of McGraw-Hill in making the many decisions required in such a venture.

MERLE C. POTTER, E. LANSING, MI
Michigan State University, 2010

Contents

Tension and Compression

1.1 Internal Effects of Forces

In this book we shall be concerned with what might be called the *internal effects* of forces acting on a body. The bodies themselves will no longer be considered to be perfectly rigid as was assumed in statics; instead, the calculation of the deformations of various bodies under a variety of loads will be one of our primary concerns in the study of strength of materials.

Axially Loaded Bar

The simplest case to consider at the start is that of an initially straight metal bar of constant cross section, loaded at its ends by a pair of oppositely directed collinear forces coinciding with the longitudinal axis of the bar and acting through the centroid of each cross section. For static equilibrium the magnitudes of the forces must be equal. If the forces are directed away from the bar, the bar is said to be in *tension*; if they are directed toward the bar, a state of *compression* exists. These two conditions are illustrated in Fig. 1-1.

Under the action of this pair of applied forces, internal resisting forces are set up within the bar and their characteristics may be studied by imagining a plane to be passed through the bar anywhere along its length and oriented perpendicular to the longitudinal axis of the bar. Such a plane is designated as *a-a* in Fig. 1-2(*a*). If for purposes of analysis the portion of the bar to the right of this plane is considered to be removed, as in Fig. 1-2(*b*), then it must be replaced by whatever effect it exerts upon the left portion. By this technique of introducing a cutting plane, the originally internal forces now become external with respect to the remaining portion of the body. For equilibrium of the portion to the left this "effect" must be a horizontal force of magnitude P. However, this force P acting normal to the cross section *a-a* is actually the resultant of distributed forces acting over this cross section in a direction normal to it.

At this point it is necessary to make some assumption regarding the manner of variation of these distributed forces, and since the applied force P acts through the centroid it is commonly assumed that they are uniform across the cross section.

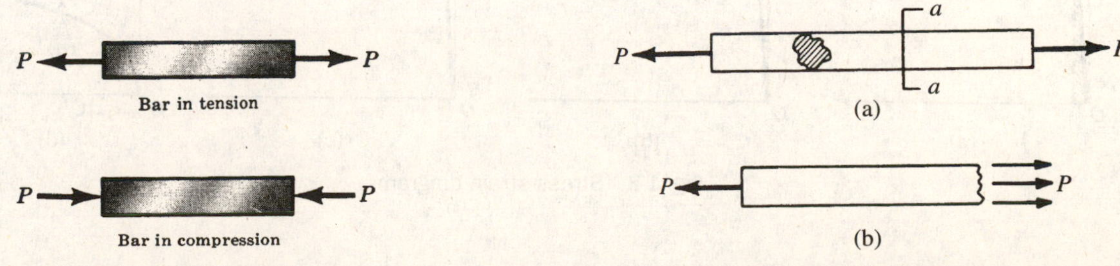

Fig. 1-1 Axially loaded bars.

Fig. 1-2 Internal force.

Normal Stress

Instead of speaking of the internal force acting on some small element of area, it is better for comparative purposes to treat the normal force acting over a *unit* area of the cross section. The intensity of normal force per unit area is termed the *normal stress* and is expressed in units of force per unit area, N/m^2. If the forces applied to the ends of the bar are such that the bar is in tension, then *tensile stresses* are set up in the bar; if the bar is in compression we have *compressive stresses*. The line of action of the applied end forces passes through the centroid of each cross section of the bar.

Normal Strain

Let us suppose that the bar of Fig. 1-1 has tensile forces gradually applied to the ends. The elongation per unit length, which is termed *normal strain* and denoted by ϵ, may be found by dividing the total elongation Δ by the length L, i.e.,

$$\epsilon = \frac{\Delta}{L} \tag{1.1}$$

The strain is usually expressed in units of meters per meter and consequently is dimensionless.

Stress-Strain Curve

As the axial load in Fig. 1-1 is gradually increased, the total elongation over the bar length is measured at each increment of load and this is continued until fracture of the specimen takes place. Knowing the original cross-sectional area of the test specimen, the *normal stress*, denoted by σ, may be obtained for any value of the axial load by the use of the relation

$$\sigma = \frac{P}{A} \tag{1.2}$$

where P denotes the axial load in newtons and A the original cross-sectional area. Having obtained numerous pairs of values of normal stress σ and normal strain ϵ, experimental data may be plotted with these quantities considered as ordinate and abscissa, respectively. This is the *stress-strain curve* or *diagram* of the material for this type of loading. Stress-strain diagrams assume widely differing forms for various materials. Figure 1-3(*a*) is the stress-strain diagram for a medium-carbon structural steel, Fig. 1-3(*b*) is for an alloy steel, and Fig. 1-3(*c*) is for hard steels and certain nonferrous alloys. For nonferrous alloys and cast iron the diagram has the form indicated in Fig. 1-3(*d*).

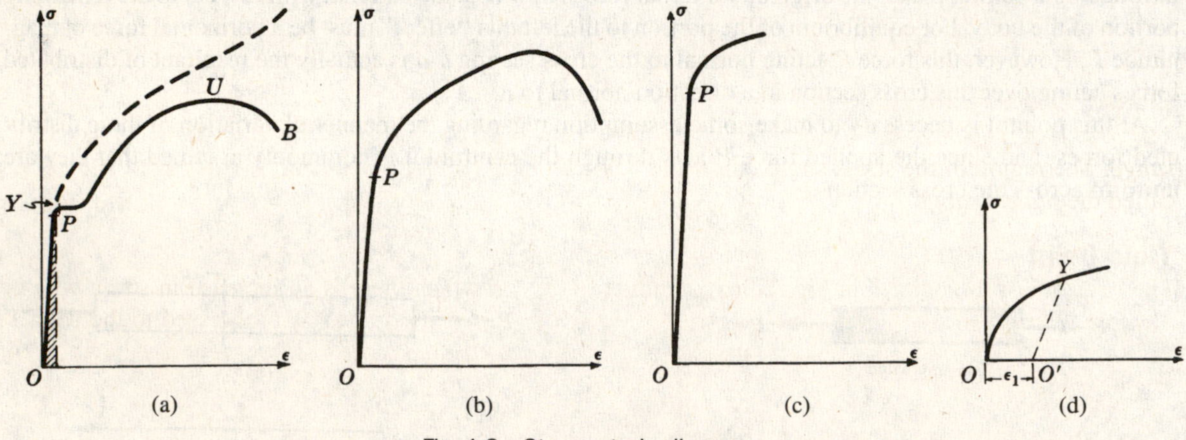

Fig. 1-3 Stress-strain diagrams.

Ductile and Brittle Materials

Metallic engineering materials are commonly classified as either *ductile* or *brittle* materials. A *ductile material* is one having a relatively large tensile strain up to the point of rupture (for example, structural steel or aluminum) whereas a *brittle material* has a relatively small strain up to this same point. An arbitrary strain of 0.05 mm/mm is frequently taken as the dividing line between these two classes of materials. Cast iron and concrete are examples of brittle materials.

Hooke's Law

For any material having a stress-strain curve of the form shown in Fig. 1-3(*a*), (*b*), or (*c*), it is evident that the relation between stress and strain is linear for comparatively small values of the strain. This linear relation between elongation and the axial force causing it is called *Hooke's law*. To describe this initial linear range of action of the material we may consequently write

$$\sigma = E\epsilon \tag{1.3}$$

where E denotes the slope of the straight-line portion OP of each of the curves in Figs.1-3(*a*), (*b*), and (*c*).

The quantity E, i.e., the ratio of the unit stress to the unit strain, is the *modulus of elasticity* of the material in tension, or, as it is often called, *Young's modulus*. Values of E for various engineering materials are tabulated in handbooks. Table 1-3 for common materials appears at the end of this chapter. Since the unit strain ϵ is a pure number (being a ratio of two lengths) it is evident that E has the same units as does the stress, N/m^2. For many common engineering materials the modulus of elasticity in compression is very nearly equal to that found in tension. *It is to be carefully noted that the behavior of materials under load as discussed in this book is restricted (unless otherwise stated) to the linear region of the stress-strain curve.*

Problems 1.1 through 1.8 illustrate Hooke's law.

1.2 Mechanical Properties of Materials

The stress-strain curve shown in Fig. 1-3(*a*) may be used to characterize several strength characteristics of the material. They are:

Proportional Limit

The ordinate of the point P is known as the *proportional limit*, i.e., the maximum stress that may be developed during a simple tension test such that the stress is a linear function of strain. For a material having the stress-strain curve shown in Fig. 1-3(*d*), there is no proportional limit.

Elastic Limit

The ordinate of a point almost coincident with P is known as the *elastic limit*, i.e., the maximum stress that may be developed during a simple tension test such that there is no permanent or residual deformation when the load is entirely removed. For many materials the numerical values of the elastic limit and the proportional limit are almost identical and the terms are sometimes used synonymously. In those cases where the distinction between the two values is evident, the elastic limit is almost always greater than the proportional limit.

Elastic and Plastic Ranges

The region of the stress-strain curve extending from the origin to the proportional limit is called the *elastic range*. The region of the stress-strain curve extending from the proportional limit to the point of rupture is called the *plastic range*.

Yield Point

The ordinate of the point Y in Fig. 1-3(*a*), denoted by σ_{yp}, at which there is an increase in strain with no increase in stress, is known as the *yield point* of the material. After loading has progressed to the point Y, yielding is said to take place. Some materials exhibit two points on the stress-strain curve at which there is an increase of strain without an increase of stress. These are called *upper* and *lower yield points*.

Ultimate Strength or Tensile Strength

The ordinate of the point U in Fig. 1-3(*a*), the maximum ordinate to the curve, is known either as the *ultimate strength* or the *tensile strength* of the material.

Breaking Strength

The ordinate of the point B in Fig. 1-3(*a*) is called the *breaking strength* of the material.

Modulus of Resilience

The work done on a unit volume of material, as a simple tensile force is gradually increased from zero to such a value that the proportional limit of the material is reached, is defined as the *modulus of resilience*.

This may be calculated as the area under the stress-strain curve from the origin up to the proportional limit and is represented as the shaded area in Fig. 1-3(*a*). The unit of this quantity is $N \cdot m/m^3$ in the SI system. Thus, resilience of a material is its ability to absorb energy in the elastic range.

Modulus of Toughness

The work done on a unit volume of material as a simple tensile force is gradually increased from zero to the value causing rupture is defined as the *modulus of toughness*. This may be calculated as the entire area under the stress-strain curve from the origin to rupture. Toughness of a material is its ability to absorb energy in the plastic range of the material.

Percentage Reduction in Area

The decrease in cross-sectional area from the original area upon fracture divided by the original area and multiplied by 100 is termed *percentage reduction in area*. It is to be noted that when tensile forces act upon a bar, the cross-sectional area decreases, but calculations for the normal stress are usually made upon the basis of the original area. This is the case for the curve shown in Fig. 1-3(*a*). As the strains become increasingly larger it is more important to consider the instantaneous values of the cross-sectional area (which are decreasing), and if this is done the *true* stress-strain curve is obtained. Such a curve has the appearance shown by the dashed line in Fig. 1-3(*a*).

Percentage Elongation

The increase in length of a bar after fracture divided by the initial length and multiplied by 100 is the *percentage elongation*. Both the percentage reduction in area and the percentage elongation are considered to be measures of the *ductility* of a material.

Working Stress

The above-mentioned strength characteristics may be used to select a *working stress*. Frequently such a stress is determined merely by dividing either the stress at yield or the ultimate stress by a number termed the *safety factor*. Selection of the safety factor is based upon the designer's judgment and experience. Specific safety factors are sometimes specified in design codes.

Strain Hardening

If a ductile material can be stressed considerably beyond the yield point without failure, it is said to *strain-harden*. This is true of many structural metals.

The nonlinear stress-strain curve of a brittle material, shown in Fig. 1-3(*d*), characterizes several other strength measures that cannot be introduced if the stress-strain curve has a linear region. They are:

Yield Strength

The ordinate to the stress-strain curve such that the material has a predetermined permanent deformation or "set" when the load is removed is called the *yield strength* of the material. The permanent set is often taken to be either 0.002 or 0.0035 mm per mm. These values are of course arbitrary. In Fig. 1-3(*d*) a set ϵ_1 is denoted on the strain axis and the line $O'Y$ is drawn parallel to the initial tangent to the curve. The ordinate of Y represents the yield strength of the material, sometimes called the *proof stress*.

Tangent Modulus

The rate of change of stress with respect to strain is known as the *tangent modulus* of the material. It is essentially an instantaneous modulus given by $E_t = d\sigma/d\epsilon$.

Coefficient of Linear Expansion

This is defined as the change of length per unit length of a straight bar subject to a temperature change of one degree and is usually denoted by α. The value of this coefficient is independent of the unit of length but does depend upon the temperature scale used. For example, from Table 1-3 at the end of this chapter the coefficient for steel is 12×10^{-6}/°C. Temperature changes in a structure give rise to internal stresses, just as do applied loads. The thermal strain due to a temperature change ΔT is

$$\epsilon_T = \alpha \, \Delta T \tag{1.4}$$

Problem 1.7 illustrates the contraction of a cable due to a temperature decrease.

Poisson's Ratio

When a bar is subjected to a simple tensile loading there is an increase in length of the bar in the direction of the load, but a decrease in the lateral dimensions perpendicular to the load. The ratio of the strain in the lateral direction to that in the axial direction is defined as *Poisson's ratio*. It is denoted by the Greek letter ν. For most metals it lies in the range 0.25 to 0.35. For cork, ν is very nearly zero.

General Form of Hooke's Law

The simple form of Hooke's law has been given for axial tension when the loading is entirely along one straight line, i.e., uniaxial. Only the deformation in the direction of the load was considered and it was given by Eq. (1.3), which is now written as

$$\epsilon = \frac{\sigma}{E} \tag{1.5}$$

Problems 1.8 and 1.9 illustrate the application of this uniaxial loading.

In the more general case an element of material is subject to three mutually perpendicular normal stresses σ_x, σ_y, σ_z, which are accompanied by the strains ϵ_x, ϵ_y, ϵ_z, respectively. By superposing the strain components arising from lateral contraction due to Poisson's effect upon the direct strains we obtain the general statement of Hooke's law:

$$\epsilon_x = \frac{1}{E}[\sigma_x - \nu(\sigma_y + \sigma_z)] \qquad \epsilon_y = \frac{1}{E}[\sigma_y - \nu(\sigma_x + \sigma_z)] \qquad \epsilon_z = \frac{1}{E}[\sigma_z - \nu(\sigma_x + \sigma_y)] \tag{1.6}$$

Problems 1.10 through 1.14 will illustrate the more general case.

1.3 Statically Indeterminate Force Systems

If the values of all the external forces which act on a body can be determined by the equations of static equilibrium alone, then the force system is *statically determinate*. Most problems considered will be of this type.

The bar shown in Fig. 1-4(*a*) is loaded by the known force P. The reactions are R_1, R_2, and R_3. The system is statically determinate because there are three equations of static equilibrium available ($\sum F_x = 0$, $\sum F_y = 0$, $\sum M = 0$) for the system and these are sufficient to determine the three unknowns.

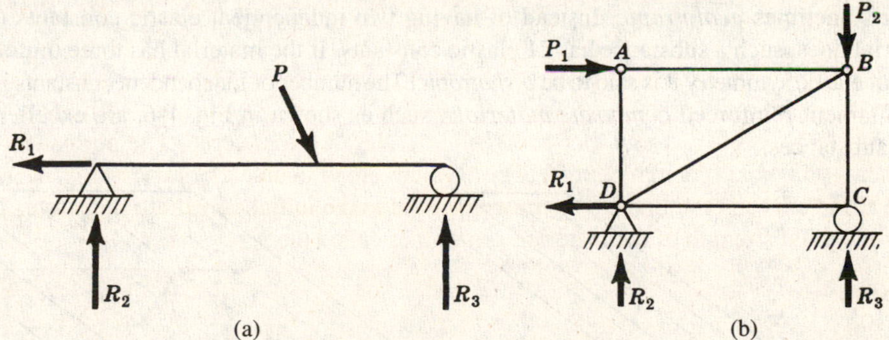

Fig. 1-4 Determinate force systems.

The truss *ABCD* shown in Fig. 1-4(*b*) is loaded by the known forces P_1 and P_2. The reactions are R_1, R_2, and R_3. Again, since there are three equations of static equilibrium available, all three unknown reactions may be determined and consequently the external force system is statically determinate.

The above two illustrations refer only to external reactions and the force systems may be defined as statically determinate *externally*.

In many cases the forces acting on a body cannot be determined by the equations of statics alone because there are more unknown forces than there are equations of equilibrium. In such a case the force system is said to be *statically indeterminate*.

The bar shown in Fig. 1-5(*a*) is loaded by the known force *P*. The reactions are R_1, R_2, R_3, and R_4. The force system is statically indeterminate because there are four unknown reactions but only three equations of static equilibrium. Such a force system is said to be *indeterminate to the first degree*.

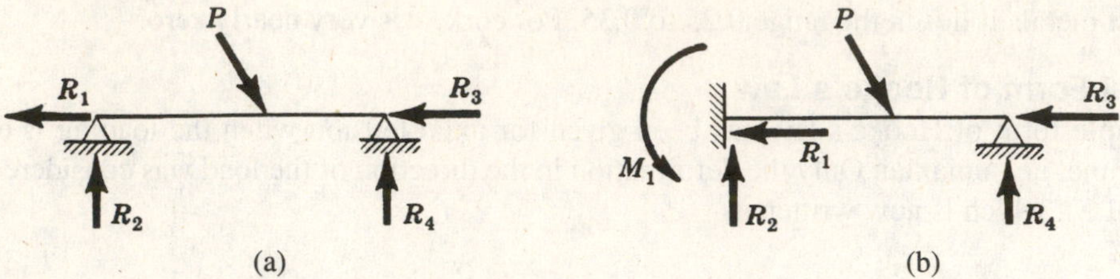

Fig. 1-5 Indeterminate force systems.

The bar shown in Fig. 1-5(*b*) is statically indeterminate to the second degree because there are five unknown reactions R_1, R_2, R_3, R_4, and M_1 but only three equations of static equilibrium. Consequently the values of all reactions cannot be determined by use of statics equations alone.

The approach that we will consider here is called the *deformation method* because it considers the deformations in the system. Briefly, the procedure to be followed in analyzing an indeterminate system is first to write all equations of static equilibrium that pertain to the system and then *supplement* these equations with additional equations based upon the deformations of the structure. Enough equations involving deformations must be written so that the total number of equations from both statics and deformations is equal to the number of unknown forces involved. Problems 1.15 through 1.19 illustrate the method.

1.4 Classification of Materials

Up to now, this entire discussion has been based upon the assumptions that two characteristics prevail in the material. They are:

- A *homogeneous material*, one with the same elastic properties (E, ν) at all points in the body.
- An *isotropic material*, one having the same elastic properties in all directions at any one point of the body.

Not all materials are isotropic. If a material does not possess any kind of elastic symmetry it is called *anisotropic*, or sometimes *aeolotropic*. Instead of having two independent elastic constants (E, ν) as an isotropic material does, such a substance has 21 elastic constants. If the material has three mutually perpendicular planes of elastic symmetry it is said to be *orthotropic*. The number of independent constants is nine in this case. Modern filament-reinforced *composite materials*, such as shown in Fig. 1-6, are excellent examples of anisotropic substances.

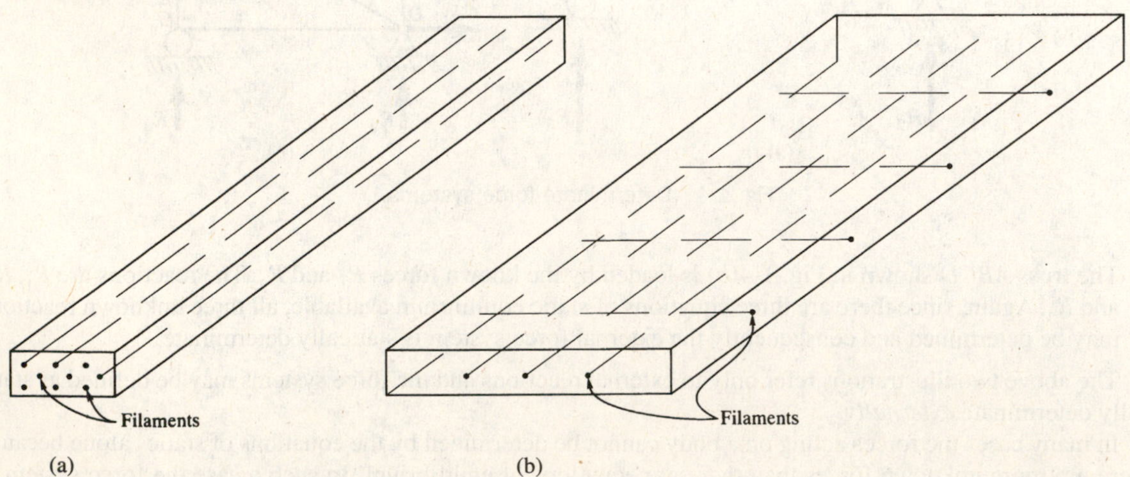

Fig. 1-6 (*a*) Epoxy bar reinforced by fine filaments in one direction; (*b*) epoxy plate reinforced by fine filaments in two directions.

Elastic Versus Plastic Material Analysis

Stresses and deformations in the plastic range of action of a material are frequently permitted in certain structures. Some building codes allow particular structural members to undergo plastic deformation, and certain components of aircraft and missile structures are deliberately designed to act in the plastic range so as to achieve weight savings. Furthermore, many metal-forming processes involve plastic action of the material. For small plastic strains of low- and medium-carbon structural steels the stress-strain curve of Fig. 1-7 is usually idealized by two straight lines, one with a slope of E, representing the elastic range, the other with zero slope representing the plastic range. This plot, shown in Fig. 1-7, represents a so-called *elastic, perfectly plastic material*. It takes no account of still larger plastic strains occurring in the strain-hardening region shown as the right portion of the stress-strain curve of Fig. 1-3(*a*). See Problem 1.20.

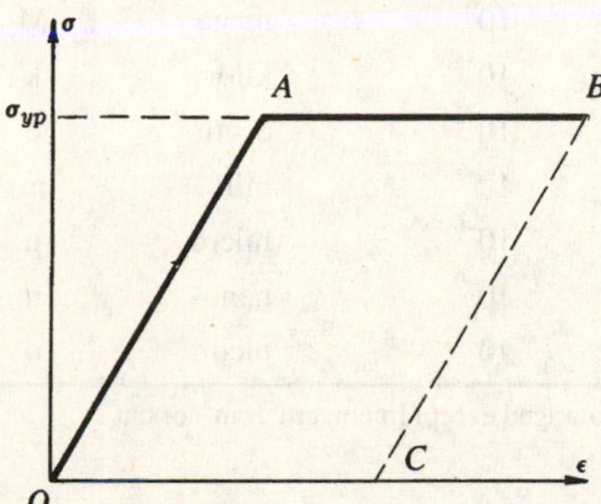

Fig. 1-7 The stress-strain curve for an elastic, perfectly plastic material.

If the load increases so as to bring about the strain corresponding to point B in Fig. 1-7, and then the load is removed, unloading takes place along the line BC so that complete removal of the load leaves a permanent "set" or elongation corresponding to the strain OC.

1.5 Units

While the student is undoubtedly comfortable using SI units, much of the data gathered and available for use in the United States are in English units. Table 1.1 lists units and conversations for many quantities of interest.

Table 1.1. Conversion Factors

Quantity	Symbol	SI Units	English Units	To Convert from English to SI Units Multiply by
Length	L	m	ft	0.3048
Mass	m	kg	lbm	0.4536
Time	t	s	sec	1
Area	A	m^2	ft^2	0.09290
Volume	V	m^3	ft^3	0.02832
Velocity	V	m/s	ft/sec	0.3048
Acceleration	a	m/s^2	ft/sec^2	0.3048
Angular velocity	ω	rad/s	rad/sec	1
		rad/s	rpm	9.55
Force, Weight	F, W	N	lbf	4.448
Density	ρ	kg/m^3	lbm/ft^3	16.02
Specific weight	γ	N/m^3	lbf/ft^3	157.1
Pressure, stress	ρ, σ, τ	kPa	psi	6.895
Work, Energy	W, E, U	J	ft-lbf	1.356
Power	W	W	ft-lbf/sec	1.356
		W	hp	746

When expressing a quantity in SI units, certain letter prefixes shown in Table 1.2 may be used to represent multiplication by a power of 10. So, rather than writing 30000 Pa (commas are not used in the SI system) or 30×10^3 Pa, we may simply write 30 kPa.

Table 1.2. Prefixes for SI Units

Multiplication Factor	Prefix	Symbol
10^{12}	tera	T
10^{9}	giga	G
10^{6}	mega	M
10^{3}	kilo	k
10^{-2}	centi*	c
10^{-3}	mili	m
10^{-6}	micro	μ
10^{-9}	nano	n
10^{-12}	pico	p

*Discouraged except in cm, cm^2, cm^3, or cm^4.

The units of various quantities are interrelated via the physical laws obeyed by the quantities. It follows that, no matter the system used, all units may be expressed as algebraic combinations of a selected set of *base units*. There are seven base units in the SI system: m, kg, s, K, mol (mole), A (ampere), cd (candela). The last one is rarely encountered in engineering mechanics. Note that N (newton) is not listed as a base unit. It is related to the other units by Newton's second law,

$$F = ma \tag{1.7}$$

If we measure F in newtons, m in kg, and a in m/s^2, we see that $N = kg \cdot m/s^2$. So, the newton is expressed in terms of the bass units.

Weight is the force of gravity; by Newton's second law,

$$W = mg \tag{1.8}$$

Since mass remains constant, the variation of W is due to the change in the acceleration of gravity g (from about 9.77 m/s^2 on the highest mountain to 9.83 m/s^2 in the deepest ocean trench, only about a 0.3% variation from 9.80 m/s^2). We will use the standard sea-level value of 9.81 m/s^2 (32.2 ft/sec^2), unless otherwise stated.

A discussion of significant digits may be in order. Answers are presented using three or four significant digits, never more. Most, if not all, engineering problems contain given information such as the radius of a shaft, the density of the material, as well as other material constants. Seldom is such information known to more than four digits, so it is not appropriate to present an answer using more digits than assumed present in the given information. However, when a number begins with "1," that digit is not counted as a significant digit. So, 1024 represents three significant digits, as does 0.0001800.

Finally, we will not include the units in most problems for which we show the steps of the solutions. If the quantities in the equations are input with units of meters (m), seconds (s), pascals (Pa), kilograms (kg), and newtons (N), the units on the answer will be predictable. There's no need, most often, to work two problems: one with the numbers and one with the units. For example, a material property of 200 GPa when used in an equation would be input as 200×10^9 Pa; a moment of inertia of 2000 mm^4 would be entered as 2000×10^{-12} m^4; and a stress of 35 MPa would be input as 35×10^6 Pa.

SOLVED PROBLEMS

1.1. In Fig. 1-8, determine an expression for the total elongation of an initially straight bar of length L, cross-sectional area A, and modulus of elasticity E if a tensile load P acts on the ends of the bar.

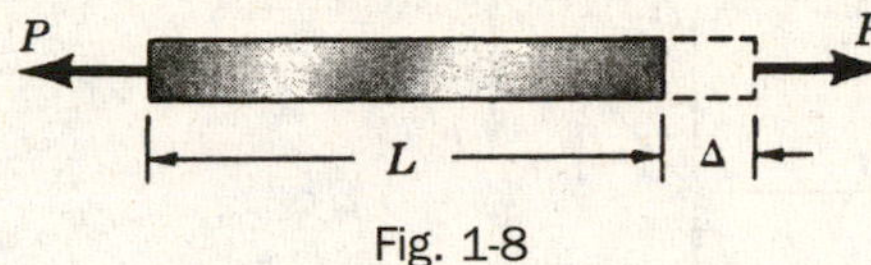

Fig. 1-8

SOLUTION: The unit stress in the direction of the force P is merely the load divided by the cross-sectional area, i.e., $\sigma = P/A$. Also the unit strain ϵ is given by the total elongation Δ divided by the original length, i.e., $\epsilon = \Delta/L$. By definition the modulus of elasticity E is the ratio of σ to ϵ, i.e.,

$$E = \frac{\sigma}{\epsilon} = \frac{P/A}{\Delta/L} = \frac{PL}{A\Delta} \qquad \text{or} \qquad \Delta = \frac{PL}{AE}$$

Note that Δ has the units of length, in meters.

1.2. A steel bar of cross section 500 mm^2 is acted upon by the forces shown in Fig. 1-9(a). Determine the total elongation of the bar. For steel, consider E = 200 GPa.

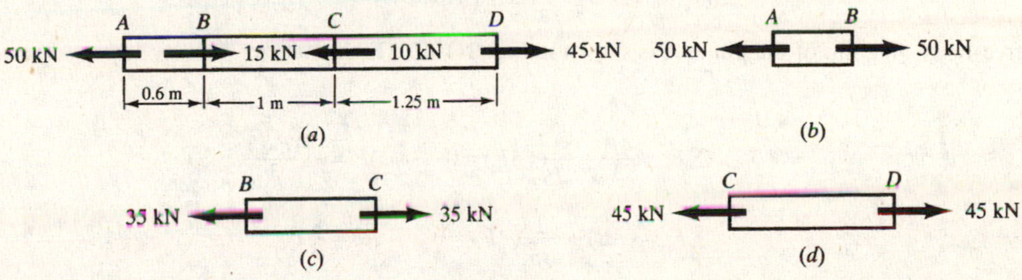

Fig. 1-9

SOLUTION: The entire bar is in equilibrium, and hence are all portions of it. The portion between A and B has a resultant force of 50 kN acting over every cross section and a free-body diagram of this 0.6-m length appears as in Fig. 1-9(b). The force at the right end of this segment must be 50 kN to maintain equilibrium with the applied load at A. The elongation of this portion is, from Problem 1.1,

$$\Delta_1 = \frac{P_1L_1}{AE} = \frac{(50\,000\,\text{N})(0.6\,\text{m})}{(500\times10^{-6}\,\text{m}^2)(200\times10^9\,\text{N/m}^2)} = 0.0003\ \text{m}$$

The force acting in the segment between B and C is found by considering the algebraic sum of the forces to the left of any section between B and C, i.e., a resultant force of 35 kN acts to the left, so that a tensile force exists. The free-body diagram of the segment between B and C is shown in Fig. 1-9(c) and the elongation of it is

$$\Delta_2 = \frac{P_2L_2}{AE} = \frac{(35000)(1)}{(500\times10^{-6})(200\times10^9)} = 0.00035\ \text{m}$$

Similarly, the force acting over any cross section between C and D must be 45 kN to maintain equilibrium with the applied load at D. The elongation of CD is

$$\Delta_3 = \frac{P_3L_3}{AE} = \frac{(45000)(1.25)}{(500\times10^{-6})(200\times10^9)} = 0.00056\ \text{m}$$

The total elongation is

$$\Delta = \Delta_1 + \Delta_2 + \Delta_3 = 0.00121\ \text{m} \qquad \text{or} \qquad 1.21\ \text{mm}$$

1.3. The pinned members shown in Fig. 1-10(a) carry the loads P and $2P$. All bars have cross-sectional area A. Determine the stresses in bars AB and AF.

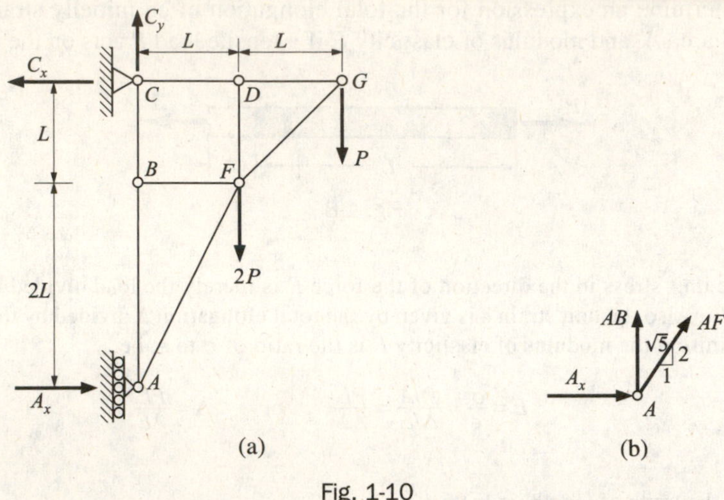

Fig. 1-10

SOLUTION: The reactions are indicated by C_x, C_y, and A_x. From statics we have

$$\Sigma M_c = -(2PL) - P(2L) + A_x(3L) = 0 \qquad A_x = \frac{4}{3}P$$

A free-body diagram of the pin at A is shown in Fig. 1-10(b). From statics

$$\Sigma F_x = \frac{4P}{3} + \frac{1}{\sqrt{5}}(AF) = 0 \qquad AF = -\frac{4P\sqrt{5}}{3}$$

$$\Sigma F_y = (AB) + \frac{2}{\sqrt{5}}(AF) = 0 \qquad AB = \frac{8}{3}P$$

The bar stresses are

$$\sigma_{AF} = -\frac{4P\sqrt{5}}{3A} \qquad \sigma_{AB} = \frac{8P}{3A}$$

1.4. In Fig. 1-11, determine the total increase of length of a bar of constant cross section hanging vertically and subject to its own weight as the only load. The bar is initially straight.

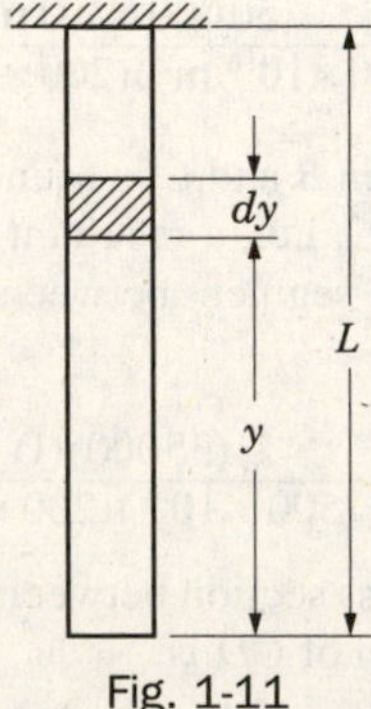

Fig. 1-11

SOLUTION: The normal stress (tensile) over any horizontal cross section is caused by the weight of the material below that section. The elongation of the element of thickness dy shown is (see Problem 1.1)

$$d\Delta = \frac{\gamma Ay\, dy}{AE}$$

where A denotes the cross-sectional area of the bar and γ its specific weight (weight/unit volume). Integrating, the total elongation of the bar is

$$\Delta = \int_0^L \frac{\gamma A y\, dy}{AE} = \frac{\gamma A}{AE}\frac{L^2}{2} = \frac{(\gamma AL)L}{2AE} = \frac{WL}{2AE}$$

where W denotes the total weight of the bar. Note that the total elongation produced by the weight of the bar is equal to that produced by a load of half its weight applied at the end.

1.5. In 1989, *Jason*, a research-type submersible with remote TV monitoring capabilities and weighing 35 200 N, was lowered to a depth of 646 m in an effort to send back to the attending surface vessel photographs of a sunken Roman ship offshore from Italy. The submersible was lowered at the end of a hollow steel cable having an area of $452 \times 10^{-6}\ \text{m}^2$ and $E = 200$ GPa. Determine the extension of the steel cable. Due to the small volume of the entire system, buoyancy may be neglected. (*Note: Jason* was the system that took the first photographs of the sunken *Titanic* in 1986.)

SOLUTION: The total cable extension is the sum of the extensions due to (*a*) the weight of *Jason*, and (*b*) the weight of the steel cable. From Problem 1.1, we have for (*a*)

$$\Delta_1 = \frac{PL}{AE} = \frac{(35200)(646)}{(452 \times 10^{-6})(200 \times 10^9)} = 0.252 \text{ m}$$

and from Problem 1.4, we have for (*b*)

$$\Delta_2 = \frac{WL}{2AE}$$

where W is the weight of the cable. W may be found from the volume of the cable, i.e.,

$$(452 \times 10^{-6})(646) = 0.292 \text{ m}^3$$

which must be multiplied by the weight of steel per unit volume which, from Table 1-3 at the end of the chapter, is 77 kN/m^3. Thus, the cable weight is

$$W = (0.292)\,(77000) = 22484 \text{ N}$$

so that the elongation due to the weight of the cable is

$$\Delta_2 = \frac{(22484)(646)}{2(452 \times 10^{-6})(200 \times 10^9)} = 0.080 \text{ m}$$

The total elongation is the sum of the effects,

$$\Delta = \Delta_1 + \Delta_2 = 0.252 + 0.080 = 0.332 \text{ m} \qquad \text{or} \qquad 33.2 \text{ cm}$$

1.6. Two prismatic bars are rigidly fastened together and support a vertical load of 45 kN, as shown in Fig. 1-12. The upper bar is steel having length 10 m and cross-sectional area 60 cm^2. The lower bar is brass having length 6 m and cross-sectional area 50 cm^2. For steel $E = 200$ GPa, for brass $E = 100$ GPa. Determine the maximum stress in each material.

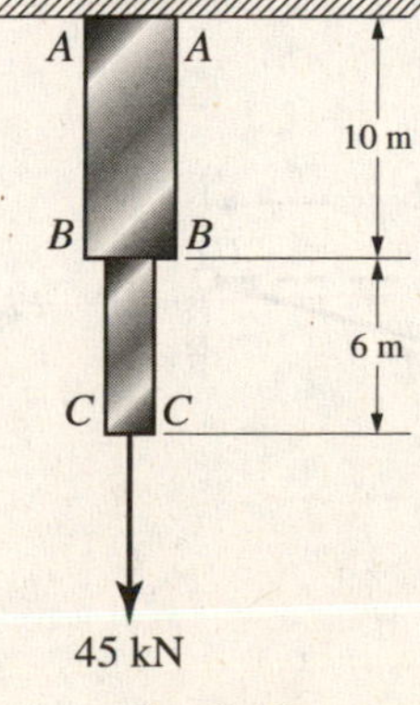

Fig. 1-12

SOLUTION: The maximum stress in the brass bar occurs just below the junction at section B-B. There, the vertical normal stress is caused by the combined effect of the load of 45 000 N together with the weight of the entire brass bar below B-B. Use specific weight in Table 1-3.

The weight of the brass bar is $W_b = 6 \times (50 \times 10^{-4}) \times 84\,000 = 2520$ N. The stress at this section is

$$\sigma = \frac{P}{A} = \frac{45000 + 2520}{50 \times 10^{-4}} = 9.5 \times 10^6 \text{ N/m}^2 \quad \text{or} \quad 9.5 \text{ MPa}$$

The maximum stress in the steel bar occurs at section A-A, the point of suspension, because there the entire weight of the steel and brass bars gives rise to normal stress. The weight of the steel bar is $W_s = 10 \times (60 \times 10^{-4}) \times 77\,000 = 4620$ N.

The stress across section A-A is

$$\sigma = \frac{P}{A} = \frac{45000 + 2520 + 4620}{60 \times 10^{-4}} = 8.69 \times 10^6 \text{ N/m}^2 \quad \text{or} \quad 8.69 \text{ MPa}$$

1.7. In 1989 a new fiber-optic cable capable of handling 40 000 telephone calls simultaneously was laid under the Pacific Ocean from California to Japan, a distance of 13 300 km. The cable was unreeled from shipboard at a mean temperature of 22°C and dropped to the ocean floor having a mean temperature of 5°C. The coefficient of linear expansion of the cable is 75×10^{-6}/°C. Determine the length of cable that must be carried on the ship to span the 13 300 km.

SOLUTION: The length of cable that must be carried on board ship consists of the 13 300 km plus an unknown length ΔL that will allow for contraction to a final length of 13 300 km when resting on the ocean floor. From the definition of the coefficient of thermal expansion [Eq. (1.4)], we have

$$\Delta L = \alpha L(\Delta T)$$

$$= (75 \times 10^{-6}/°\text{C})\,[13\,300 \text{ km} + \Delta L](22 - 5)°\text{C} \qquad (1)$$

Solving, we find

$$\Delta L = 16.96 \text{ km}$$

The percent change of length is thus

$$\frac{\Delta L}{L} \times 100 = \frac{16.96}{13300 + 16.96} \times 100 = 0.13\%$$

so that the ΔL term in (1) is of minor consequence. Thus, the required length of cable at shipboard temperature is approximately 13 317 km.

1.8. Consider two thin rods as shown in Fig. 1-13(a), which are pinned at A, B, and C and are initially horizontal and of length L when no load is applied. The weight of each rod is negligible. A force Q is then applied (gradually) at point B. Determine the magnitude of Q so as to produce a prescribed vertical deflection δ at point B.

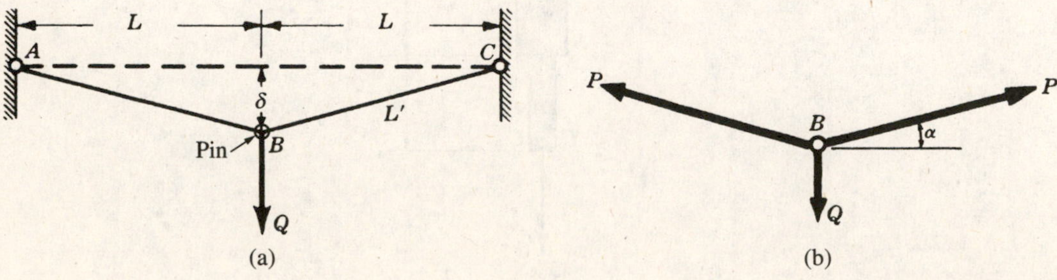

Fig. 1-13

SOLUTION: This is an interesting example of a system in which the elongations of the individual members satisfy Hooke's law and yet for geometric reasons deflection is *not* proportional to force. Each bar obeys the relation $\Delta = PL/AE$ developed in Problem 1.1. Initially each bar is of length L and after the load Q has been applied the length is L'. Thus

$$\Delta = L' - L = \frac{PL}{AE} \tag{1}$$

The free-body diagram of the pin at B is shown in Fig. 1-13(*b*). From statics,

$$\Sigma F_y = 2P\sin\alpha - Q = 0 \qquad \text{or} \qquad Q = 2P\left(\frac{\delta}{L'}\right)$$

Using (1),

$$Q = 2\frac{(L'-L)AE}{L}\frac{\delta}{L'} = \frac{2\delta AE}{L}\left(1-\frac{L}{L'}\right) \tag{2}$$

But

$$(L')^2 = L^2 + \delta^2 \tag{3}$$

Consequently

$$Q = \frac{2\delta AE}{L}\left(1-\frac{L}{\sqrt{L^2+\delta^2}}\right) \tag{4}$$

This can be simplified using the binomial theorem:

$$\sqrt{L^2+\delta^2} = L\left(1+\frac{\delta^2}{L^2}\right)^{1/2} = L\left(1+\frac{1}{2}\frac{\delta^2}{L^2}+\cdots\right) \tag{5}$$

and thus

$$1-\frac{L}{L\left(1+\frac{1}{2}\frac{\delta^2}{L^2}\right)} \approx 1-\left(1-\frac{1}{2}\frac{\delta^2}{L^2}\right) = \frac{1}{2}\frac{\delta^2}{L^2} \tag{6}$$

From this we have the approximate relation between force and displacement,

$$Q \approx \frac{2AE\delta}{K}\frac{\delta^2}{2L^2} = \frac{AE\delta^3}{L^3} \tag{7}$$

Thus the displacement δ is *not* proportional to the force Q even though Hooke's law holds for each bar individually.

1.9. For the system discussed in Problem 1.8, let us consider rods each of initial length 2 m and cross-sectional area 0.6 cm^2. For a load Q of 85 N determine the central deflection δ by both the exact and the approximate relations. Use $E = 200$ GPa.

SOLUTION: The exact expression relating force and deflection is $Q = \frac{2\delta AE}{L}\left(1-\frac{L}{\sqrt{L^2+\delta^2}}\right)$. Substituting the given numerical values,

$$85 = \frac{2\delta(6\times10^{-5})(200\times10^9)}{2}\left(1-\frac{2}{\sqrt{2^2+\delta^2}}\right)$$

Solving by trial and error we find $\delta = 0.0385$ m.

The approximate relation between force and deflection is $Q \approx \frac{AE\delta^3}{L^3}$. Substituting,

$$85 = \frac{(6\times10^{-5})(200\times10^9)\delta^3}{2^3} \qquad \therefore \delta = 0.038 \text{ m} \quad \text{or} \quad 3.8 \text{ cm}$$

The approximate relation has an error of less than 1.5%.

1.10. A square steel bar 50 mm on a side and 1 m long is subject to an axial tensile force of 250 kN. Determine the decrease Δt in the lateral dimension due to this load. Use $E = 200$ GPa and $\nu = 0.3$.

SOLUTION: The loading is axial, hence the stress in the direction of the load is given by

$$\sigma = \frac{P}{A} = \frac{(250 \times 10^3)}{(0.05)(0.05)} = 100 \times 10^6 \text{ Pa} \quad \text{or} \quad 100 \text{ MPa}$$

The simple form of Hooke's law for uniaxial loading states that $E = \sigma/\epsilon$. The strain ϵ in the direction of the load is thus $(100 \times 10^6)/(200 \times 10^9) = 5 \times 10^{-4}$.

The ratio of the lateral strain to the axial strain is denoted as Poisson's ratio, i.e.,

$$\nu = \frac{\text{lateral strain}}{\text{axial strain}} \quad \text{or} \quad 0.3 = \frac{\epsilon_L}{5 \times 10^{-4}} \quad \therefore \epsilon_L = 1.5 \times 10^{-4} \text{ m/m}$$

The change in a 50 mm length is

$$\Delta t = 0.05 \times 1.5 \times 10^{-4} = 7.5 \times 10^{-6} \text{ m} \quad \text{or} \quad 0.0075 \text{ mm}$$

which represents the decrease in the lateral dimension of the bar.

1.11. Consider an elemental block subject to uniaxial tension (see Fig. 1-14). Derive approximate expressions for the change of volume per unit volume due to this loading.

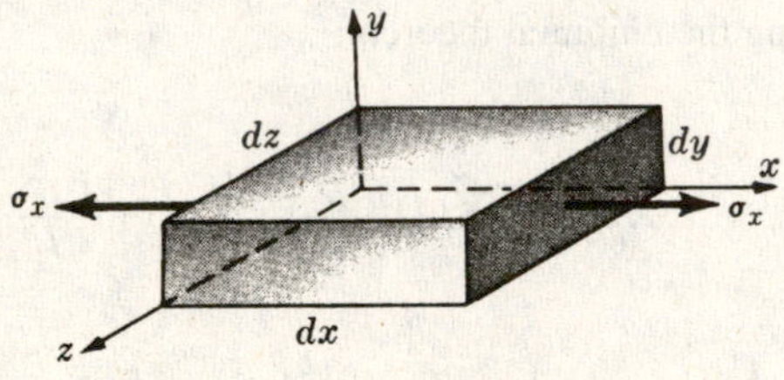

Fig. 1-14

SOLUTION: The strain in the direction of the forces may be denoted by ϵ_x. The strains in the other two orthogonal directions are then each $\nu\epsilon_x$. Consequently, if the initial dimensions of the element are dx, dy, and dz then the final dimensions are

$$(1 + \epsilon_x)dx \qquad (1 - \nu\epsilon_x)dy \qquad (1 - \nu\epsilon_x)dz$$

and the volume after deformation is

$$\begin{aligned} V' &= [(1 + \epsilon_x)dx][(1 - \nu\epsilon_x)dy][(1 - \nu\epsilon_x)dz] \\ &= (1 + \epsilon_x)(1 - 2\nu\epsilon_x)dx\, dy\, dz \\ &= (1 - 2\nu\epsilon_x + \epsilon_x)dx\, dy\, dz \end{aligned}$$

since the deformations are so small that the *squares* and *products* of strains may be neglected. Since the initial volume was $dx\, dy\, dz$, the change of volume per unit volume is

$$\frac{\Delta V}{V} = (1 - 2\nu)\epsilon_x$$

Hence, for a tensile force the volume increases slightly, for a compressive force it decreases.

Also, the cross-sectional area of the element in a plane normal to the direction of the applied force is given approximately by

$$A = (1 - \nu\epsilon_x)^2\, dy\, dz = (1 - 2\nu\epsilon_x)\, dy\, dz$$

1.12. A square bar of aluminum 50 mm on a side and 250 mm long is loaded by axial tensile forces at the ends. Experimentally, it is found that the strain in the direction of the load is 0.001. Determine the volume of the bar when the load is acting. Consider $\nu = 0.33$.

SOLUTION: From Problem 1.11 the change of volume per unit volume is given by

$$\frac{\Delta V}{V} = \epsilon(1-2\nu) = 0.001(1-0.66) = 0.00034$$

Consequently, the change of volume of the entire bar is given by

$$\Delta V = (50)\,(50)\,(250)\,(0.00034) = 212.5\ \text{mm}^3$$

The original volume of the bar in the unstrained state is 6.25×10^5 mm^3. Since a tensile force increases the volume, the final volume under load is 6.252125×10^5 mm^3. Measurements made with the aid of lasers do permit determination of the final volume under load to the indicated accuracy of seven significant figures. Ordinary methods of measurement do not of course lead to such accuracy.

1.13. The general three-dimensional form of Hooke's law in which strain components are expressed as functions of stress components has already been presented in Eq. (1.6). Occasionally it is necessary to express the stress components as functions of the strain components. Derive these expressions.

SOLUTION: Given the previous expressions

$$\epsilon_x = \frac{1}{E}[\sigma_x - \nu(\sigma_y + \sigma_z)] \tag{1}$$

$$\epsilon_y = \frac{1}{E}[\sigma_y - \nu(\sigma_x + \sigma_z)] \tag{2}$$

$$\epsilon_z = \frac{1}{E}[\sigma_z - \nu(\sigma_x + \sigma_y)] \tag{3}$$

let us introduce the notation

$$e = \epsilon_x + \epsilon_y + \epsilon_z \tag{4}$$

$$\theta = \sigma_x + \sigma_y + \sigma_z \tag{5}$$

With this notation, (1), (2), and (3) may be readily solved by determinants for the unknowns σ_x, σ_y, σ_z to yield

$$\sigma_x = \frac{\nu E}{(1+\nu)(1-2\nu)}e + \frac{E}{1+\nu}e_x \tag{6}$$

$$\sigma_y = \frac{\nu E}{(1+\nu)(1-2\nu)}e + \frac{E}{1+\nu}e_y \tag{7}$$

$$\sigma_z = \frac{\nu E}{(1+\nu)(1-2\nu)}e + \frac{E}{1+\nu}e_z \tag{8}$$

These are the desired expressions.

Further information may also be obtained from (1) through (5). If (1), (2), and (3) are added and the symbols e and θ introduced, we have

$$\epsilon = \frac{1}{E}(1-2\nu)\theta \tag{9}$$

For the special case of a solid subjected to uniform hydrostatic pressure p, $\sigma_x = \sigma_y = \sigma_z = -p$. Hence

$$\epsilon = \frac{-3(1-2\nu)p}{E} \quad \text{or} \quad \frac{p}{\epsilon} = -\frac{E}{3(1-2\nu)} \tag{10}$$

The quantity $E/3(1-2\nu)$ is often denoted by K and is called the *bulk modulus* or *modulus of volume expansion* of the material. Physically, the bulk modulus K is a measure of the resistance of a material to change of volume.

We see that the final volume of an element having sides dx, dy, dz prior to loading and subject to strains ϵ_x, ϵ_y, ϵ_z is $(1+\epsilon_x)\,dx\,(1+\epsilon_y)\,dy\,(1+\epsilon_z)\,dz \sim (1+\epsilon_x+\epsilon_y+\epsilon_z)\,dx\,dy\,dz$.

Thus the ratio of the increase in volume to the original volume is given approximately by

$$\epsilon = \epsilon_x + \epsilon_y + \epsilon_z$$

This change of volume per unit volume, e, is defined as the *dilatation*.

1.14. A steel cube is subjected to a hydrostatic pressure of 1.5 MPa. Because of this pressure the volume decreases to give a dilatation of -10^{-5}. The Young's modulus of the material is 200 GPa. Determine Poisson's ratio of the material and also the bulk modulus.

SOLUTION: From Problem 1.13 for hydrostatic loading the dilatation e is given by Eq. (10)

$$e = \frac{-3(1-2\nu)p}{E}$$

Substituting the given numerical values, we have

$$-10^{-5} = \frac{-3(1-2\nu)(1.5\times 10^6)}{200\times 10^9}$$

from which $\nu = 0.278$. Also from Problem 1.13 the bulk modulus is

$$K = \frac{E}{3(1-2\nu)}$$

which becomes

$$K = \frac{200\times 10^9}{3(1-0.556)} = 150\times 10^9 \text{ Pa} \quad \text{or} \quad 150 \text{ GPa}$$

1.15. Consider a steel tube surrounding a solid aluminum cylinder, the assembly being compressed between rigid cover plates by centrally applied forces as shown in Fig. 1-15(*a*). The aluminum cylinder is 8 cm in diameter and the outside diameter of the steel tube is 9.2 cm. If $P = 200$ kN, find the stress in the steel and also in the aluminum. For steel, $E = 200$ GPa and for aluminum $E = 80$ GPa.

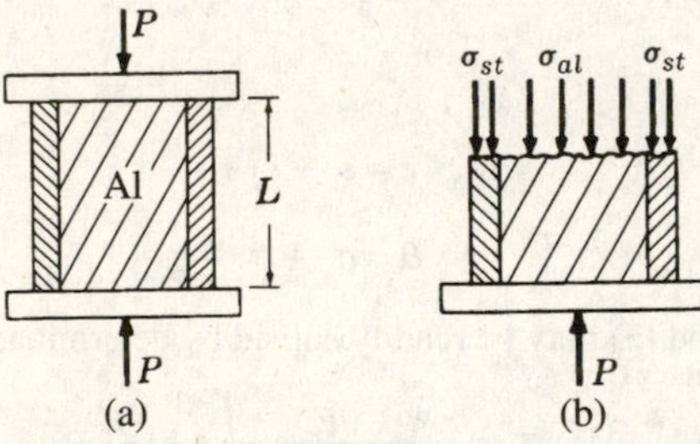

Fig. 1-15

SOLUTION: Pass a horizontal plane through the assembly at any elevation except in the immediate vicinity of the cover plates and then remove one portion or the other, say the upper portion. In that event the portion that we have removed must be replaced by the effect it exerted upon the remaining portion and that effect consists of vertical normal stresses distributed over the two materials. The free-body diagram of the portion of the assembly below this cutting plane is shown in Fig. 1-15(*b*) where σ_{st} and σ_{al} denote the normal stresses existing in the steel and aluminum, respectively.

Let us denote the resultant force carried by the steel by P_{st} and that carried by the aluminum by P_{al}. Then $P_{st} = A_{st}\,\sigma_{st}$ and $P_{al} = A_{al}\,\sigma_{al}$ where A_{st} and A_{al} denote the cross-sectional areas of the steel tube and the aluminum cylinder, respectively. There is only one equation of static equilibrium available for such a force system and it takes the form

$$\Sigma F_y = P - P_{st} - P_{al} = 0$$

Thus, we have one equation in two unknowns, P_{st} and P_{al}, and hence the problem is statically indeterminate. In that event we must supplement the available statics equation by an equation derived from the deformations of the structure. Such an equation is readily obtained because the rigid cover plates force the axial deformations of the two metals to be identical.

The deformation due to axial loading is given by $\Delta = PL/AE$. Equating axial deformations of the steel and the aluminum we have

$$\frac{P_{st}L}{A_{st}E_{st}} = \frac{P_{al}L}{A_{al}E_{al}}$$

or

$$\frac{P_{st}L}{(\pi/4)(0.092^2 - 0.08^2)(200\times 10^9)} = \frac{P_{al}L}{(\pi/4)0.08^2(80\times 10^9)} \qquad \text{from which} \quad P_{st} = 0.806P_{al}$$

This equation is now solved simultaneously with the statics equation, $P = P_{st} + P_{al}$, and we find $P_{al} = 0.554P$, $P_{st} = 0.446P$.

For a load of $P = 200$ kN this becomes $P_{al} = 111$ kN and $P_{st} = 89$ kN. The desired stresses are found by dividing the resultant force in each material by its cross-sectional area:

$$\sigma_{al} = \frac{111}{\pi \times 0.08^2/4} = 22\,100 \text{ kPa} \qquad \sigma_{st} = \frac{89}{\pi(0.092^2 - 0.08^2)/4} = 54\,900 \text{ kPa}$$

1.16. The three-bar assembly shown in Fig. 1-16 supports the vertical load P. Bars AB and BD are identical, each of length L and cross-sectional area A_1. The vertical bar BC is also of length L but of area A_2. All bars have the same modulus E and are pinned at A, B, C, and D. Determine the axial force in each of the bars.

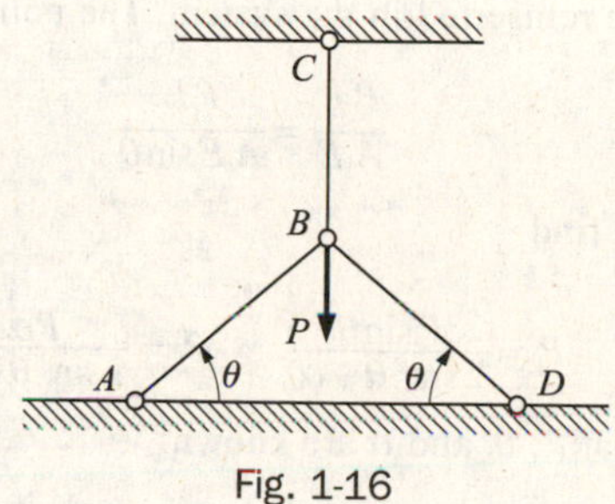

Fig. 1-16

SOLUTION: First, we draw a free-body diagram of the pin at B. The forces in each of the bars are represented by P_1 and P_2 as shown in Fig. 1-17. For vertical equilibrium we find

$$\Sigma F_y = 2P_1 \sin\theta + P_2 - P = 0 \tag{1}$$

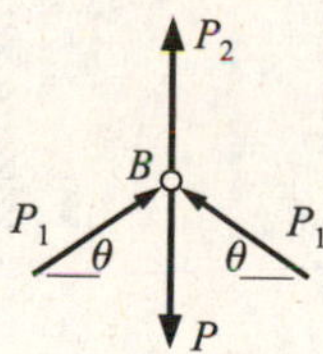

Fig. 1-17

We assume, temporarily, that the pin at B is removed. Next we examine deformations. Under the action of the axial force P_2 the vertical bar extends downward an amount

$$\Delta_1 = \frac{P_2 L}{A_2 E} \tag{2}$$

so that the lower end (originally at B) moves to B' as shown in Fig. 1-18.

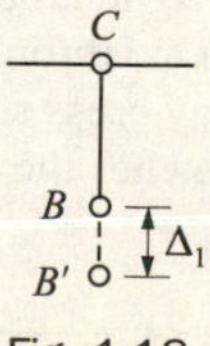

Fig. 1-18

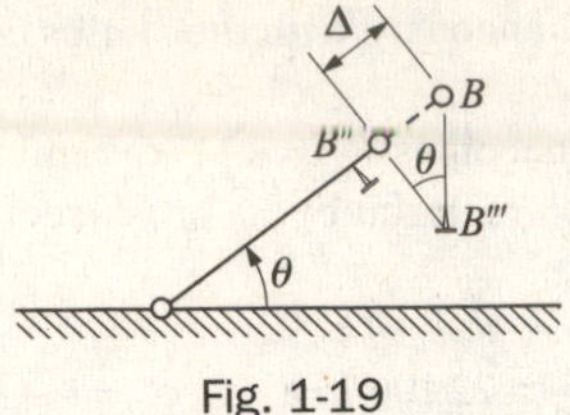

Fig. 1-19

The compressive force in AB causes it to shorten an amount Δ shown as BB'' in Fig. 1-19. The bar AB then rotates about A as a rigid body so that B'' moves to B''' directly below point C. From Fig. 1-19 the vertical component of Δ is

$$BB''' = \frac{P_1 L}{A_1 E \sin\theta}$$

Next, we consider the pin to be reinserted in the system. The points B' and B''' must coincide so that

$$\frac{P_2 L}{A_2 E} = \frac{P_1 L}{A_1 E \sin\theta} \tag{3}$$

Substituting Eq. (3) in Eq. (1) we find

$$P_1 = \frac{P\sin\theta}{2\sin^2\theta + \alpha} \qquad P_2 = \frac{P\alpha}{2\sin^2\theta + \alpha}$$

where $\alpha = A_2/A_1$. It is assumed that P, θ, and α are known.

1.17. The composite bar shown in Fig. 1-20(a) is rigidly attached to the two supports. The left portion of the bar is copper, of uniform cross-sectional area 80 cm^2 and length 30 cm. The right portion is aluminum, of uniform cross-sectional area 20 cm^2 and length 20 cm. At a temperature of 26°C the entire assembly is stress free. The temperature of the structure drops and during this process the right support yields 0.025 mm in the direction of the contracting metal. Determine the minimum temperature to which the assembly may be subjected in order that the stress in the aluminum does not exceed 160 MPa. For copper $E = 100$ GPa, $\alpha = 17 \times 10^{-6}$/°C, and for aluminum $E = 80$ GPa, $\alpha = 23 \times 10^{-6}$/°C.

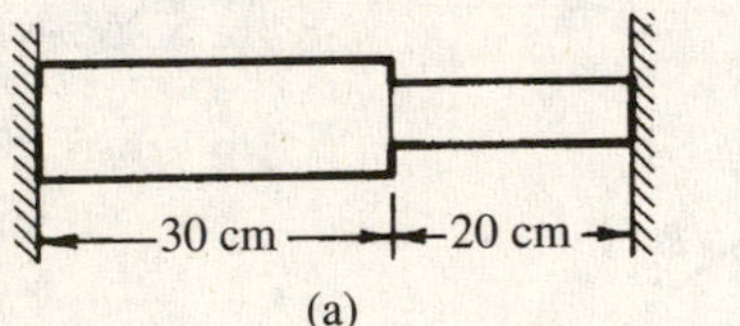

(a)

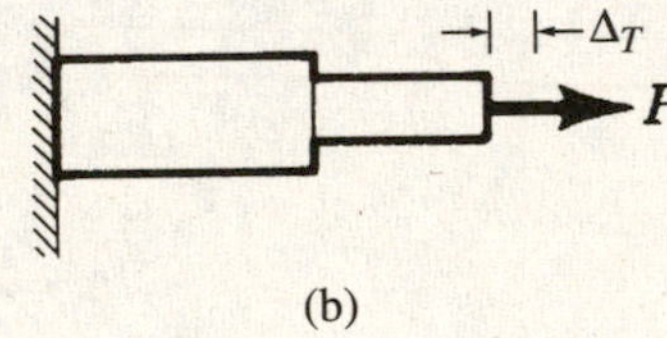

(b)

Fig. 1-20

SOLUTION: It is perhaps simplest to consider that the bar is cut just to the left of the supporting wall at the right and is then free to contract due to the temperature drop ΔT. The total shortening of the composite bar is given by

$$\Delta_T = (\alpha L \Delta t)_{cu} + (\alpha L \Delta T)_{al}$$
$$= (17\times10^{-6})(0.30)\Delta T + (23\times10^{-6})(0.20)\Delta T$$

according to the definition of the coefficient of linear expansion [Eq. (1.4)]. It is to be noted that the shape of the cross section has no influence upon the change in length of the bar due to a temperature change.

Even though the bar has contracted this amount, it is still stress free. However, this is not the complete analysis because the reaction of the wall at the right has been neglected by cutting the bar there. Consequently, we must represent the action of the wall by an axial force P applied to the bar, as shown in Fig. 1-20(b). For equilibrium, the resultant force acting over any cross section of either the copper or the aluminum must be equal to P. The application of the force P stretches the composite bar by an amount

$$\Delta = \frac{P(0.30)}{(80\times10^{-4})(100\times10^{9})} + \frac{P(0.20)}{(20\times10^{-4})(80\times10^{9})} = 16.25\times10^{-10}P$$

If the right support were unyielding, we would equate the last expression to the expression giving the total shortening due to the temperature drop. Actually, the right support yields 0.025 cm and consequently we may write

$$16.25 \times 10^{-10} P = 9.7 \times 10^{-6} \Delta T - 0.025 \times 10^{-3}$$

The stress in the aluminum is not to exceed 160 MPa, and since it is given by the formula $\sigma = P/A$, the maximum force P becomes $P = A\sigma = 20 \times 10^{-4}\ (160 \times 10^6) = 320\ 000$ N. Substituting this value of P in the above equation relating deformations, we find $\Delta T = 56.2°C$. Therefore the temperature may drop 56.2°C from the original 26°C. The final temperature would be –30.2°C.

1.18. A hollow steel cylinder surrounds a solid copper cylinder and the assembly is subjected to an axial loading of 200 kN as shown in Fig. 1-21(*a*). The cross-sectional area of the steel is 20 cm^2, while that of the copper is 60 cm^2. Both cylinders are the same length before the load is applied. Determine the temperature rise of the entire system required to place all of the load on the copper cylinder. The cover plate at the top of the assembly is rigid. For copper $E = 100$ GPa, $\alpha = 1.7 \times 10^{-6}$/°C, while for steel $E = 200$ GPa, $\alpha = 12 \times 10^{-6}$/°C.

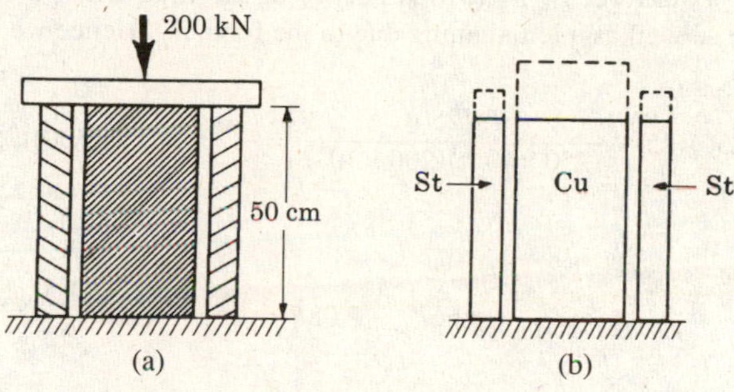

Fig. 1-21

SOLUTION: One method of analyzing this problem is to assume that the load as well as the upper cover plate are removed and that the system is allowed to freely expand vertically because of a temperature rise ΔT. In that event the upper ends of the cylinders assume the positions shown by the dashed lines in Fig. 1-21(*b*).

The copper cylinder naturally expands upward more than the steel one because the coefficient of linear expansion of copper is greater than that of steel. The upward expansion of the steel cylinder is $(12 \times 10^{-6})(0.50)\Delta T$, while that of the copper is $(17 \times 10^{-6})(0.50)\Delta T$.

This is not of course the true situation because the load of 200 kN has not as yet been considered. If all of this axial load is carried by the copper then only the copper will be compressed and the compression of the copper is given by

$$\Delta_{cu} = \frac{PL}{AE} = \frac{200\,000 \times 0.50}{(60 \times 10^{-4}) \times (100 \times 10^9)} = 1.667 \times 10^{-4} \text{ m}$$

The condition of the problem states that the temperature rise ΔT is just sufficient so that all of the load is carried by the copper. Thus, the expanded length of the copper indicated by the dashed lines in the sketch will be decreased by the action of the force. The net expansion of the copper is the expansion caused by the rise of temperature minus the compression due to the load. The change of length of the steel is due only to the temperature rise. Consequently, we may write

$$(17 \times 10^{-6})(0.50)\Delta T - 1.667 \times 10^{-4} = (12 \times 10^{-6})(0.50)\Delta T \qquad \therefore \Delta T = 66.7°C$$

1.19. The rigid bar AD is pinned at A and attached to the bars BC and ED, as shown in Fig.1-22(*a*). The entire system is initially stress free and the weights of all bars are negligible. The temperature of bar BC is lowered 25°C and that of bar ED is raised 25°C. Find the normal stresses in bars BC and ED. For BC, which is brass, assume $E = 90$ GPa, $\alpha = 20 \times 10^{-6}$/°C, and for ED, which is steel, take $E = 200$ GPa and $\alpha = 12 \times 10^{-6}$/°C. The cross-sectional area of BC is 500 mm^2 and of ED is 250 mm^2.

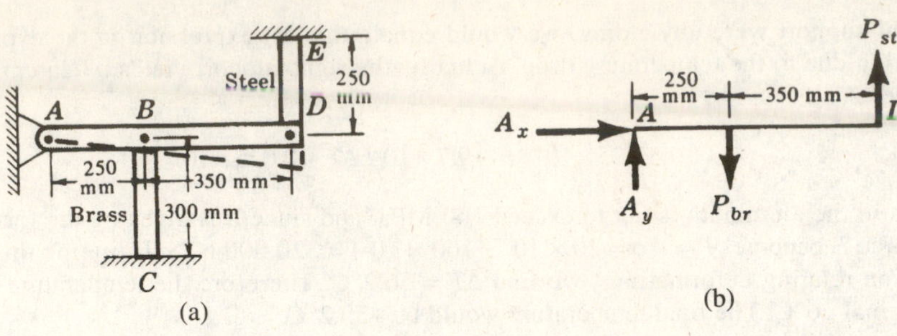

Fig. 1-22

SOLUTION: Let us denote the forces on *AD* by P_{st} and P_{br} acting in the assumed directions shown in the free-body diagram, Fig. 1-22(*b*). Since *AD* rotates as a rigid body about *A* (as shown by the dashed line) we have $\Delta_{br}/250 = \Delta_{st}/600$ where Δ_{br} and Δ_{st} denote the axial compression of *BC* and the axial elongation of *DE*, respectively.

The total change of length of *BC* is composed of a shortening due to the temperature drop as well as a lengthening due to the axial force P_{br}. The total change of length of *DE* is composed of a lengthening due to the temperature rise as well as a lengthening due to the force P_{st}. Hence we have

$$\left(\frac{250}{600}\right)\left[(12\times10^{-6})(250)(25)+\frac{P_{st}(250)}{(250\times10^{-6})(200\times10^{9})}\right]=-(20\times10^{-6})(300)(25)+\frac{P_{br}(300)}{(500\times10^{-6})(90\times10^{9})}$$

or

$$6.66P_{br} - 2.08P_{st} = 153.0\times10^{3}$$

From statics,

$$\Sigma M_A = 250P_{br} - 600P_{st} = 0$$

Solving these equations simultaneously, $P_{st} = 10.99$ kN and $P_{br} = 26.3$ kN. Using $\sigma = P/A$ for each bar, we obtain $\sigma_{st} = 43.9$ MPa and $\sigma_{br} = 52.6$ kN.

1.20. Consider a low-carbon square steel bar 20 mm on a side and 1.7 m long having a material yield point of 275 MPa and $E = 200$ GPa. An applied axial load gradually builds up from zero to a value such that the elongation of the bar is 15 mm, after which the load is removed. Determine the permanent elongation of the bar after removal of the load. Assume elastic, perfectly plastic behavior as shown in Fig. 1-23.

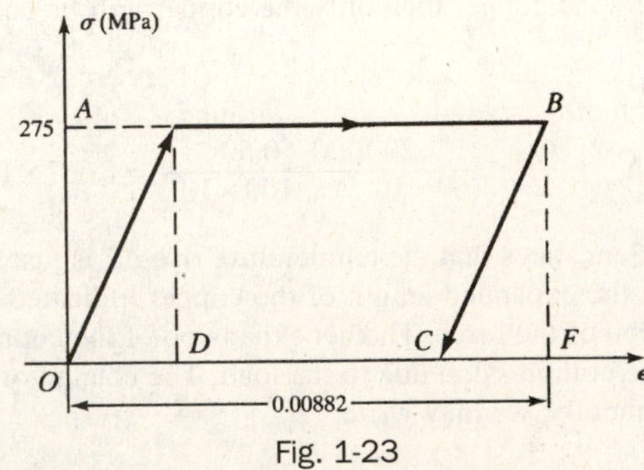

Fig. 1-23

SOLUTION: Yield begins when the applied load reaches a value of

$$P = \sigma_{yp}(\text{area})$$

$$= (275\times10^{6}\ \text{N/m}^2)(0.020\ \text{m})^2$$

$$= 110\,000\ \text{N}$$

which corresponds to point A. Note that in that figure the ordinate is stress and the abscissa is strain. However, values on each of these axes differ only by constants from those on a force-elongation plot.

When the elongation is 15 mm, corresponding to point B, unloading begins and the axial strain at the initiation of unloading is

$$\epsilon = \frac{15\ \text{mm}}{1700\ \text{mm}} = 0.00882$$

Unloading follows along line BC (parallel to AO) until the horizontal axis is reached, so that OC corresponds to the strain after complete removal of the load. We next find the strain CF—but this is readily found from using the similar triangles OAD and CBF to be

$$E = \frac{\sigma}{\epsilon} \qquad \epsilon = \frac{275 \times 10^6\ \text{Pa}}{200 \times 10^9\ \text{Pa}} = 1.375 \times 10^{-3}$$

Thus, after load removal the residual strain is

$$\begin{aligned} OC &= OF - CF \\ &= 0.00882 - 0.00138 = 0.00744 \end{aligned}$$

The elongation of the 1.7-m long bar is consequently

$$(1.7\ \text{m})\,(0.00744) = 0.0126\ \text{m} \qquad \text{or} \qquad 12.6\ \text{mm}$$

SUPPLEMENTARY PROBLEMS

1.21. Laboratory tests on human teeth indicate that the area effective during chewing is approximately 0.25 cm^2 and that the tooth length is about 1.1 cm. If the applied load in the vertical direction is 880 N and the measured shortening is 0.004 cm, determine Young's modulus. *Ans.* $8.8 \times 10^9\ \text{N/m}^2$ or 8.8 GPa

1.22. A hollow right-circular cylinder is made of cast iron and has an outside diameter of 75 mm and an inside diameter of 60 mm. If the cylinder is loaded by an axial compressive force of 50 kN, determine the total shortening in a 600 mm length. Also determine the normal stress under this load. Take the modulus of elasticity to be 100 GPa. *Ans.* $\Delta = 0.188$ mm, $\sigma = 31.45$ MPa

1.23. A solid circular steel rod 6 mm in diameter and 500 mm long is rigidly fastened to the end of a square brass bar 25 mm on a side and 400 mm long, the geometric axes of the bars lying along the same line. An axial tensile force of 5 kN is applied at each of the extreme ends. Determine the total elongation of the assembly. For steel, $E = 200$ GPa and for brass, $E = 90$ GPa. *Ans.* 0.477 mm

1.24. A high-performance jet aircraft cruises at three times the speed of sound at an altitude of 25 000 m. It has a long, slender titanium body reinforced by titanium ribs. The length of the aircraft is 30 m and the coefficient of thermal expansion of the titanium is 10×10^{-6}/°C. Determine the increase of overall length of the aircraft at cruise altitude over its length on the ground if the temperature while cruising is 500°C above ground temperature. (*Note:* This change of length is of importance since the designer must account for it because it changes the performance characteristics of the system.) *Ans.* 0.150 m

1.25. A material that can be used as a superconductor is composed of yttrium (a rare earth metal), barium, copper, and oxygen. This material acts as a superconductor (i.e., transmits electricity with essentially no resistance) at temperatures down to –178°C. If the temperature is then raised from –178°C to 67°C, and the coefficient of thermal expansion is 11.0×10^{-6}/°C, determine the elongation of a 100-m long segment due to this temperature differential. *Ans.* 0.27 m

1.26. A Z-shaped rigid bar $ABCD$, shown in Fig. 1-24, is suspended by a pin at B, and loaded by a vertical force P. At A a 10-mm-diameter steel tie rod AF connects the section to a firm ground support at F. Use $E = 200$ GPa. Determine the vertical deflection at D. *Ans.* 4.53 mm

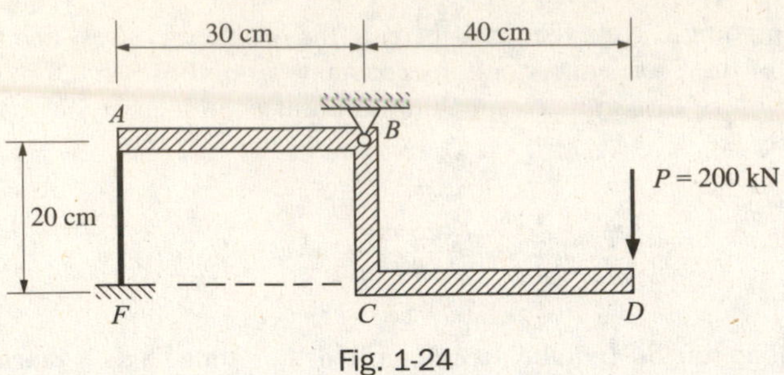

Fig. 1-24

1.27. The rigid bar *ABC* is pinned at *B* and at *A* attached to a vertical steel bar *AD* which in turn is attached to a larger steel bar *DF* which is firmly attached to a rigid foundation. The geometry of the system is shown in Fig. 1-25. If a vertical force *P* of magnitude 40 kN is applied at *C*, determine the vertical displacement of point *C*. *Ans.* 9.17 mm

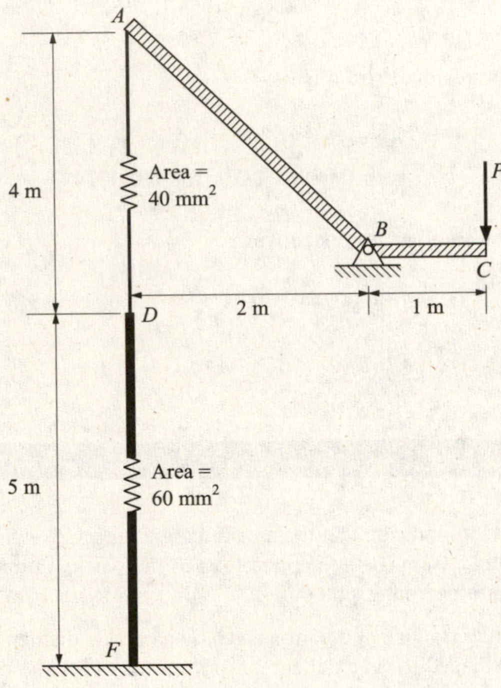

Fig. 1-25

1.28. The steel bars *AB* and *BC* are pinned at each end and support the load of 200 kN, as shown in Fig. 1-26. The material is structural steel, having a yield point of 200 MPa, and safety factors of 2 and 3.5 for tension and compression, respectively. Determine the size of each bar and also the horizontal and vertical components of displacement of point *B*. Take E = 200 GPa. Neglect any possibility of lateral buckling of bar *BC*. *Ans.* d_{AB} = 47.0 mm, d_{BC} = 47.2 mm, Δ_x = 0.37 mm (to right), Δ_y = 1.78 mm (downward)

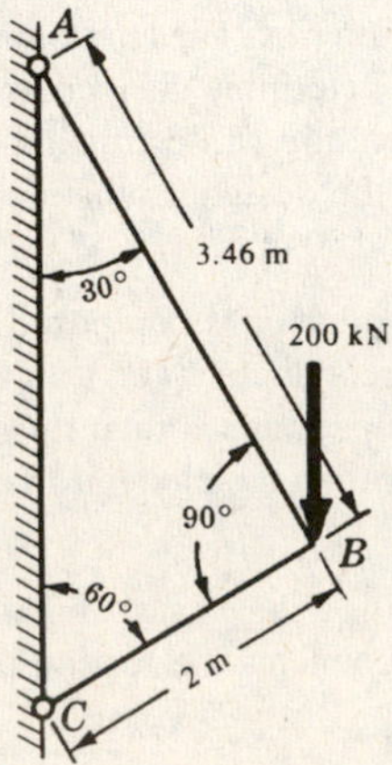

Fig. 1-26

1.29. The two bars *AB* and *CB* shown in Fig. 1-27 are pinned at each end and subject to a single vertical force *P*. The geometric and elastic constants of each are as indicated. Determine the horizontal and vertical components of displacement of pin *B*.

Ans. $\Delta_x = -\dfrac{PL_1}{\sqrt{3}\,A_1E_1} + \dfrac{PL_2}{\sqrt{3}\,A_2E_2}$, $\Delta_y = \dfrac{PL_1}{3A_1E_1} + \dfrac{PL_2}{3A_2E_2}$

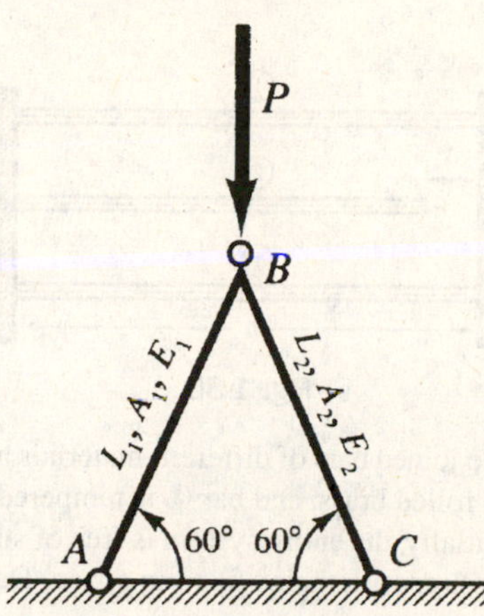

Fig. 1-27

1.30. In Problem 1-29, the bar *AB* is titanium, having an area of 1000 mm^2, length of 2.4 m, and $E_1 = 110$ GPa. Bar *CB* is steel having an area of 400 mm^2, length of 2.4 m, and $E_2 = 200$ GPa. What are the horizontal and vertical components of displacement of the pin *B* if $P = 600$ kN? *Ans.* $\Delta_x = 2.83$ mm, $\Delta_y = 10.4$ mm

1.31. Two bars are joined together and attached to supports as in Fig. 1-28. The left bar is brass for which $E = 90$ GPa, $\alpha = 20 \times 10^{-6}$/°C, and the right bar is aluminum for which $E = 70$ GPa, $\alpha = 25 \times 10^{-6}$/°C. The cross-sectional area of the brass bar is 500 mm^2, and that of the aluminum bar is 750 mm^2. Let us suppose that the system is initially stress free and that the temperature then drops 20°C.

(*a*) If the supports are unyielding, find the normal stress in each bar.

(*b*) If the right support yields 0.1 mm, find the normal stress in each bar.

Ans. (*a*) $\sigma_{br} = 41$ MPa, $\sigma_{al} = 27.33$ MPa; (*b*) $\sigma_{br} = 28.4$ MPa, $\sigma_{al} = 19$ MPa

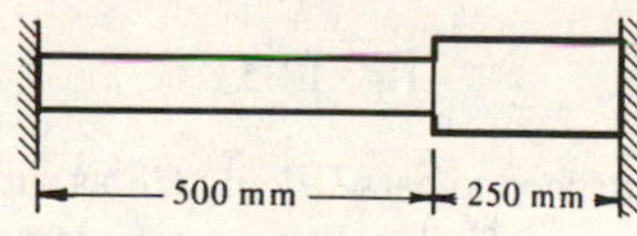

Fig. 1-28

1.32. An aluminum right-circular cylinder surrounds a steel cylinder as shown in Fig. 1-29. The axial compressive load of 200 kN is applied through the rigid cover plate shown. If the aluminum cylinder is originally 0.25 mm longer than the steel before any load is applied, find the normal stress in each when the temperature has dropped 20 K and the entire load is acting. For steel take $E = 200$ GPa, $\alpha = 12 \times 10^{-6}$/°C, and for aluminum assume $E = 70$ GPa, $\alpha = 25 \times 10^{-6}$/°C.

Ans. $\sigma_{st} = 9$ MPa, $\sigma_{at} = 15.5$ MPa

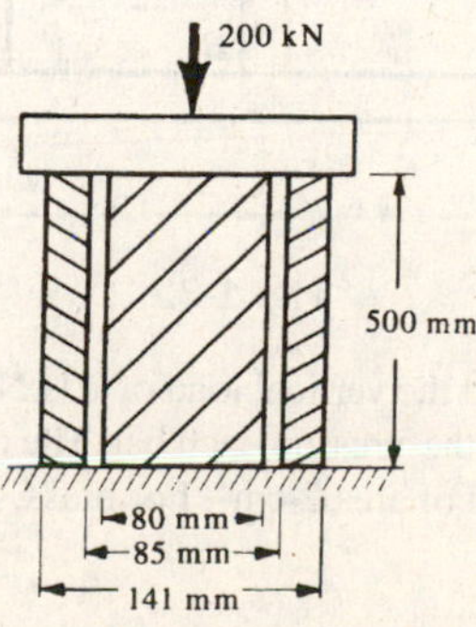

Fig. 1-29

1.33. A system consists of two rigid end-plates, tied together by three horizontal bars, as shown in Fig. 1-30. Through a fabrication error, the central bar ② is $0.0005L$ too short. All bars are of identical cross section and of steel having $E = 210$ GPa. Find the stress in each bar after bar ② has been physically attached to the end plate. Any external force is removed.

Ans. $\sigma_1 = -35$ MPa

$\sigma_2 = 70$ MPa

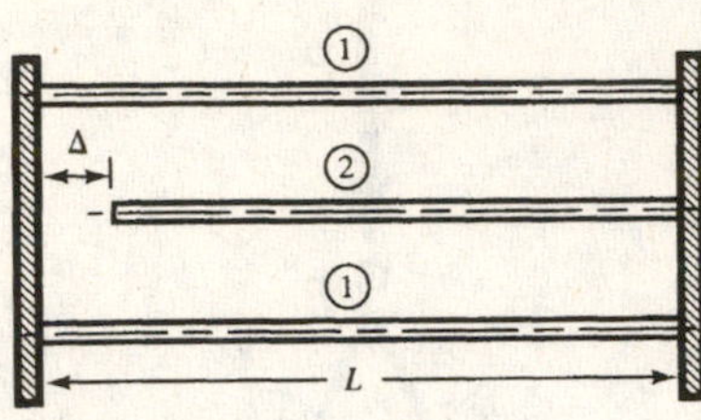

Fig. 1-30

1.34. A structural system consists of three joined bars of different materials and geometries, as shown in Fig. 1-31. Bar ① is aluminum alloy, bar ② is cold rolled brass, and bar ③ is tempered alloy steel. Properties and dimensions of all three are shown in the figure. Initially, the entire system is free of stresses, but then the right support is moved 3 mm to the right whereas the left support remains fixed in space. Determine the stress in each bar due to this 3 mm displacement.

Ans. $\sigma_1 = 223$ MPa

$\sigma_2 = 178$ MPa

$\sigma_3 = 446$ MPa

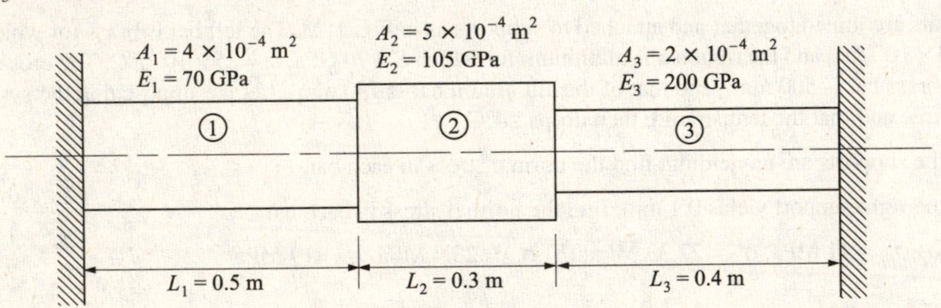

Fig. 1-31

1.35. The rigid bar *AC* is pinned at *A* and attached to bars *DB* and *CE* as shown in Fig. 1-32. The weight of *AC* is 50 kN and the weights of the other two bars are negligible. Consider the temperature of both bars *DB* and *CE* to be raised 35°C. Find the resulting normal stresses in these two bars. *DB* is copper for which $E = 90$ GPa, $\alpha = 18 \times 10^{-6}$/°C, and the cross-sectional area is 1000 mm^2, while *CE* is steel for which $E = 200$ GPa, $\alpha = 12 \times 10^{-6}$/°C, and the cross section is 500 mm^2. Neglect any possibility of lateral buckling of the bars. *Ans.* $\sigma_{st} = 72$ MPa, $\sigma_{cu} = -21.7$ MPa

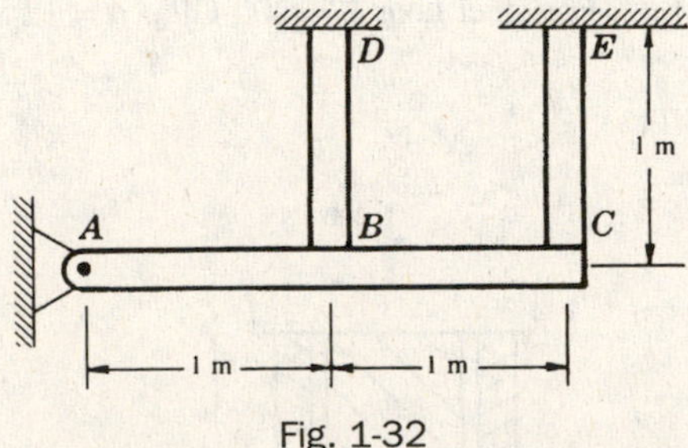

Fig. 1-32

1.36. The three bars shown in Fig. 1-33 support the vertical load of 20 kN. The bars are all stress free and joined by the pin at *A* before the load is applied. Calculate the stress in each bar. The outer bars are each of brass and of cross-sectional area 2.5 cm^2. The central bar is steel and of area 2 cm^2. For brass, $E = 85$ GPa and for steel, $E = 200$ GPa.

Ans. $\sigma_{br} = 16.8$ MPa, $\sigma_{st} = 79$ MPa

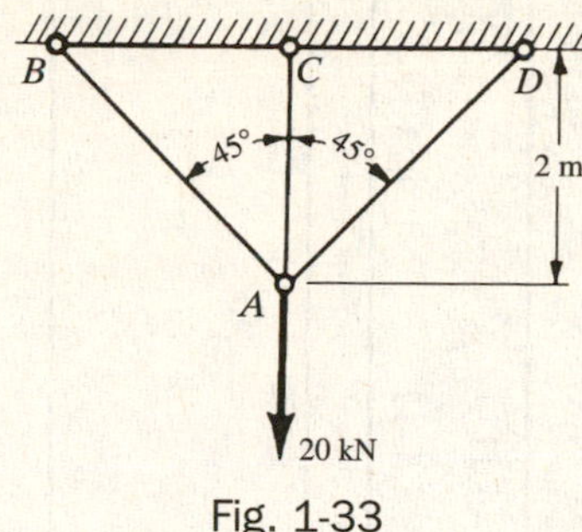

Fig. 1-33

1.37. The rigid bar AD in Fig. 1-34 is pinned at A and supported by a steel rod at D together with a linear spring at B. The bar carries a vertical load of 30 kN applied at C. Determine the vertical displacement of point D. *Ans.* 0.8 mm

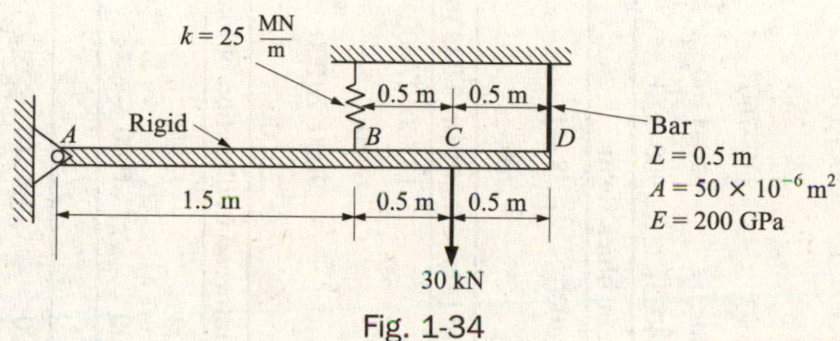

Fig. 1-34

1.38. A bar of uniform cross section is subjected to uniaxial tension and develops a strain in the direction of the force of 1/800. Calculate the change of volume per unit volume. Assume $\nu = 1/3$. *Ans.* 1/2400 (increase)

1.39. A square steel bar is 50 mm on a side and 250 mm long. It is loaded by an axial tensile force of 200 kN. If $E = 200$ GPa and $\nu = 0.3$, determine the change of volume per unit volume. *Ans.* 0.00016

1.40. Consider a steel square steel bar 2 cm on a side and 150 cm long having a material yield point of 270 MPa and a Young's modulus of 200 GPa. An axial tensile load gradually builds up from zero to a value such that the elongation of the bar is 18 mm, after which the load is removed. Determine the permanent elongation of the bar. Assume that the material is elastic, perfectly plastic. *Ans.* 1.27 cm

Table 1-3. Properties of Common Engineering Materials at 20°C (68°F)

Material	Specific weight		Young's modulus		Ultimate stress		Coefficient of linear thermal expansion		Poisson's ratio
	lb/in³	kN/m³	lb/in²	GPa	lb/in²	kPa	10*e*–6/°F	10*e*–6/°C	
I. Metals in slab, bar, or block form									
Aluminum alloy	0.0984	27	10–12*e*6	70–79	45–80*e*3	310–550	13	23	0.33
Brass	0.307	84	14–16*e*6	96–110	43–85*e*3	300–590	11	20	0.34
Copper	0.322	87	16–18*e*6	112–120	33–55*e*3	230–380	9.5	17	0.33
Nickel	0.318	87	30*e*6	210	45–110*e*3	310–760	7.2	13	0.31
Steel	0.283	77	28–30*e*6	195–210	80–200*e*3	550–1400	6.5	12	0.30
Titanium alloy	0.162	44	15–17*e*6	105–120	130–140*e*3	900–970	4.5–5.5	8–10	0.33
II. Nonmetallics in slab, bar, or block form									
Concrete (composite)	0.0868	24	3.6*e*6	25	4000–6000	28–41	6	11	
Glass	0.0955	26	7–12*e*6	48–83	10 000	70	3–6	5–11	0.23
III. Materials in filamentary (whisker) form: [dia. < 0.001 in (0.025 mm)]									
Aluminum oxide	0.141	38	100–350*e*6	690–2410	2–4*e*6	13 800–27 600			
Barium carbide	0.090	25	65*e*6	450	1*e*6	6900			
Glass			50*e*6	345	1–3*e*6	7000–20 000			
Graphite	0.081	22	142*e*6	980	3*e*6	20 000			
IV. Composite materials (unidirectionally reinforced in direction of loading)									
Boron epoxy	0.071	19	31*e*6	210	198 000	1365	2.5	4.5	
S-glass-reinforced epoxy	0.0766	21	9.6*e*6	66.2	275 000	1900			
V. Others									
Graphite-reinforced epoxy	0.054	15	15*e*6	104	190 000	1310			
Kevlar-49 epoxy*	0.050	13.7	12.5*e*6	86	220 000	1520			

*Tradename of E. I. duPont Co.

Note: 12*e*–6°/C = 12×10^{-6}/°C

Shear Stresses

2.1 Shear Force and Shear Stress

If a plane is passed through a body, a force acting along this plane is called a *shear force* or *shearing force*. It will be denoted by F_s. The shear force, divided by the area over which it acts, is called the *shear stress* or *shearing stress*. It is denoted by τ. Thus

$$\tau = \frac{F_s}{A} \tag{2.1}$$

Let us consider a bar cut by a plane *a-a* perpendicular to its axis, as shown in Fig. 2-1. A normal stress σ is perpendicular to this plane. This type of stress was considered in Chapter 1. The shear stress acts *along* the plane, is shown by the stress τ, whereas normal stress acts *normal* to the plane.

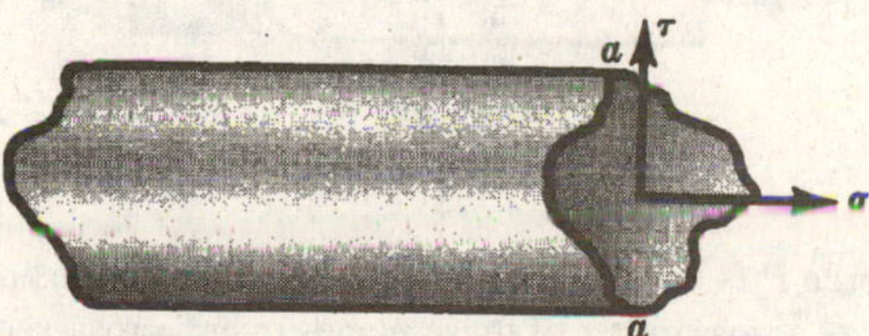

Fig. 2-1 Normal and shear stresses acting on a plane area.

The shear stress is taken to be uniform in all problems discussed in this chapter. Thus the expression $\tau = F_s/A$ indicates an average shear stress over the area.

Punching operations (Problem 2.2), wood test specimens (Problem 2.3), riveted joints (Problem 2.5), and welded joints (Problems 2.6 and 2.10) are common examples of systems involving shear stresses.

2.2 Deformations Due to Shear Stresses

Let us consider the deformation of a plane rectangular element cut from a solid where the forces acting on the element are known to be shearing stresses τ in the directions shown in Fig. 2-2(*a*). The faces of the element parallel to the plane of the paper are assumed to be load free. Since there are no normal stresses acting on the element, the lengths of the sides of the rectangular element will not change when the shearing stresses assume the value τ. However, there will be a distortion of the originally right *angles* of the element, and after this distortion due to the shearing stresses the element assumes the configuration shown by the dashed lines in Fig. 2-2(*b*).

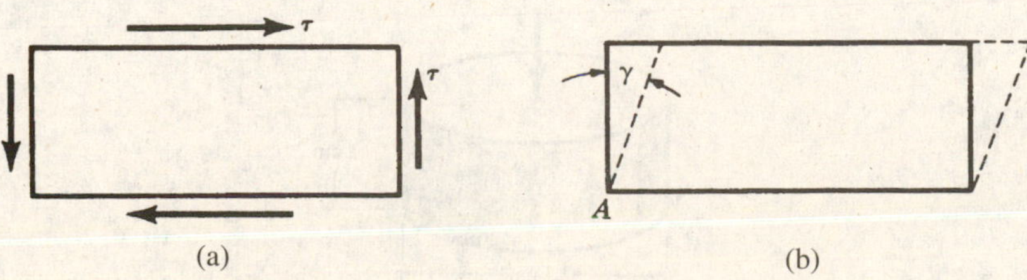

Fig. 2-2 Shearing stresses acting on an element.

2.3 Shear Strain

The change of angle at the corner of an originally rectangular element is defined as the *shear strain*. It must be expressed in radians and is denoted by γ, as shown.

Shear Modulus

The ratio of the shear stress τ to the shear strain γ is called the *shear modulus* and is denoted by G. Thus

$$G = \frac{\tau}{\gamma} \tag{2.2}$$

G is also known as the *modulus of rigidity*.

The units of G are the same as those of the shear stress, i.e., N/m^2, since the shear strain is dimensionless. Stress-strain diagrams for various materials may be drawn for shearing loads, just as they were drawn for normal loads in Chapter 1. They have the same general appearance as those sketched in Chapter 1 but the numerical values associated with the plots are of course different.

SOLVED PROBLEMS

2.1. Consider the bolted joint shown in Fig. 2-3. The force P is 30 kN and the diameter of the bolt is 10 mm. Determine the average value of the shearing stress existing across either of the planes *a-a* or *b-b*.

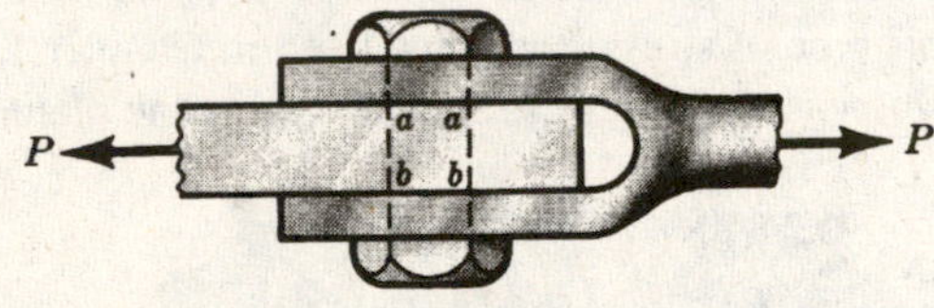

Fig. 2-3

SOLUTION: Assume that force P is equally divided between the sections *a-a* and *b-b*. Consequently, a force of $\frac{1}{2}(30 \times 10^3) = 15 \times 10^3$ N acts across either of these planes over a cross-sectional area

$$A = \frac{1}{4}\pi(10)^2 = 78.6 \text{ mm}^2$$

Thus the average shearing stress across either plane is

$$\tau = \frac{1}{2}\frac{P}{A} = \frac{15 \times 10^3}{78.6 \times 10^{-6}} = 191 \times 10^6 \text{ Pa} \qquad \text{or} \qquad 191 \text{ MPa}$$

2.2. Low-carbon structural steel has a shearing ultimate strength of approximately 300 MPa. Determine the force P necessary to punch a 2.5-cm-diameter hole through a plate of this steel 1 cm thick. If the modulus of elasticity in shear for this material is 82 GPa, find the shear strain at the edge of this hole when the shear stress is 143 MPa.

SOLUTION: Let us assume uniform shearing on a cylindrical surface is 2.5 cm in diameter and 1 cm thick, as shown in Fig. 2-4. For equilibrium the force P is

$$P = \tau A = 0.025\pi \times 0.01 \times 300 \times 10^6 = 236\,000 \text{ N} \qquad \text{or} \qquad 236 \text{ kN}$$

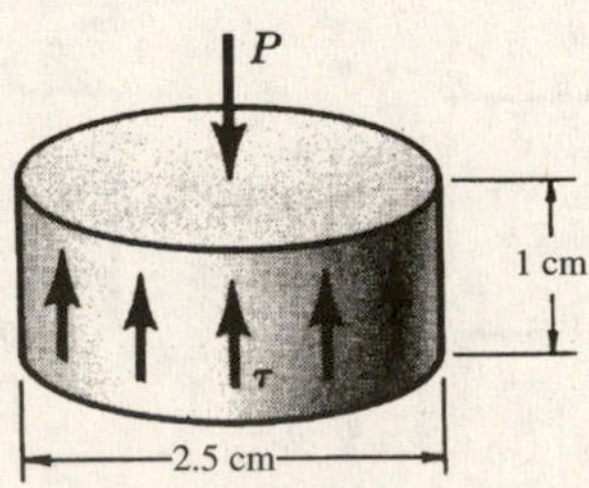

Fig. 2-4

To determine the shear strain γ when the shear stress τ is 143 MPa, we employ the definition $G = \tau/\gamma$ to obtain

$$\gamma = \frac{\tau}{G} = \frac{143 \times 10^6}{82 \times 109} = 0.00174 \text{ radians}$$

2.3. In the wood industries, inclined blocks of wood are sometimes used to determine the *compression-shear* strength of glued joints. Consider the pair of glued blocks A and B of Fig. 2-5(a) which are 38 mm deep in a direction perpendicular to the plane of the paper. Determine the shearing ultimate strength of the glue if a vertical force of 40 kN is required to cause rupture of the joint. It is to be noted that a good glue causes a large proportion of the failure to occur in the wood.

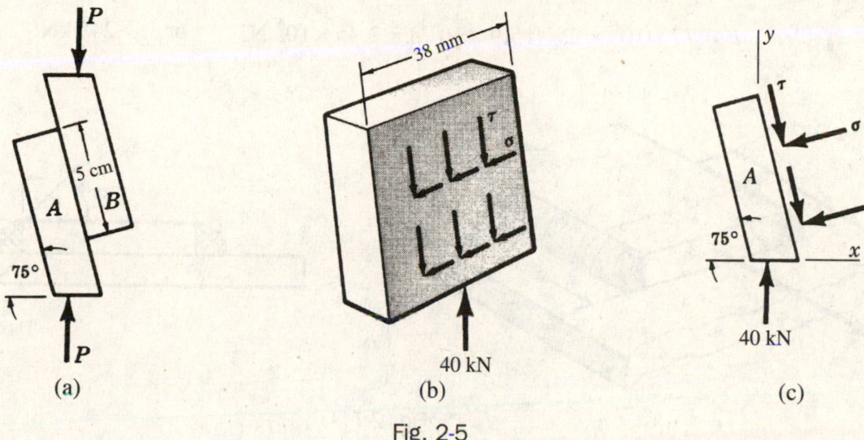

Fig. 2-5

SOLUTION: Let us consider the equilibrium of the lower block A in Fig. 2-5(b). The reactions of the upper block B upon the lower one consist of both normal and shearing forces appearing as in the perspective and orthogonal views of Figs. 2-5(b) and 2-5(c).

Referring to Fig. 2-5(c), equilibrium in the x-direction requires

$$\Sigma F_x = \tau(0.05)(0.038)\cos 75° - \sigma(0.05)(0.038)\cos 15° = 0 \qquad \text{or} \qquad \sigma = 0.268\tau$$

For equilibrium in the vertical direction we have

$$\Sigma F_y = 40000 - \tau(0.05)(0.038)\sin 75° - \sigma(0.05)(0.038)\sin 15° = 0$$

Substituting $\sigma = 0.268\tau$ and solving, we find $\tau = 20.3$ MPa.

2.4. The shearing stress in a piece of structural steel is 100 MPa. If the modulus of rigidity G is 85 GPa, find the shearing strain γ.

SOLUTION: By definition, $G = \tau/\gamma$. Then the shearing strain

$$\gamma = \frac{\tau}{G} = \frac{100 \times 10^6}{85 \times 10^9} = 0.00117 \text{ rad}$$

2.5. A single rivet is used to join two plates as shown in Fig. 2-6. If the diameter of the rivet is 20 mm and the load P is 30 kN, what is the average shearing stress developed in the rivet?

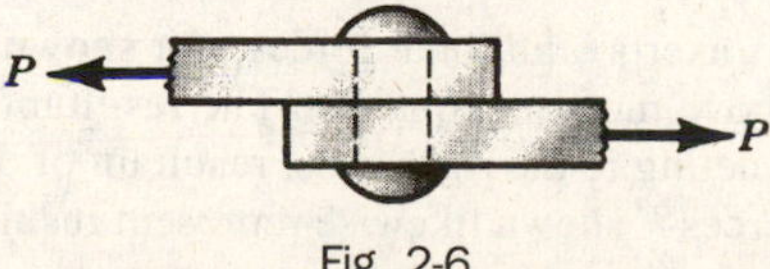

Fig. 2-6

SOLUTION: Here the average shear stress in the rivet is P/A where A is the cross-sectional area of the rivet. Hence the shearing stress is given by

$$\tau = \frac{30000 \text{ N}}{(\pi/4)[0.020 \text{ m}]^2} = 9.55 \times 10^7 \text{ N/m}^2 \qquad \text{or} \qquad 95.5 \text{ MPa}$$

2.6. One common type of weld for joining two plates is the *fillet weld.* This weld undergoes shear as well as tension or compression and frequently bending in addition. For the two plates shown in Fig. 2-7, determine the allowable tensile force P that may be applied using an allowable working stress of 77 MPa for shear loading. Consider only shearing stresses in the weld. The load is applied midway between the two welds.

SOLUTIONS: The minimum dimension of the weld cross section is termed the *throat*, which in this case is 1.25 sin 45° = 0.884 cm. The effective weld area that resists shearing is given by the length of the weld times the throat dimension, or weld area = 18(0.884) = 15.9 cm^2 for each of the two welds. Thus the allowable tensile load P is given by the product of the working stress in shear times the area resisting shear, or

$$P = 77 \times 10^6 \times 2 \times (15.9 \times 10^{-4}) = 2.45 \times 10^5 \text{ N} \qquad \text{or} \qquad 245 \text{ kN}$$

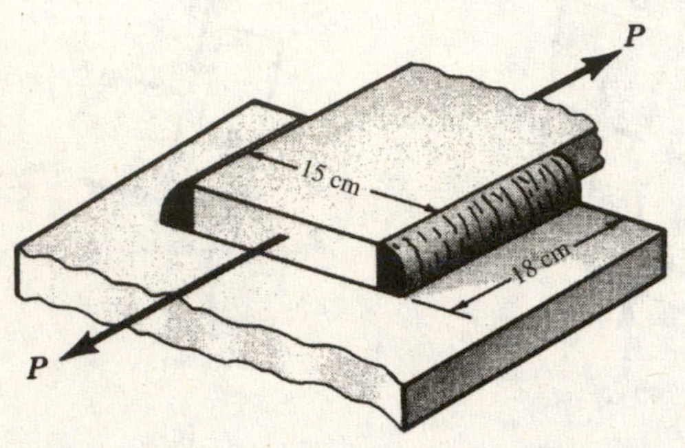

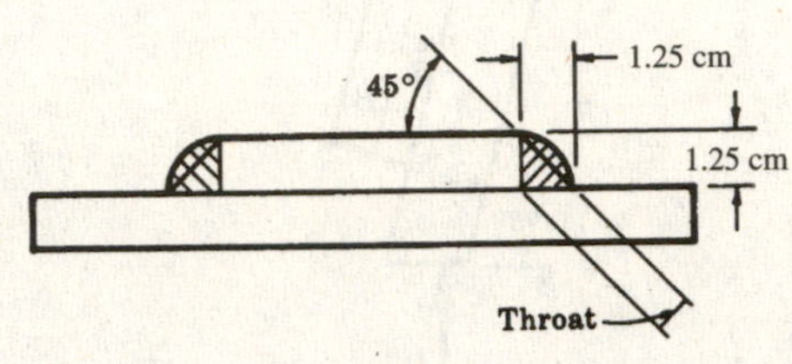

Fig. 2-7

2.7. Shafts and pulleys are usually fastened together by means of a key, as shown in Fig. 2-8(a). Consider a pulley subject to a turning moment T of 1000 N·m keyed by a 1 × 1 × 8 cm key to the shaft. The shaft is 5 cm in diameter. Determine the shear stress on a horizontal plane through the key.

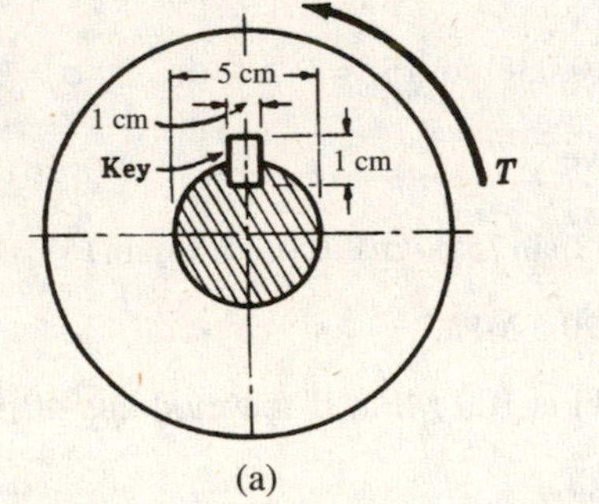

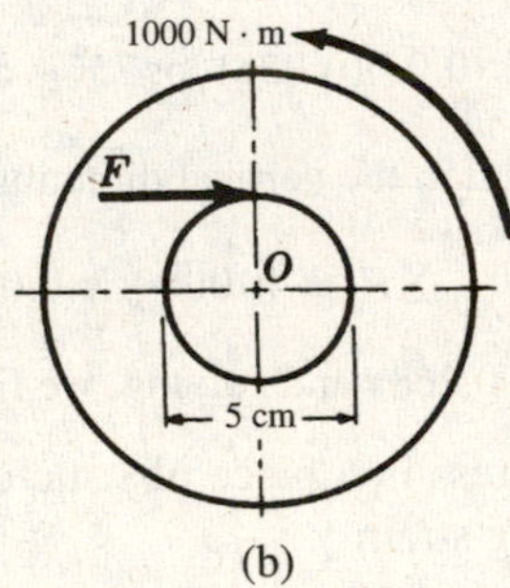

Fig. 2-8

SOLUTION: Drawing a free-body diagram of the pulley alone, as shown in Fig. 2-8(b), we see that the applied turning moment must be resisted by a horizontal tangential force F exerted on the pulley by the key. For equilibrium of moments about the center of the pulley we have

$$\Sigma M_0 = 1000 - F(0.025) = 0 \qquad \text{or} \qquad F = 40000 \text{ N}$$

It is to be noted that the shaft exerts additional forces, not shown, on the pulley. These act through the center O and do not enter the above moment equation. The resultant forces acting on the key appear as in Fig. 2-9(a). Actually the force F acting to the right is the resultant of distributed forces acting over the lower half of the left face. The other forces F shown likewise represent resultants of distributed force systems. The exact nature of the force distribution is not known.

The free-body diagram of the portion of the key below a horizontal plane a-a through its midsection is shown in Fig. 2-9(b). For equilibrium in the horizontal direction we have

$$\Sigma F_x = 40000 - \tau(0.01)(0.08) = 0 \qquad \text{or} \qquad \tau = 50 \times 10^6 \text{Pa} \qquad \text{or} \qquad 50 \text{ MPa}$$

This is the horizontal shear stress in the key.

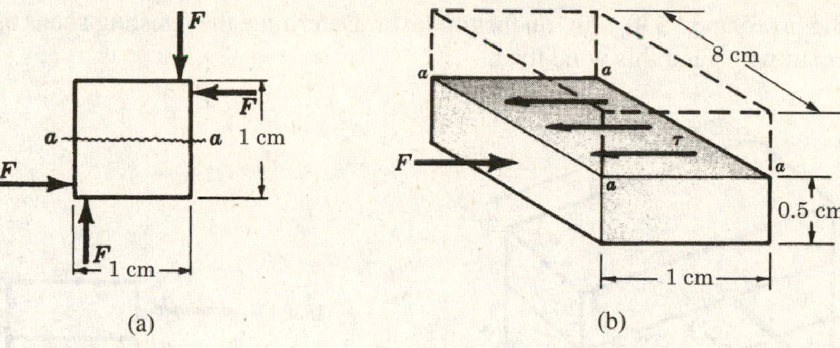

Fig. 2-9

2.8. A lifeboat on a seagoing cruise ship is supported at each end by a stranded steel cable passing over a pulley on a davit anchored to the top deck. The cable at each end carries a tension of 4000 N and the cable as well as the pulley are located in a vertical plane as shown in Fig. 2-10. The pulley may rotate freely about the horizontal circular axle indicated. Determine the diameter d of this axle if the allowable transverse shearing stress is 50 MPa.

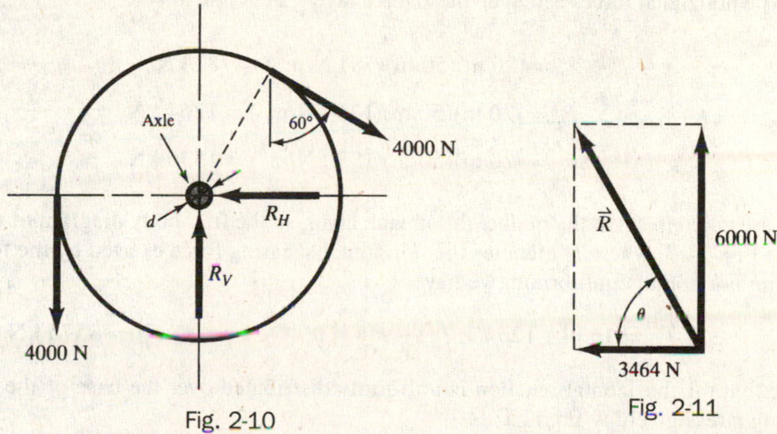

Fig. 2-10

Fig. 2-11

SOLUTION: The free-body diagram of the pulley shows not only the cable tensions but also the forces R_H and R_V exerted on the pulley by the circular axle. From statics we have

$$\Sigma F_H = -R_H + 4000\sin 60° = 0 \qquad \therefore\ R_H = 3464\ \text{N}(\leftarrow)$$

$$\Sigma F_V = R_V - 4000 - 4000\cos 60° = 0 \qquad \therefore\ R_V = 6000\ \text{N}(\uparrow)$$

The resultant of R_H and R_V is $R = \sqrt{(3464)^2 + (6000)^2} = 6930$ N oriented at an angle θ from the horizontal given by

$$\tan\theta = \frac{6000}{3464} \qquad \therefore\ \theta = 60°$$

The force exerted by the pulley upon the axle is equal and opposite to that shown in Fig. 2-11. If we assume that the resultant force of 6930 N is uniformly distributed over the cross section of the axle, the transverse shearing stress is

$$50 \times 10^6 = \frac{6930}{\pi d^2/4} \qquad \therefore\ d = 0.0133\ \text{m} \quad \text{or} \quad 13.3\ \text{mm}$$

2.9. A building that is 60 m tall has essentially the rectangular configuration shown in Fig. 2-12. Horizontal wind loads will act on the building exerting pressures on the vertical face that may be approximated as uniform within each of the three "layers" as shown. From empirical expressions for wind pressures at the midpoint of each of the three layers, we have a pressure of 781 N/m^2 on the lower layer, 1264 N/m^2

on the middle layer, and 1530 N/m^2 on the top layer. Determine the resisting shear that the foundation must develop to withstand this wind load.

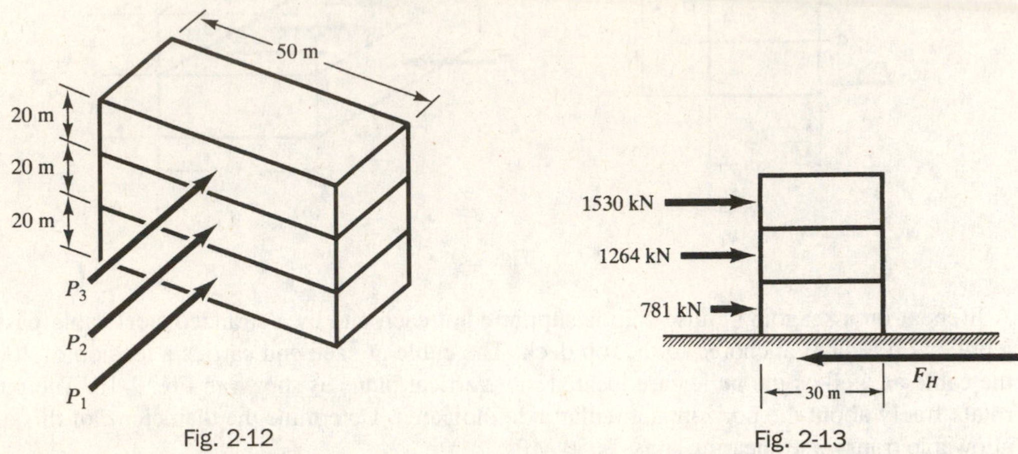

Fig. 2-12 Fig. 2-13

SOLUTION: The horizontal forces acting on these three layers are found to be

$$P_1 = (20\text{ m})(50\text{ m})(781\text{ N/m}^2) = 781\text{ kN}$$

$$P_2 = (20\text{ m})(50\text{ m})(1264\text{ N/m}^2) = 1264\text{ kN}$$

$$P_3 = (20\text{ m})(50\text{ m})(1530\text{ N/m}^2) = 1530\text{ kN}$$

These forces are taken to act at the midheight of each layer, so the free-body diagram of the building has the appearance of Fig. 2-13, where F_H denotes the horizontal shearing force exerted by the foundation upon the structure. From horizontal equilibrium, we have

$$\Sigma F_H = 1530 + 1264 + 781 - F_H = 0 \qquad \therefore\ F_H = 3575\text{ kN}$$

If we assume that this horizontal reaction is uniformly distributed over the base of the structure, the horizontal shearing stress given by Eq. (2.1) is

$$\tau = \frac{3575}{(30)(50)} = 2.38\text{ kN/m}^2$$

2.10. Two 1.5-mm-thick strips of titanium alloy 45 mm wide are joined by a 45° laser weld as shown in Fig. 2-14. A 100 kW carbon dioxide laser system is employed to form the joint. If the allowable shearing stress in the alloy is 440 MPa and the joint is assumed to be 100 percent efficient, determine the maximum allowable force P that may be applied.

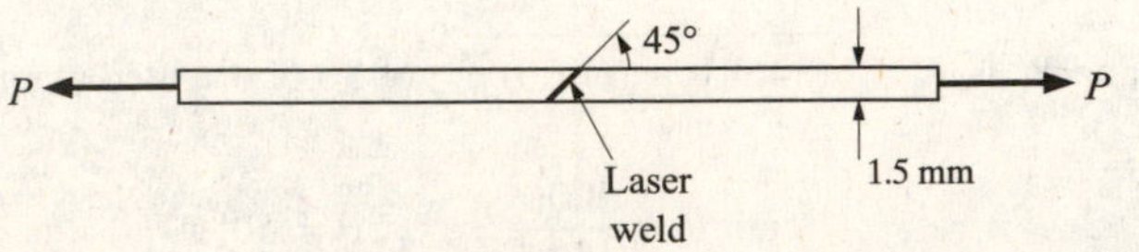

P 45° P
Laser weld
1.5 mm

Fig. 2-14

SOLUTION: A free-body diagram of the right strip has the form shown in Fig. 2-15. There, σ denotes normal stress in the weld on the 45° plane and τ the shearing stress. These are, of course, forces per unit area on the 45°

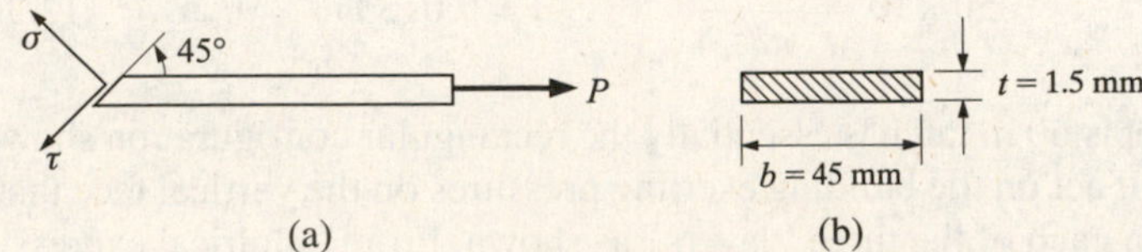

Fig. 2-15

plane and these must be multiplied by the area of the 45° plane which is $bt/\cos 45°$, where t denotes strip thickness and b the width. For horizontal equilibrium we have

$$\Sigma F_t = \tau\left(\frac{bt}{\cos 45°}\right) - P\cos 45° = 0 \qquad \therefore\ \tau = \frac{P\cos 45°}{bt}$$

$$440 \times 10^6 = \frac{P \times 0.707^2}{0.045 \times 0.0015} \qquad \therefore\ P = 59400\ \text{N}$$

SUPPLEMENTARY PROBLEMS

2.11. In Problem 2.1, if the maximum allowable working stress in shear is 93 MPa, determine the required diameter of the bolt in order that this value is not exceeded. *Ans.* $d = 2.03$ cm

2.12. A circular punch 20 mm in diameter is used to punch a hole through a steel plate 10 mm thick. If the force necessary to drive the punch through the metal is 250 kN, determine the maximum shearing stress developed in the material. *Ans.* $\tau = 400$ MPa

2.13. In structural practice, steel clip angles are commonly used to transfer loads from horizontal girders to vertical columns. If the reaction of the girder upon the angle is a downward force of 45 kN as shown in Fig. 2-16 and if two 22-mm-diameter rivets resist this force, find the average shearing stress in each of the rivets. As in Problem 2.5, assume that the rivet fills the hole. *Ans.* 59 MPa

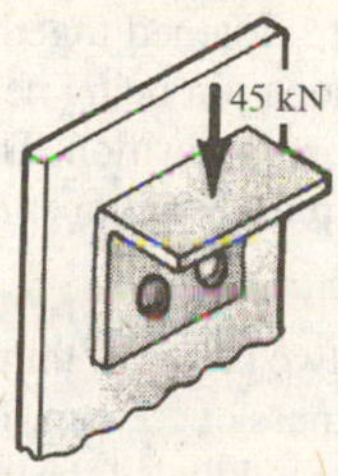

Fig. 2-16

2.14. A pulley is keyed to a 60-mm-diameter shaft. The unequal belt pulls, T_1 and T_2, on the two sides of the pulley give rise to a net turning moment of 120 N · m. The key is 10 mm × 15 mm in cross section and 75 mm long, as shown in Fig. 2-17. Determine the average shearing stress acting on a horizontal plane through the key. *Ans.* $\tau = 5.33$ MPa

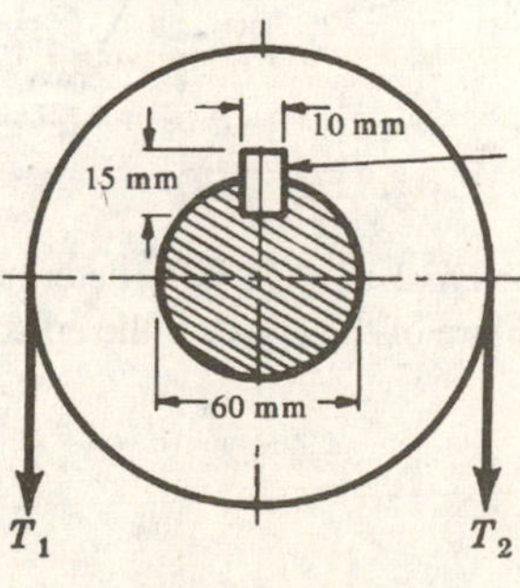

Fig. 2-17

2.15. Consider the balcony-type structure shown in Fig. 2-18. The horizontal balcony is loaded by a total load of 80 kN distributed in a radially symmetric fashion. The central support is a shaft 500 mm in diameter and the balcony is welded at both the upper and lower surfaces to this shaft by welds 10 mm on a side (or leg) as shown in the enlarged view. Determine the average shearing stress existing between the shaft and the weld. *Ans.* 2.5 MPa

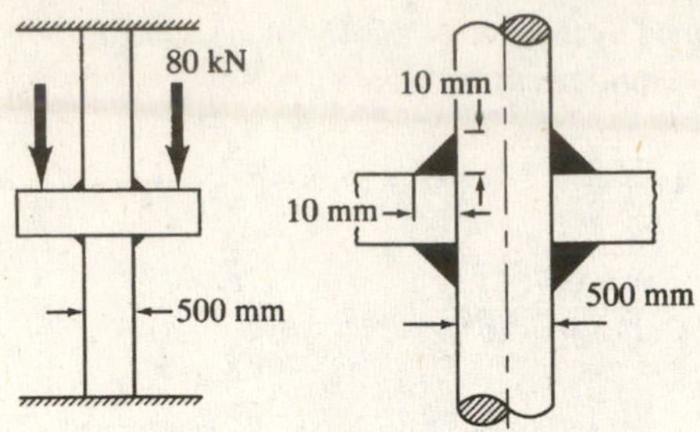

Fig. 2-18

2.16. Consider the two plates of equal thickness joined by two fillet welds as indicated in Fig. 2-19. Determine an expression for the maximum shearing stress in the welds. *Ans.* $\tau = 0.707\ P/ab$

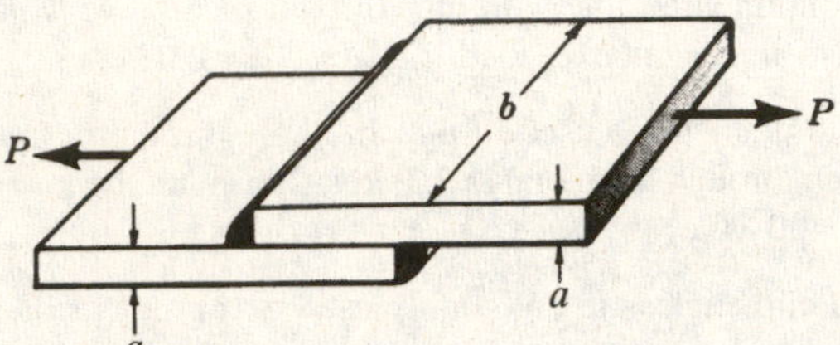

Fig. 2-19

2.17. A copper tube 55 mm in outside diameter and of wall thickness 5 mm fits loosely over a solid steel circular bar 40 mm in diameter. The two members are fastened together by two metal pins each 8 mm in diameter and passing transversely through both members, one pin being near each end of the assembly. At room temperature the assembly is just stress free when the pins are in position. The temperature of the entire assembly is then raised 40°C. Calculate the average shear stress in the pins. For copper $E = 90$ GPa, $\alpha = 18 \times 10^{-6}/°C$; for steel $E = 200$ GPa, $\alpha = 12 \times 10^{-6}/°C$. *Ans.* $\tau = 132$ MPa

2.18. In automotive as well as aircraft applications, two pieces of thin metal are often joined by a single lap shear joint, as shown in Fig. 2-20. Here, the metal has a thickness of 2.2 mm. The ultimate shearing strength of the epoxy adhesive joining the metals is 2.57×10^4 kPa, the shear modulus of the epoxy is 2.8 GPa, and the epoxy is effective over the 12.7×25.4-mm overlapping area. Determine the maximum axial load P the joint can carry. *Ans.* 8290 N

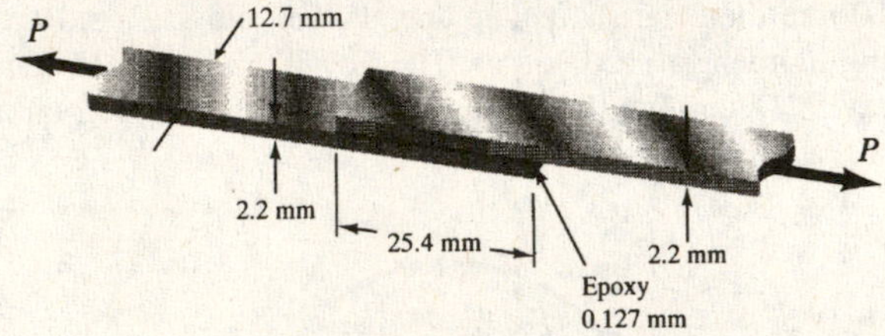

Fig. 2-20

2.19. If the shear modulus of the epoxy in Problem 2.18 is 2.8 GPa, determine the axial displacement of one piece of metal with respect to the other just prior to failure of the epoxy if the epoxy is 0.127 mm thick. *Ans.* 0.0017 mm

CHAPTER 3

Combined Stresses

3.1 Introduction

In Chapter 1, we have considered normal stresses arising in bars subject to axial loading as well as several cases involving shearing stresses in Chapter 2. It is to be noted that we have considered a bar, for example, to be subject to only *one* loading at a time. But frequently such bars are simultaneously subject to several kinds of loadings, and it is required to determine the state of stress under these conditions. Since normal and shearing stress are tensor quantities, considerable care must be exercised in combining the stresses given by the expressions for single loadings as derived in Chapters 1 and 2. It is the purpose of this chapter to investigate the state of stress on an arbitrary plane through an element in a body subject to both normal and shearing stresses.

3.2 General Case of Plane Stress

In general if a plane element is removed from a body it will be subject to the normal stresses σ_x and σ_y together with the shearing stress τ_{xy}, as shown in Fig. 3-1. We assume that no stresses act on the element in the z-direction.

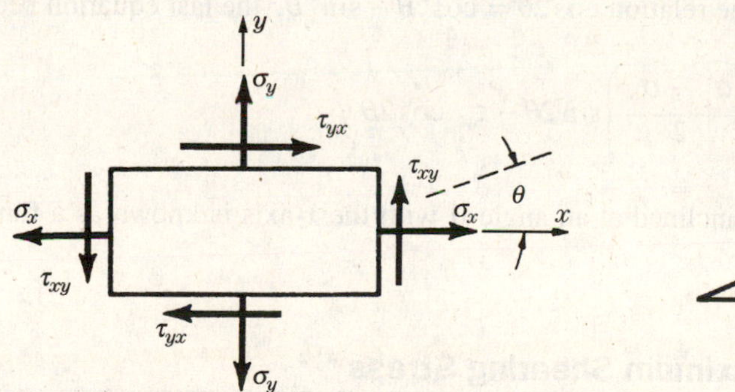

Fig. 3-1 Stresses on a plane element.

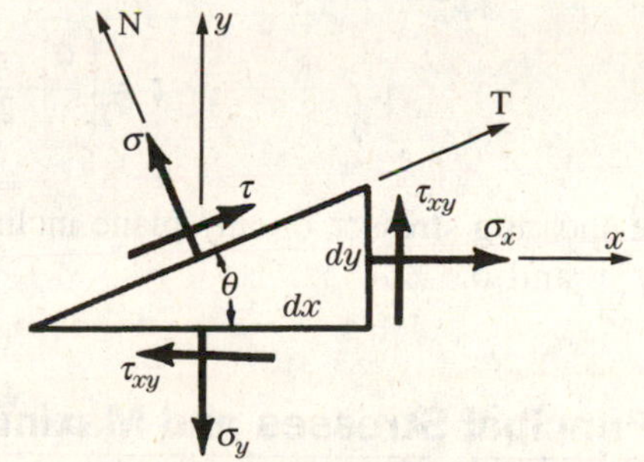

Fig. 3-2 Normal and shearing stresses on an inclined plane.

For normal stresses, tensile stress is considered to be positive, compressive stress negative. For shearing stresses, the positive sense* is that illustrated in Fig. 3-1. If moments are taken about the center of the face shown, we would see that $\tau_{xy} = \tau_{yx}$ since the normal stresses pass through the center.

We shall assume that the stresses σ_x, σ_y, and τ_{xy} are known. Frequently it is desirable to investigate the state of stress on a plane inclined at an angle θ, measured counterclockwise to the x-axis, as shown in Fig. 3-1. The normal and shearing stresses on such a plane are denoted by σ and τ and appear as in Fig. 3-2.

*A face is positive if its normal unit vector acts in a positive direction. It is a negative face if the unit vector acts in the negative direction. A positive normal stress or shear stress on a negative face acts in the negative direction.

We shall introduce the N- and T-axes normal and tangential to the inclined plane, as shown. Let t denote the thickness of the element perpendicular to the plane of the page. Let us begin by summing forces in the N-direction. For equilibrium we have

$$\Sigma F_N = \sigma t\, ds - \sigma_x t\, dy \sin\theta + \tau_{xy} t\, dy \cos\theta - \sigma_y t\, dx \cos\theta + \tau_{xy} t\, dx \sin\theta = 0$$

Substituting $dy = ds \sin\theta$, $dx = ds \cos\theta$ in the equilibrium equation,

$$\sigma\, ds = \sigma_x ds \sin^2\theta + \sigma_y ds \cos^2\theta - 2\tau_{xy} ds \sin\theta\cos\theta$$

Introducing the identities $\sin^2\theta = \frac{1}{2}(1 - \cos 2\theta)$, $\cos^2\theta = \frac{1}{2}(1 + \cos 2\theta)$, $\sin 2\theta = 2\sin\theta\cos\theta$, we find

or
$$\sigma = \left(\frac{\sigma_x + \sigma_y}{2}\right) - \left(\frac{\sigma_x - \sigma_y}{2}\right)\cos 2\theta - \tau_{xy} \sin 2\theta \qquad (3.1)$$

Thus the normal stress σ on any plane inclined at an angle θ with the x-axis is known as a function of σ_x, σ_y, τ_{xy}, and θ.

Next, summing forces acting on the element in the T-direction, we find

$$\Sigma F_T = \tau t\, ds - \sigma_x t\, dy \cos\theta - \tau_{xy} t\, dy \sin\theta + \tau_{xy} t\, dx \cos\theta + \sigma_y t\, dx \sin\theta = 0$$

Substituting for dx and dy as before, we get

$$\tau ds = \sigma_x ds \sin\theta\cos\theta + \tau_{xy} ds \sin^2\theta - \tau_{xy} ds \cos^2\theta - \sigma_y ds \sin\theta\cos\theta$$

Introducing the previous identities and the relation $\cos 2\theta = \cos^2\theta - \sin^2\theta$, the last equation becomes

$$\tau = \left(\frac{\sigma_x - \sigma_y}{2}\right)\sin 2\theta - \tau_{xy} \cos 2\theta \qquad (3.2)$$

Thus the shearing stress τ on any plane inclined at an angle θ with the x-axis is known as a function of $\sigma_x, \sigma_y, \tau_{xy}$, and θ.

3.3 Principal Stresses and Maximum Shearing Stress

To determine the maximum value that the normal stress σ may assume as the angle θ varies, we shall differentiate Eq. (3.1) with respect to θ and set this derivative equal to zero. Thus

$$\frac{d\sigma}{d\theta} = (\sigma_x - \sigma_y)\sin 2\theta - 2\tau_{xy} \cos 2\theta = 0 \qquad (3.3)$$

Hence the values of θ leading to maximum and minimum values of the normal stress are given by

$$\tan 2\theta_p = \frac{\tau_{xy}}{\frac{1}{2}(\sigma_x - \sigma_y)} \qquad (3.4)$$

The two planes defined by the angles θ_p are called *principal planes*. The normal stresses that exist on these planes are designated as *principal stresses*. They are the maximum and minimum values of the normal stress in the element under consideration.

The values of $\sin 2\theta_p$ and $\cos 2\theta_p$ may now be substituted in Eq. (3.1) to yield the maximum and minimum values of the normal stresses. From Eq. (3.4) we can write

$$\sin 2\theta_p = \frac{\pm\tau_{xy}}{\sqrt{[\frac{1}{2}(\sigma_x - \sigma_y)]^2 + (\tau_{xy})^2}} \qquad \cos 2\theta_p = \frac{\pm\frac{1}{2}(\sigma_x - \sigma_y)}{\sqrt{[\frac{1}{2}(\sigma_x - \sigma_y)]^2 + (\tau_{xy})^2}} \tag{3.5}$$

From Eq. (3.1) we obtain

$$\sigma = \left(\frac{\sigma_x + \sigma_y}{2}\right) \pm \sqrt{[\tfrac{1}{2}(\sigma_x - \sigma_y)]^2 + (\tau_{xy})^2} \tag{3.6}$$

The maximum normal stress is

$$\sigma_{\text{max}} = \left(\frac{\sigma_x + \sigma_y}{2}\right) + \sqrt{[\tfrac{1}{2}(\sigma_x - \sigma_y)]^2 + (\tau_{xy})^2} \tag{3.7}$$

The minimum normal stress is

$$\sigma_{\text{min}} = \left(\frac{\sigma_x + \sigma_y}{2}\right) - \sqrt{[\tfrac{1}{2}(\sigma_x - \sigma_y)]^2 + (\tau_{xy})^2} \tag{3.8}$$

These two principal stresses occur on the principal planes defined by Eq. (3.4). By substituting one of the values of θ_p from Eq. (3.4) into Eq. (3.1), one may determine which of the two principal stresses is acting on that plane. The other principal stress naturally acts on the other principal plane.

By substituting the values of the angle $2\theta_p$ as given by Eq. (3.4) into Eq. (3.2), it is seen that the shearing stresses τ on the principal planes are zero.

To determine the maximum value that the shearing stress τ may assume as the angle θ varies, we shall differentiate Eq. (3.2) with respect to θ and set this derivative equal to zero. Thus

$$\frac{d\tau}{d\theta} = (\sigma_x - \sigma_y)\cos 2\theta + 2\tau_{xy}\sin 2\theta = 0 \tag{3.9}$$

The values of θ leading to the maximum values of the shearing stress are thus

$$\tan 2\theta_s = -\frac{\frac{1}{2}(\sigma_x - \sigma_y)}{\tau_{xy}} \tag{3.10}$$

The planes defined by the two solutions to this equation are the planes of maximum shearing stress. Trigonometry provides

$$\sin 2\theta_s = \frac{\pm\frac{1}{2}(\sigma_x - \sigma_y)}{\sqrt{[\frac{1}{2}(\sigma_x - \sigma_y)]^2 + (\tau_{xy})^2}} \qquad \cos 2\theta_s = \frac{\pm\tau_{xy}}{\sqrt{[\frac{1}{2}(\sigma_x - \sigma_y)]^2 + (\tau_{xy})^2}}$$

Substituting these values in Eq. (3.2) we find

$$\tau_{\substack{\text{max}\\\text{min}}} = \pm\sqrt{[\tfrac{1}{2}(\sigma_x - \sigma_y)]^2 + (\tau_{xy})^2} \tag{3.11}$$

Here the positive sign represents the maximum shearing stress, the negative sign the minimum shearing stress.

If we compare Eqs. (3.4) and (3.10), it is evident that the angles $2\theta_p$ and $2\theta_s$ differ by 90°, since the tangents of these angles are the negative reciprocals of one another. Hence the planes defined by the angles

θ_p and θ_s differ by 45°; that is, the planes of maximum shearing stress are oriented 45° from the planes of maximum normal stress.

It is also of interest to determine the normal stresses on the planes of maximum shearing stress. These planes are defined by Eq. (3.10). If we now substitute the values of sin $2\theta_s$ and cos $2\theta_s$ in Eq. (3.1) for normal stress, we find

$$\sigma = \frac{1}{2}(\sigma_x + \sigma_y) \tag{3.12}$$

Thus on each of the planes of maximum shearing stress is a normal stress of magnitude $\frac{1}{2}(\sigma_x + \sigma_y)$.

3.4 Mohr's Circle

All the information contained in the above equations may be presented in a convenient graphical form known as *Mohr's circle*. In this representation normal stresses are plotted along the horizontal axis and shearing stresses along the vertical axis. The stresses σ_x, σ_y, and τ_{xy} are plotted and a circle is drawn through these points having its center on the horizontal axis. Figure 3-3 shows Mohr's circle for an element subject to the general case of plane stress, shown in Fig. 3-1. The ends of diameter *BCD* are the points (σ_y, τ_{xy}) at *B* and $(\sigma_x, -\tau_{xy})$ at *D*. The center *C* is at $[(\sigma_x + \sigma_y)/2, 0)]$. For applications see Problems 3.4, 3.5, 3.6, 3.7, 3.9, and 3.11.

Tensile stresses are considered to be positive and compressive stresses negative. Thus tensile stresses are plotted to the right of the origin in Fig. 3-3 and compressive stresses to the left.

When Mohr's circle has been drawn as in Fig. 3-3, the principal stresses are represented by the line segments *OG* and *OH*.

To determine the normal and shearing stresses on a plane whose normal vector is inclined at a counterclockwise angle θ with the *x*-axis, we measure a counterclockwise angle equal to 2θ from the diameter *BD* of Mohr's circle. The endpoints of this diameter *BD* represent the stress conditions in the original *x-y* directions; i.e., they represent the stresses σ_x, σ_y, and τ_{xy}. The angle 2θ corresponds to the diameter *EF*. The coordinates of point *F* represent the normal and shearing stresses on the plane at an angle θ to the *x*-axis. That is, the normal stress σ is represented by *ON* and the shearing stress is represented by *NF*.

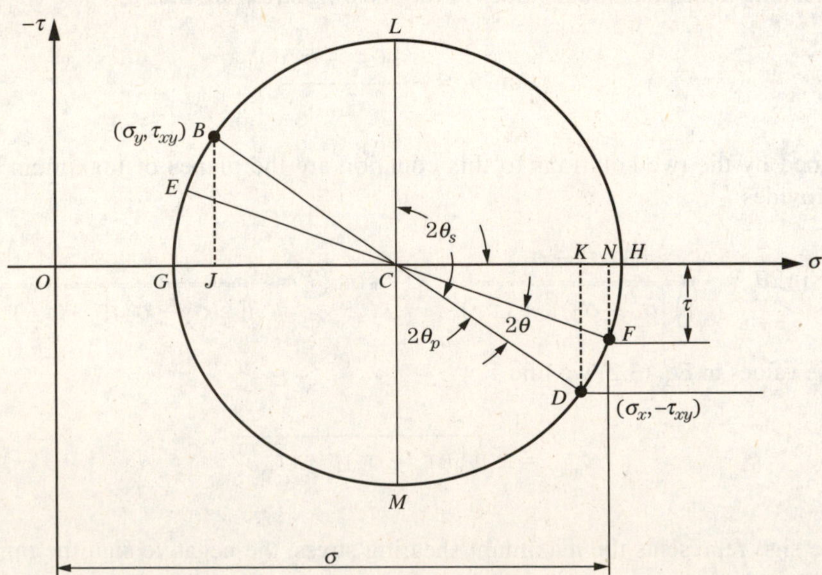

Fig. 3-3 Mohr's circle for plane stress (assume $\sigma_x > \sigma_y$).

SOLVED PROBLEMS

3.1. Let us consider a straight bar of uniform cross section loaded in axial tension. Determine the normal and shearing stress intensities on a plane inclined at an angle θ to the axis of the bar. Also, determine the magnitude and direction of the maximum shearing stress in the bar.

SOLUTION: This is the same elastic body that was considered in Chapter 1, but there the stresses studied were normal stresses in the direction of the axial force acting on the bar. In Fig. 3-4(a), P denotes the axial force acting on the bar, A the area of the cross section perpendicular to the axis of the bar; the normal stress σ_x is given by $\sigma_x = P/A$.

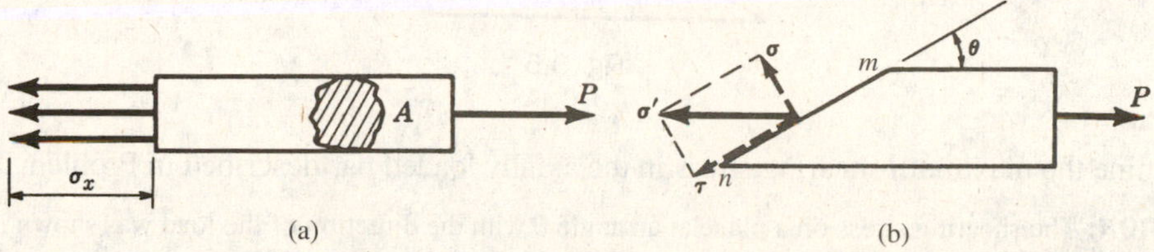

Fig. 3.4

Suppose now that instead of using a cutting plane which is perpendicular to the axis of the bar, we pass a plane through the bar at an angle θ with the axis of the bar. Such a plane mn is shown in Fig. 3-4(b).

In Fig. 3-4(b), we consider only a single stress vector σ' and resolve it into two components: a normal stress σ and a shearing stress τ.

Since the angle between σ' and τ is θ, we immediately have the relations

$$\tau = \sigma' \cos\theta \qquad \text{and} \qquad \sigma = \sigma' \sin\theta$$

But $P = \sigma' A/\sin\theta$. Substituting for σ' in the above equations, we obtain

$$\tau = \frac{P\sin\theta\cos\theta}{A} \qquad \text{and} \qquad \sigma = \frac{P\sin^2\theta}{A}$$

But $\sigma_x = P/A$. Hence we may write these in the form

$$\tau = \sigma_x \sin\theta\cos\theta \qquad \text{and} \qquad \sigma = \sigma_x \sin^2\theta$$

Now, employing the trigonometric identities

$$\sin 2\theta = 2\sin\theta\cos\theta \qquad \text{and} \qquad \sin^2\theta = \frac{1-\cos 2\theta}{2}$$

we may write

$$\tau = \frac{1}{2}\sigma_x \sin 2\theta \qquad \sigma = \frac{1}{2}\sigma_x(1-\cos 2\theta)$$

These expressions give the normal and shearing stresses on a plane inclined at an angle θ to the axis of the bar.

3.2. A bar of cross section 850 mm^2 is acted upon by axial tensile forces of 60 kN applied at each end of the bar. Determine the normal and shearing stresses on a plane inclined at 30° to the direction of loading.

SOLUTION: From Problem 3.1, the normal stress on a cross section perpendicular to the axis of the bar is

$$\sigma_x = \frac{P}{A} = \frac{60\times 10^3}{850\times 10^{-6}} = 70\times 10^6 \text{ Pa} \qquad \text{or} \qquad 70.6 \text{ MPa}$$

The normal stress on a plane at an angle θ with the direction of loading was found in Problem 3.1 to be $\sigma = \frac{1}{2}\sigma_x(1-\cos 2\theta)$. For $\theta = 30°$ this becomes

$$\sigma = \frac{1}{2}(70.6)(1-\cos 60°) = 17.65 \text{ MPa}$$

The shearing stress on a plane at an angle θ with the direction of loading was found in Problem 3.1 to be $\tau = \frac{1}{2}\sigma_x \sin 2\theta$. For $\theta = 30°$ this becomes

$$\tau = \frac{1}{2}(70.6)(\sin 60°) = 30.6 \text{ MPa}$$

These stresses together with the axial load of 60 kN are represented in Fig. 3-5.

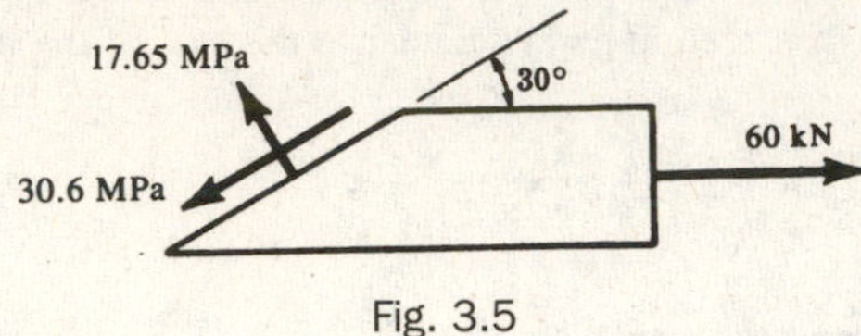

Fig. 3.5

3.3. Determine the maximum shearing stress in the axially loaded bar described in Problem 3.2.

SOLUTION: The shearing stress on a plane at an angle θ with the direction of the load was shown in Problem 3.1 to be $\tau = \frac{1}{2}\sigma_x \sin 2\theta$. This is maximum when $2\theta = 90°$, that is, when $\theta = 45°$. For this loading we have $\sigma_x = 70.6$ MPa and when $\theta = 45°$ the shear stress is

$$\tau = \frac{1}{2}(70.6)\sin 90° = 35.3 \text{ MPa}$$

That is, the maximum shearing stress is equal to one-half of the maximum normal stress.

The normal stress on this 45° plane may be found from the expression

$$\sigma = \frac{1}{2}\sigma_x(1 - \cos 2\theta) = \frac{1}{2}(70.6)(1 - \cos 90°) = 35.3 \text{ MPa}$$

3.4. Consider again the axially loaded bar discussed in Problem 3.2. Use Mohr's circle to determine the normal and shearing stresses on the 30° plane.

SOLUTION: In Fig. 3-6(*a*), the normal stress of 70.6 MPa is shown on an element with $\sigma_y = 0$ and $\tau_{xy} = 0$. The ends of diameter *OH* in Fig. 3-6(*b*) are at (70.6, 0) and (0, 0). The plane of interest is 60° clockwise from the vertical plane of the element. So, point *D* on Mohr's circle is 120° clockwise from point *H*, as shown. The coordinates of the point *D* are

$$\overline{KD} = \tau = -\frac{1}{2}(70.6)\sin 60° = -30.6 \text{ MPa}$$

$$\overline{OK} = \sigma = \overline{OC} - \overline{KC} = \frac{1}{2}(70.6) - \frac{1}{2}(70.6)\cos 60° = 17.65 \text{ MPa}$$

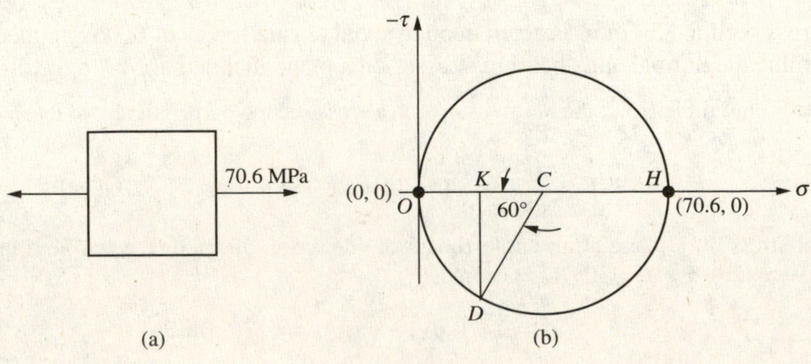

Fig. 3.6

3.5. A bar of cross section 10 cm^2 is acted upon by axial compressive forces of 60 kN applied to each end of the bar. Using Mohr's circle, find the normal and shearing stresses on a plane inclined at 30° to the direction of loading.

SOLUTION: The normal stress on a cross-section perpendicular to the axis of the bar is

$$\sigma_x = \frac{P}{A} = \frac{-60}{10 \times 10^{-4}} = -60\,000 \text{ kPa}$$

An element in Fig. 3-7(*a*) shows the 60 MPa compressive stress and zero shear stress. It is plotted as point *G* on Mohr's circle in Fig. 3-7(*b*). The origin *O* represents the face with zero normal stress. The desired area is 60° clockwise (or counterclockwise) from the area shown. Hence, the point of interest is 120° clockwise at *D*. There we have

$$\overline{KD} = \tau = \frac{1}{2}(60)\sin 60° = 52 \text{ MPa}$$

$$\overline{OK} = \sigma = -\frac{1}{2}(60) + \frac{1}{2}(60)\cos 60° = -15 \text{ MPa}$$

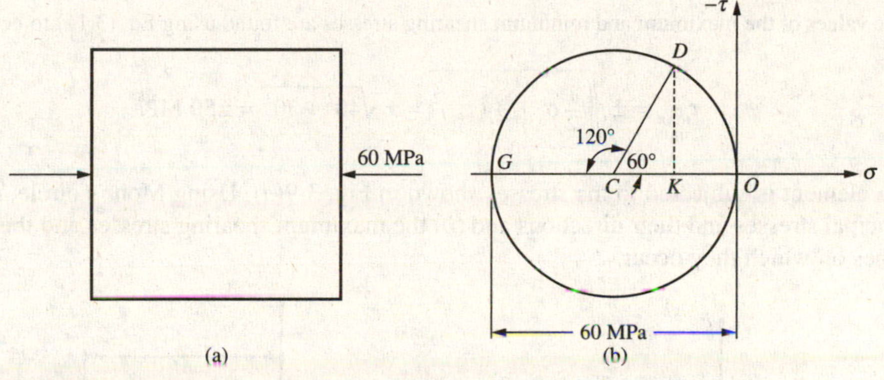

Fig. 3-7

Note: We could have rotated the top face of the element 30° counterclockwise, from *O* to *D*, and obtained the same result.

3.6. A plane element in a body is subjected to a normal stress in the *x*-direction of 80 MPa, as well as a shearing stress of 30 MPa, as shown in Fig. 3-8. (*a*) Determine the normal and shearing stress intensities on a plane inclined at an angle of 30° to the normal stress. (*b*) Determine the maximum and minimum values of the normal stress that may exist on inclined planes and the directions of these stresses. (*c*) Determine the magnitude of the maximum shearing stress that may exist on an inclined plane.

SOLUTION: (*a*) We have $\sigma_x = 80$ MPa and $\tau_{xy} = 30$ MPa. From Eq. (3.1), the normal stress on a plane inclined at 30° to the *x*-axis is

$$\sigma = \frac{1}{2}\sigma_x - \frac{1}{2}\sigma_x \cos 2\theta - \tau_{xy} \sin 2\theta$$
$$= \frac{1}{2}(80) - \frac{1}{2}(80)\cos 60° - 30\sin 60° = -6 \text{ MPa}$$

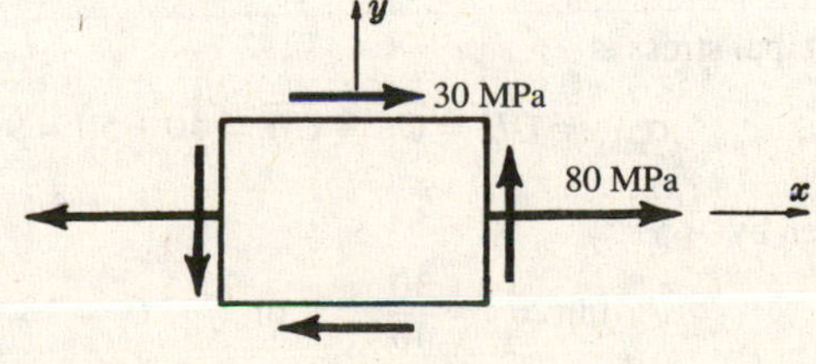

Fig. 3-8

From Eq. (3.2), the shearing stress on a plane inclined at 30° to the x-axis is

$$\tau = \frac{1}{2}\sigma_x \sin 2\theta - \tau_{xy} \cos 2\theta = \frac{1}{2}(80)\sin 60° - 30\cos 60° = 19.6 \text{ MPa}$$

(*b*) The values of the principal stresses, that is, the maximum and minimum values of the normal stresses existing in this element, were given by Eqs. (3.7) and (3.8). They provide

$$\sigma_{max} = \frac{1}{2}\sigma_x + \sqrt{\left(\frac{1}{2}\sigma_x\right)^2 + (\tau_{xy})^2} = 40 + \sqrt{40^2 + 30^2} = 90 \text{ MPa}$$

$$\sigma_{min} = \frac{1}{2}\sigma_x - \sqrt{\left(\frac{1}{2}\sigma_x\right)^2 + (\tau_{xy})^2} = 40 - \sqrt{40^2 + 30^2} = -10 \text{ MPa}$$

The directions of the planes on which these principal stresses occur are found using Eq. (3.4) to be

$$\tan 2\theta_p = \frac{\tau_{xy}}{\frac{1}{2}\sigma_x} = \frac{30}{40} \qquad \therefore 2\theta_p = 36.9°,\ 216.9°$$

Hence, $\theta_p = 18.4°$, 108°.

(*c*) The values of the maximum and minimum shearing stresses are found using Eq. (3.11) to be

$$\tau_{\substack{max \\ min}} = \pm\sqrt{\left(\frac{1}{2}\sigma_x\right)^2 + (\tau_{xy})^2} = \pm\sqrt{40^2 + 30^2} = \pm 50 \text{ MPa}$$

3.7. **A plane element is subjected to the stresses shown in Fig. 3-9(*a*). Using Mohr's circle, determine (*a*) the principal stresses and their directions and (*b*) the maximum shearing stresses and the directions of the planes on which they occur.**

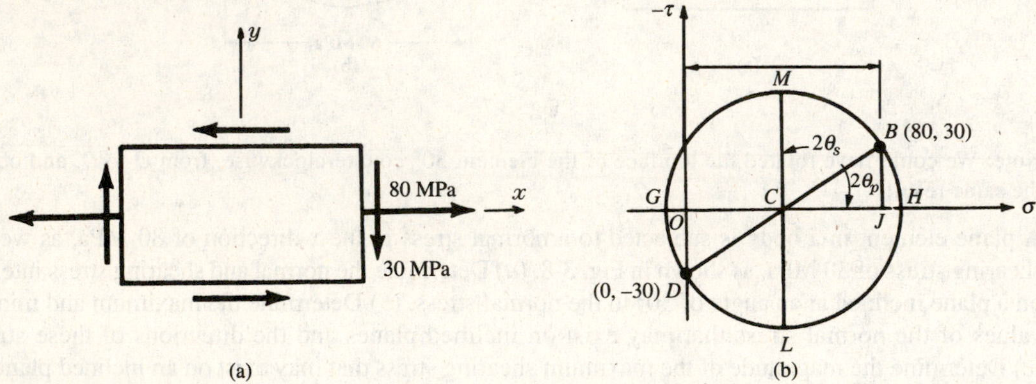

Fig. 3.9

SOLUTION: Mohr's circle is sketched in Fig. 3-9(*b*). The ends of diameter $\overline{BD}$ are at (80, 30) and (0,−30).

(*a*) The principal stresses are represented by points *G* and *H*. The principal stresses may be determined by realizing that the coordinate of *C* is 40, and that $\overline{CD} = \sqrt{40^2 + 30^2} = 50$. Therefore the minimum principal stress is

$$\sigma_{min} = \overline{OG} = \overline{OC} - \overline{CG} = 40 - 50 = -10 \text{ MPa}$$

The maximum principal stress is

$$\sigma_{max} = \overline{OH} = \overline{OC} + \overline{CH} = 40 + 50 = 90 \text{ MPa}$$

The angle $2\theta_p$ is given by

$$\tan 2\theta_p = \frac{30}{40} \qquad \text{or} \qquad \theta_p = 18.43°$$

From this it is readily seen that the principal stress represented by point H acts on a plane oriented 18.43° clockwise from the original plane. The principal stresses thus appear as in Fig. 3-10(*a*). It is evident from Mohr's

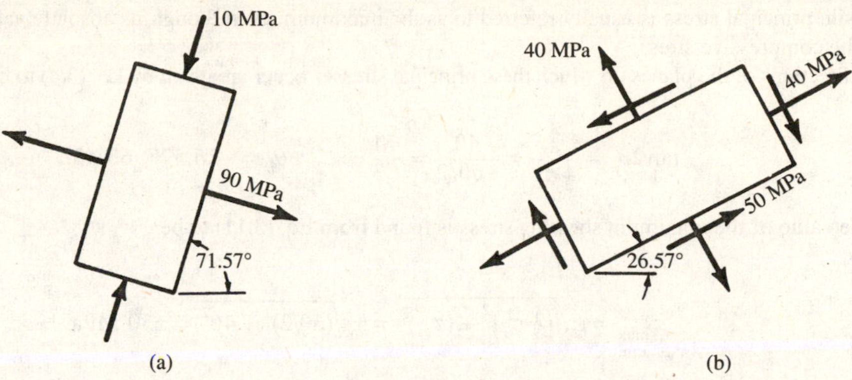

Fig. 3-10

circle that the shearing stresses on these planes are zero, since points G and H lie on the horizontal axis of Mohr's circle.

(*b*) The maximum shearing stress is represented by $\overline{CL}$ in Mohr's circle. This radius has already been found to be equal to 50 MPa. The angle $2\theta_s$ may be found by subtracting 90° from the angle $2\theta_p$, which has already been determined. This leads to $2\theta_s = 53.14°$ and $\theta_s = 26.57°$. Also, from Mohr's circle the abscissa of point L is 40 MPa and this represents the normal stress occurring on the planes of maximum shearing stress. The maximum shearing stresses thus appear as in Fig. 3-10 (*b*).

3.8. A plane element in a body is subject to a normal compressive stress in the x-direction of 60 MPa as well as a shearing stress of 40 MPa, as shown in Fig. 3-11. (*a*) Determine the normal and shearing stress intensities on a plane inclined at an angle of 30° to the normal stress. (*b*) Determine the maximum and minimum values of the normal stress that may exist on inclined planes and the direction of these stresses. (*c*) Find the magnitude and direction of the maximum shearing stress that may exist on an inclined plane. Use the equations.

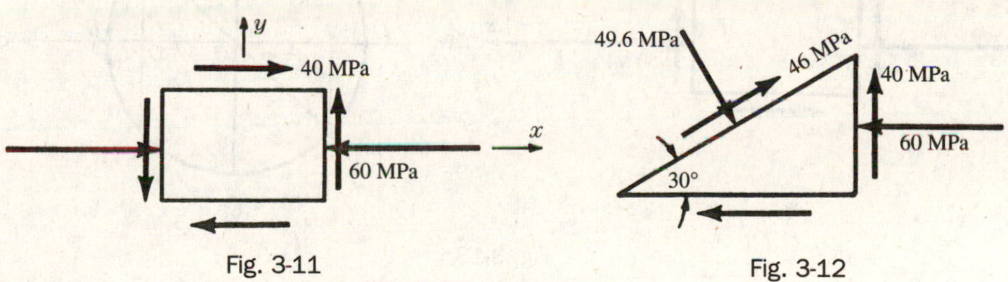

Fig. 3-11 Fig. 3-12

SOLUTION:

(*a*) By the sign convention for normal and shearing stresses, we have here $\sigma_x = -60$ MPa, $\tau_{xy} = 40$ MPa. From Eq. (3.1), the normal stress on the 30° plane is

$$\sigma = -\frac{60}{2} + \frac{60}{2}\cos 60° - 40\sin 60° = -49.6 \text{ MPa}$$

From Eq. (3.2), the shearing stress on the 30° plane is

$$\tau = -\frac{60}{2}\sin 60° - 40\cos 60° = -46 \text{ MPa}$$

The stresses on the 30° plane appear as in Fig. 3-12.

(*b*) The values of the principal stresses are given by Eqs. (3.7) and (3.8):

$$\sigma_{\max} = -\frac{60}{2} + \sqrt{30^2 + 40^2} = 20 \text{ MPa}$$

$$\sigma_{\min} = -\frac{60}{2} - \sqrt{30^2 + 40^2} = -80 \text{ MPa}$$

The tensile principal stress is usually referred to as the maximum, even though its absolute value is smaller than that of the compressive stress.

The directions of the planes on which these principal stresses occur are given by Eq. (3.4) to be

$$\tan 2\theta_p = \frac{\tau_{xy}}{\frac{1}{2}\sigma_x} = \frac{40}{-60/2} = -\frac{4}{3} \qquad \therefore \theta_p = -26.57°,\ 63.43°$$

(*c*) The value of the maximum shearing stress is found from Eq. (3.11) to be

$$\tau_{\substack{max\\min}} = \pm\sqrt{\left(\tfrac{1}{2}\sigma_x\right)^2 + (\tau_{xy})^2} = \pm\sqrt{(60/2)^2 + 40^2} = \pm 50 \text{ MPa}$$

The directions of the planes on which these shearing stresses occur are found using Eq. (3.10) to be

$$\tan 2\theta_s = \frac{-\frac{1}{2}\sigma_x}{\tau_{xy}} = \frac{60/2}{40} = \frac{3}{4} \qquad \therefore \theta_s = 18.43°,\ 108.4°$$

3.9. **A plane element is subject to the stresses shown in Fig. 3-13(*b*). Using Mohr's circle, determine (*a*) the principal stresses and their directions and (*b*) the maximum shearing stresses and the directions of the planes on which they occur.**

SOLUTION: Mohr's circle is sketched in Fig. 3-13(*b*). The ends of diameter $\overline{BD}$ are at (−60, −40) and (0, 40).

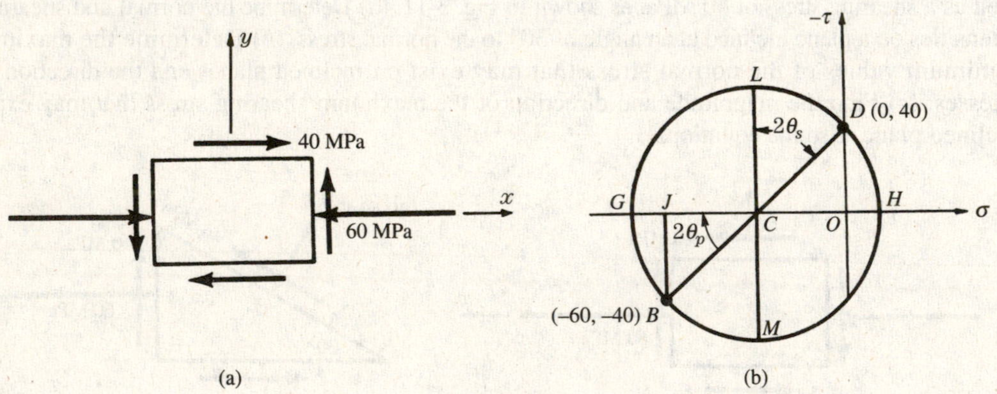

Fig. 3-13

(*a*) The principal stresses are represented by points *G* and *H*. They may be determined by realizing that the coordinate of *C* is –30, and that $\overline{CD} = \sqrt{30^2 + 40^2} = 50$. Thus the minimum principal stress is

$$\sigma_{min} = \overline{OG} = +(\overline{OC} + \overline{CG}) = -30 - 50 = -80 \text{ MPa}$$

The maximum principal stress is

$$\sigma_{max} = \overline{OH} = \overline{CH} - \overline{CO} = 50 - 30 = 20 \text{ MPa}$$

The angle $2\theta_p$ is given by

$$\tan 2\theta_p = \frac{40}{30} \qquad \therefore \theta_p = 26.6°$$

The principal stresses thus appear as in Fig. 3-14. It is evident from Mohr's circle Fig. 3-13(*b*) that the shearing stresses on these planes are zero, since points *G* and *H* lie on the horizontal axis of Mohr's circle.

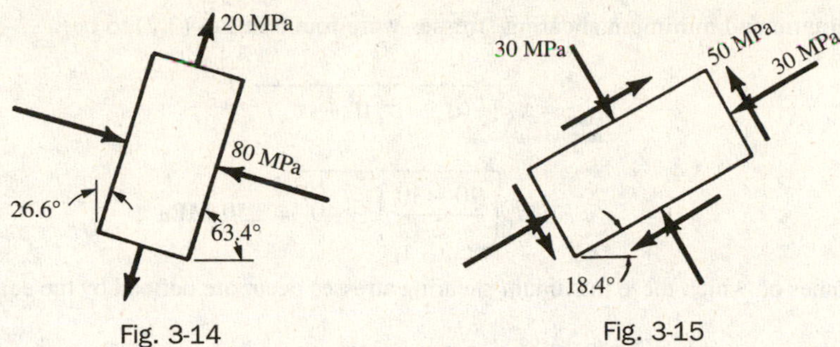

Fig. 3-14 Fig. 3-15

(*b*) The maximum shearing stress is represented by $\overline{CL}$ in Mohr's circle. This radius has already been found to be equal to 50 MPa. The angle $2\theta_s$ may be found by subtracting 90° from the above value of $2\theta_p$. This leads to θ_s = 18.4°. Also, from Mohr's circle the abscissa of point *L* is –30 MPa which represents the normal stress occurring on the planes of maximum shearing stresses, as shown in Fig. 3-15.

3.10. A plane element is subject to the stresses shown in Fig. 3-16. Determine (*a*) the principal stresses and their directions and (*b*) the maximum shearing stresses and the directions of the planes on which they occur. Use the equations.

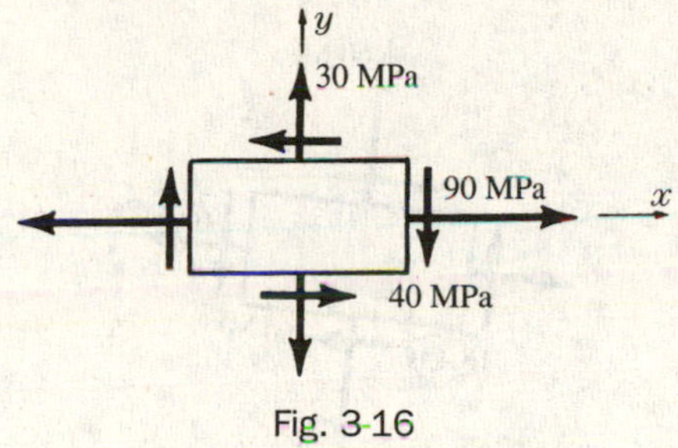

Fig. 3-16

SOLUTION:

(*a*) We have $\sigma_x = 90$ MPa, $\sigma_y = 30$ MPa, and $\tau_{xy} = 40$ MPa. The maximum normal stress, is given by Eq. (3.7) to be

$$\sigma_{\max} = \left(\frac{\sigma_x + \sigma_y}{2}\right) + \sqrt{[\tfrac{1}{2}(\sigma_x - \sigma_y)]^2 + (\tau_{xy})^2}$$

$$= \frac{90+30}{2} + \sqrt{\left(\frac{90-30}{2}\right)^2 + 40^2} = 110 \text{ MPa}$$

The minimum normal stress is given by Eq. (3.8) to be

$$\sigma_{\min} = \left(\frac{\sigma_x + \sigma_y}{2}\right) - \sqrt{[\tfrac{1}{2}(\sigma_x - \sigma_y)]^2 + (\tau_{xy})^2} = \frac{90+30}{2} - \sqrt{30^2 + 40^2} = 10 \text{ MPa}$$

The directions of the principal planes on which these stresses occur are given by

$$\tan 2\theta_p = \frac{\tau_{xy}}{\frac{1}{2}(\sigma_x - \sigma_y)} = \frac{-40}{\frac{1}{2}(90-30)} = -\frac{4}{3} \qquad \therefore \theta_p = -26.6°, 63.4°$$

The element displaying the principal stresses is sketched in Fig. 3-17.

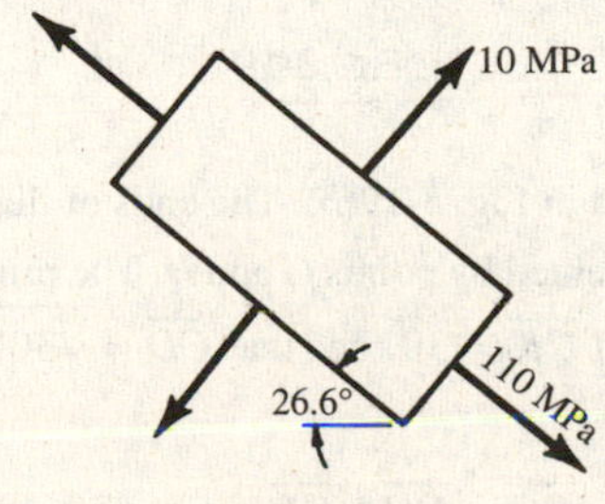

Fig. 3-17

(*b*) The maximum and minimum shearing stresses were found in Eq. (3.2) to be

$$\tau_{\substack{\max\\ \min}} = \pm\sqrt{[\tfrac{1}{2}(\sigma_x - \sigma_y)]^2 + (\tau_{xy})^2}$$

$$= \pm\sqrt{\left(\frac{90-30}{2}\right)^2 + 40^2} = \pm 50 \text{ MPa}$$

The planes on which these maximum shearing stresses occur are defined by the equation

$$\tan 2\theta_s = -\frac{\frac{1}{2}(\sigma_x - \sigma_y)}{\tau_{xy}} = -\frac{(90-30)/2}{-40} = \frac{3}{4} \qquad \therefore \theta_s = 18.4°,\ 108.4°$$

Finally, the normal stresses on these planes of maximum shearing stress are found from Eq. (3.12) to be

$$\sigma = \frac{1}{2}(\sigma_x + \sigma_y) = \frac{90+30}{2} = 60 \text{ MPa}$$

The orientation of the element for which the shearing stresses are maximum is shown in Fig. 3-18.

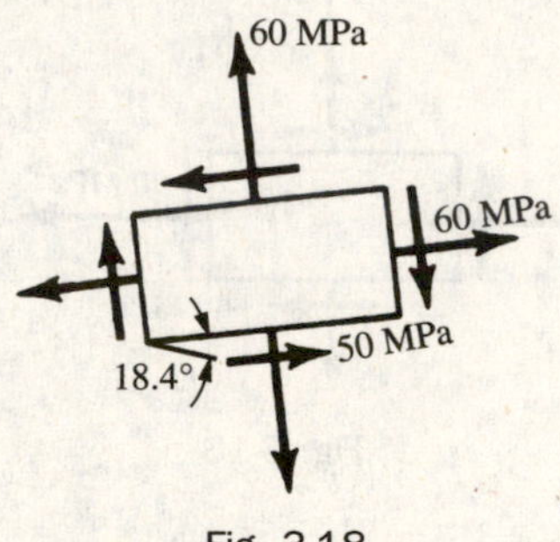

Fig. 3-18

3.11. A plane element is subject to the stresses shown in Fig. 3-19(*a*). Using Mohr's circle, determine (*a*) the principal stresses and their directions and (*b*) the maximum shearing stresses and the directions of the planes on which they occur.

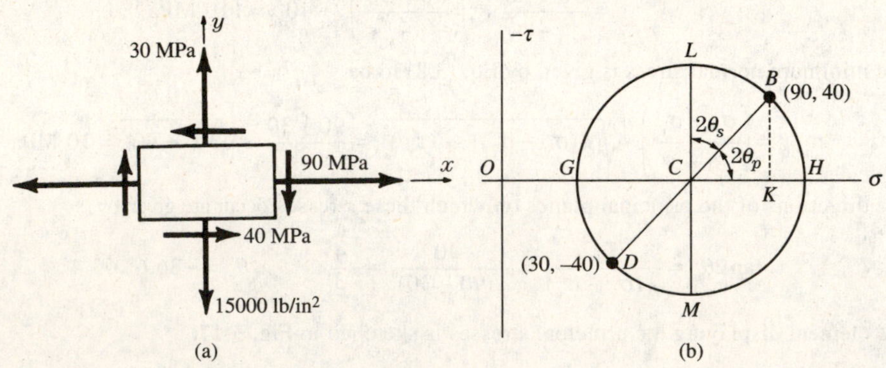

Fig. 3-19

SOLUTION: Mohr's circle is sketched in Fig. 3-19(*b*). The ends of diameter $\overline{BD}$ are at (90, 40) and (30, –40).

(*a*) The principal stresses are represented by points *G* and *H*. The principal stress may be determined by realizing that the coordinate of *C* is 60, that $\overline{CK} = 30$, and that $\overline{CD} = \sqrt{30^2 + 40^2} = 50$. Thus the minimum principal stress is

$$\sigma_{\min} = \overline{OG} = \overline{OC} - \overline{CG} = 60 - 50 = 10 \text{ MPa}$$

Also, the maximum principal stress is

$$\sigma_{max} = \overline{OH} = \overline{OC} + \overline{CH} = 60 + 50 = 110 \text{ MPa}$$

The angle $2\theta_p$ is given by

$$\tan 2\theta_p = \frac{40}{30} \qquad \therefore \theta_p = 26.6°$$

The principal stresses thus appear as in Fig. 3-17.

(*b*) The maximum shearing stress is represented by $\overline{CL}$ in Mohr's circle. This radius has already been found to represent 50 MPa. The angle $2\theta_s$ may be found by adding 90° to the angle $2\theta_p$, which leads to $\theta_s = 18.4°$.
From Mohr's circle the abscissa of point L is 60 MPa and this represents the normal stress occurring on the planes of maximum shearing stress. The maximum shearing stress thus appears as in Fig. 3-18.

SUPPLEMENTARY PROBLEMS

3.12. A bar of uniform cross section 50 mm × 75 mm is subject to an axial tensile force of 500 kN applied at each end of the bar. Determine the maximum shearing stress existing in the bar. *Ans.* 66.7 MPa

3.13. In Problem 3.12 determine the normal and shearing stresses acting on a plane inclined at 11° to the line of action of the axial loads. *Ans.* 4.87 MPa, 24.97 MPa

3.14. A square steel bar 2 cm on a side is subject to an axial compressive load of 40 kN. Determine the normal and shearing stresses acting on a plane inclined at 30° to the line of action of the axial loads. Use the equations. *Ans.* $\sigma = -25$ MPa, $\tau = -43.3$ MPa

3.15. Rework Problem 3.14 by use of Mohr's circle. *Ans.* $\sigma = -25$ MPa, $\tau = -43.3$ MPa

3.16. A plane element in a body is subject to the stresses $\sigma_x = 20$ MPa, $\sigma_y = 0$, and $\tau_{xy} = 30$ MPa. Determine analytically the normal and shearing stresses existing on a plane inclined at 45° to the x-axis. *Ans.* $\sigma = -25$ MPa, $\tau = -43.3$ MPa

3.17. A plane element is subject to the stresses $\sigma_x = 50$ MPa, $\sigma_y = 50$ MPa, and $\tau_{xy} = 0$. Determine analytically the maximum shearing stress existing in the element. *Ans.* 0

3.18. A plane element is subject to the stresses $\sigma_x = 60$ MPa, $\sigma_y = -60$ MPa, and $\tau_{xy} = 0$. Determine analytically the maximum shearing stress existing in the element. What is the direction of the planes on which the maximum shearing stresses occur? *Ans.* 60 MPa at 45°

3.19. For the element described in Problem 3.18 determine analytically the normal and shearing stresses acting on a plane inclined at 30° to the x-axis. *Ans.* $\sigma = -30$ MPa, $\tau = 52$ MPa

3.20. A plane element is subject to the stresses $\sigma_x = 50$ MPa and $\sigma_y = -50$ MPa. From Mohr's circle determine the stresses acting on a plane inclined at 20° to the x-axis. *Ans.* $\sigma = -38.3$ MPa, $\tau = -32.1$ MPa

3.21. A plane element removed from a thin-walled cylindrical shell loaded in torsion is subject to the shearing stresses shown in Fig. 3-20. Determine the principal stresses existing in this element and the directions of the planes on which they occur. *Ans.* 50 MPa at 45°

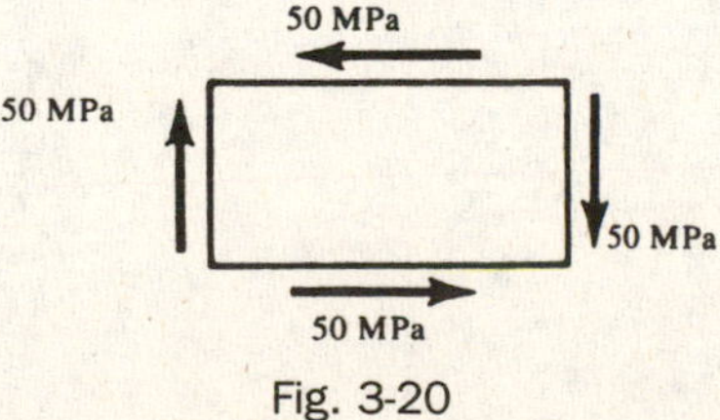

Fig. 3-20

3.22. A plane element is subject to the stresses shown in Fig. 3-21. Determine analytically (*a*) the principal stresses and their directions and (*b*) the minimum shearing stresses and the directions of the planes on which they act.

Ans. (*a*) $\sigma_{max} = 1.2$ MPa at 50.7°, $\sigma_{min} = -126.2$ MPa at 140.7°; (*b*) $\tau_{max} = 63.7$ MPa at 5.67°

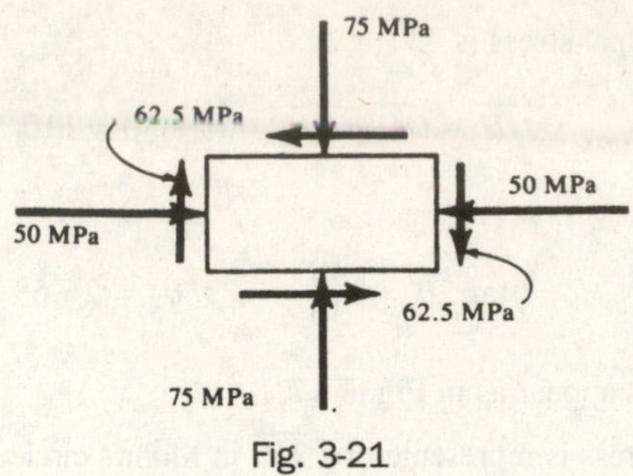

Fig. 3-21

3.23. Rework Problem 3.22 by the use of Mohr's circle. *Ans.* Same as for Problem 3.22.

3.24. A plane element is subject to the stresses indicated in Fig. 3.22. Determine principal stresses together with their orientation. *Ans.* 198.1 MPa @ 24.8°, 66.9 MPa

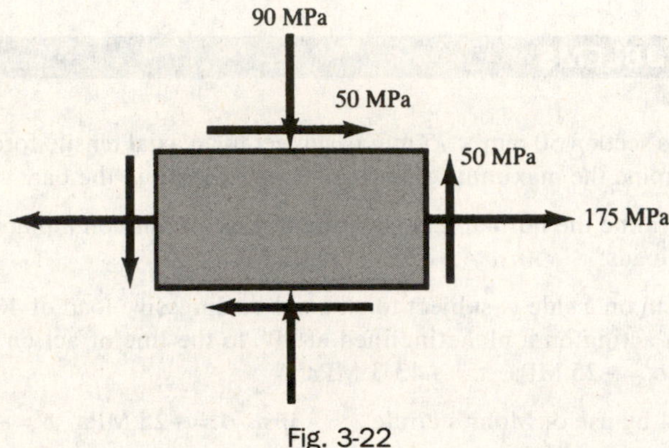

Fig. 3-22

3.25. A plane element is subject to the stresses indicated in Fig. 3-23. Determine principal stresses and the maximum shearing stress together with their orientation.

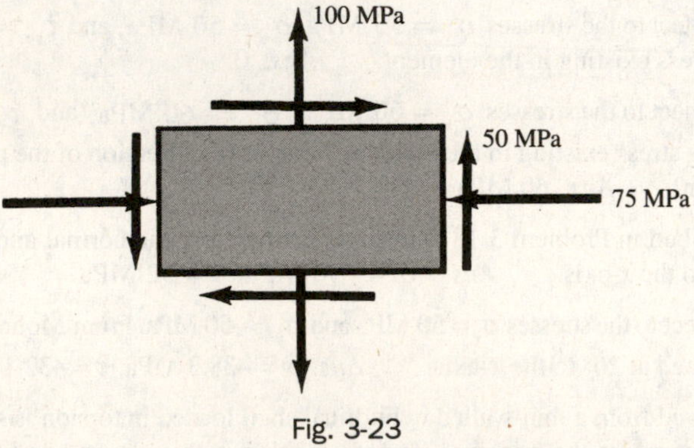

Fig. 3-23

Ans. $\sigma_{max} = 113.3$ MPa @ 75.1°, $\sigma_{min} = 88.3$ MPa, $\tau_{max} = 12.5$ MPa @ −59.9°

Thin-Walled Pressure Vessels

4.1 Introduction

In Chapter 1 we examined various cases involving uniform normal stresses acting in bars. Another application of uniformly distributed normal stresses occurs in the approximate analysis of thin-walled pressure vessels, such as cylindrical, spherical, conical, or toroidal shells subject to internal or external pressure from a gas or a liquid. In this chapter we will treat only thin shells of revolution and restrict ourselves to axisymmetric deformations of these shells.

Limitations

Liquid and gas storage tanks and containers, water pipes, boilers, submarine hulls, and certain air plane components are common examples of thin-walled pressure vessels.

Considering thin-walled vessels, the ratio of the wall thickness to the radius of the vessel should not exceed approximately 0.10. Also there must be no discontinuities in the structure. The simplified treatment presented here does not permit the consideration of reinforcing rings on a cylindrical shell, as shown in Fig. 4-1, nor does it give an accurate indication of the stresses and deformations in the vicinity of end closure plates on cylindrical pressure vessels. Even so, the treatment is satisfactory in many design problems.

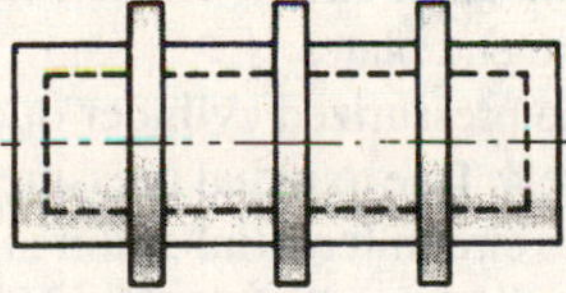

Fig. 4-1 Reinforcing rings on a cylinder.

We will be concerned with the stresses arising from a uniform internal pressure acting on a thin shell of revolution. The formulas for the various stresses will be correct if the sense of the pressure is reversed, i.e., if external pressure acts on the container. However, it is to be noted that an additional consideration, beyond the scope of this book, must then be taken into account; another study of an entirely different nature must be carried out to determine the load at which the shell will *buckle* due to the compression. A buckling or instability failure may take place even though the peak stress is far below the maximum allowable working stress of the material.

4.2 Cylindrical Pressure Vessels

Consider the thin-walled cylinder of Fig. 4-2 closed at both ends by cover plates and subject to a uniform internal pressure p. The wall thickness is h and the inner radius r. Neglecting the restraining effects of the end-plates, let us calculate the *axial* (*longitudinal, meridional*) and *circumferential* (*hoop, tangential*) normal stresses existing in the walls due to this loading. To determine the hoop stress σ_h let us consider a section of the cylinder of length L to be removed from the vessel. The free-body diagram of half of this section appears as in Fig. 4-2(a). Note that the body has been cut in such a way that the originally *internal* effect (σ_h) now appears as an *external* effect to this free body. Figure 4-2(b) shows the pressure and the hoop stress acting on a cross section.

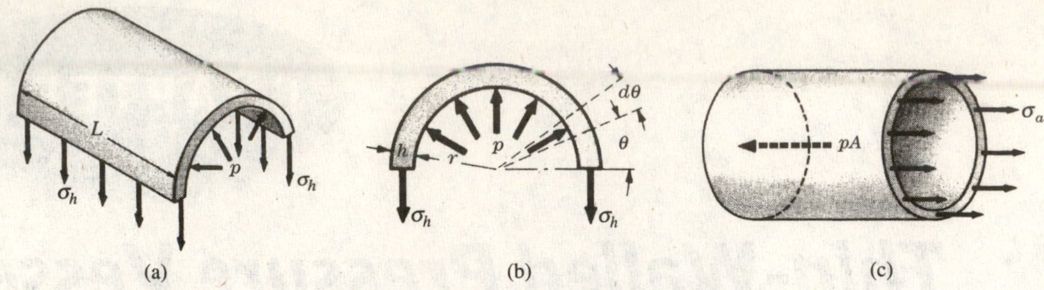

Fig. 4-2 A cylindrical pressure vessel.

The horizontal components of the radial pressures cancel one another by virtue of symmetry about the vertical centerline. In the vertical direction we have the equilibrium equation

$$\Sigma F_v = -2\sigma_h hL + \int_0^\pi pr(d\theta)L(\sin\theta) = 0 \tag{4.1}$$

Integrating,

$$2\sigma_h hL = -prL[\cos\theta]_0^\pi \qquad \text{or} \qquad \sigma_h = \frac{pr}{h} \tag{4.2}$$

Note that the resultant vertical force due to the pressure p could have been obtained by multiplying the pressure by the horizontal *projected area* $2rL$ upon which the pressure acts.

To determine the axial stress σ_a consider a section to be passed through the cylinder normal to its axis. The free-body diagram of the remaining portion of the cylinder is shown in Fig. 4-2(*c*). For equilibrium

$$\Sigma F_h = -p\pi r^2 + 2\pi rh\sigma_a = 0 \qquad \text{or} \qquad \sigma_a = \frac{pr}{2h} \tag{4.3}$$

Consequently, the circumferential stress is twice the axial stress. These rather simple expressions for stresses are not accurate in the immediate vicinity of the end plates.

Now, let us calculate the increase in the radius of the pressurized cylinder due to the internal pressure p. Consider the axial and circumferential loadings separately. Due to radial pressure p only, the circumferential stress is given by $\sigma_h = pr/h$, and because $\sigma = E\epsilon$ the circumferential strain is given by $\epsilon_h = pr/Eh$. The length over which ϵ_h acts is the circumference of the cylinder, which is $2\pi r$. Hence the total elongation of the circumference is

$$\Delta = \epsilon_h(2\pi r) = \frac{2\pi pr^2}{Eh} \tag{4.4}$$

The final length of the circumference is thus $2\pi r + 2\pi pr^2/Eh$. Dividing this circumference by 2π we find the radius of the deformed cylinder to be $r + pr^2/Eh$, so that the increase in radius is pr^2/Eh.

Due to the axial pressure p only, axial stresses $\sigma_a = pr/2h$ are set up. These axial stresses give rise to axial strains $\epsilon_a = pr/2Eh$. As in Chapter 1, an extension in the direction of loading, which is the axial direction here, is accompanied by a decrease in the dimension perpendicular to the load. Thus here the circumferential dimension decreases. The ratio of the strain in the lateral direction to that in the direction of loading was defined in Chapter 1 to be Poisson's ratio, denoted by ν. Consequently the above strain ϵ_a induces a circumferential strain equal to $-\nu\epsilon_a$ and if this strain is denoted ϵ'_h we have $\epsilon'_h = -\nu pr/2Eh$, which tends to decrease the radius of the cylinder, as shown by the negative sign.

In a manner exactly analogous to the treatment of the increase of radius due to radial loading only, the decrease of radius corresponding to the strain ϵ'_h is given by $\nu pr^2/2Eh$. The resultant increase of radius due to the internal pressure p is thus

$$\Delta r = \frac{pr^2}{Eh} - \frac{\nu pr^2}{2Eh} = \frac{pr^2}{Eh}\left(1 - \frac{\nu}{2}\right) \tag{4.5}$$

4.3 Spherical Pressure Vessel

Consider a closed thin-walled spherical shell subject to a uniform internal pressure p. The inside radius of the shell is r and its wall thickness is h. Let us derive an expression for the tensile stress existing in the wall. For a free-body diagram, let us consider exactly half of the entire sphere. This body is acted upon by the applied internal pressure p as well as the forces that the other half of the sphere, which has been removed, exerts upon the half under consideration. Because of the symmetry of loading and deformation, these forces may be represented by axial tensile stresses σ_a as shown in Fig. 4-3.

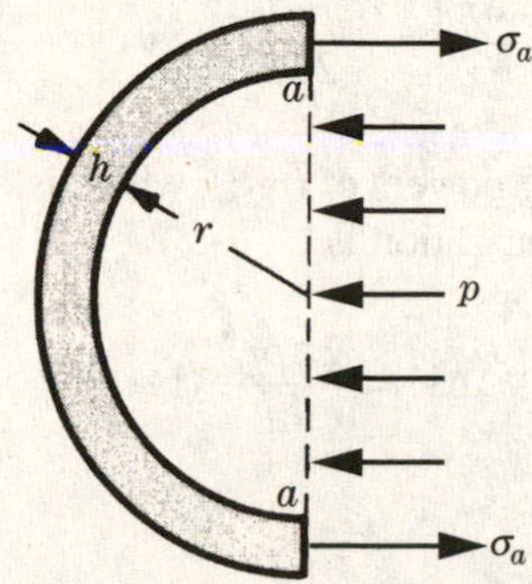

Fig. 4-3 A spherical pressure vessel.

This free-body diagram represents the forces acting on the hemisphere. Actually the pressure p acts over the entire inside surface of the hemisphere and in a direction perpendicular to the surface at every point. However, it is permissible to consider the force exerted by this same pressure p upon the projection of this area, which in this case is the vertical circular area denoted by *a-a*. This is possible because the hemisphere is symmetric about the horizontal axis and the vertical components of the pressure annul one another. Only the horizontal components produce the axial stress σ_a. For equilibrium we have

$$\Sigma F_x = \sigma_a 2\pi rh - p\pi r^2 = 0 \quad \text{or} \quad \sigma_a = \frac{pr}{2h} \tag{4.6}$$

From symmetry this axial stress is the same in all directions at any point in the wall of the sphere.

Let us find the increase in volume of a thin-walled spherical shell subject to a uniform internal pressure p. From Eq. (4.6) we know that the axial stress is assumed to be constant throughout the shell thickness and is given by

$$\sigma_a = \frac{pr}{2h} \tag{4.7}$$

in all directions at any point in the shell. From the two-dimensional form of Hooke's law (see Chap. 1), we have the axial strain as

$$\epsilon_a = \frac{1}{E}[\sigma_a - \nu\sigma_a] = \frac{pr}{2Eh}[1 - \nu] \tag{4.8}$$

So the increase in the circumference is the strain multiplied by the circumference:

$$\frac{pr}{2Eh}[1 - \nu] \times 2\pi r$$

The radius of the spherical shell subject to internal pressure p is now found by dividing the circumference of the pressurized shell by the factor 2π. Thus the final radius is

$$r + \Delta r = \left[2\pi r + (2\pi r) \times \frac{pr}{2Eh}(1 - \nu)\right] \Big/ 2\pi = \left[r + \frac{pr^2}{2Eh}(1 - \nu)\right] \tag{4.9}$$

and the volume of the pressurized sphere is

$$V + \Delta V = \frac{4}{3}\pi\left[r + \frac{pr^2}{2Eh}(1-\nu)\right]^3 \tag{4.10}$$

The desired increase of volume due to pressurization is found by subtracting the initial volume from Eq. (4.10):

$$\Delta V = \frac{4\pi}{3}\left[r + \frac{pr^2}{2Eh}(1-\nu)\right]^3 - \frac{4}{3}\pi r^3 \tag{4.11}$$

Expanding and dropping terms involving powers of (p/E), which is ordinarily of the order of 1/1000, we see that the increase of volume due to pressurization is

$$\Delta V = \frac{2\pi p r^4}{Eh}(1-\nu) \tag{4.12}$$

SOLVED PROBLEMS

4.1 A space simulator consists of a 9-m-diameter cylindrical vessel which is 28 m high. It is made of cold-rolled stainless steel having a proportional limit of 1.1 GPa. The minimum operating pressure of the chamber is 10^{-6} torr, where 1 torr = 1/760 of a standard atmosphere. Determine the required wall thickness so that a working stress based upon the proportional limit together with a safety factor of 2.5 will not be exceeded.

SOLUTION: The significant stress is the circumferential stress, given by Eq. (4.2). The pressure to be used for design is essentially the atmospheric pressure acting on the outside of the shell, which is 100 kPa since the internal pressure of 10^{-6} torr is negligible* compared to 100 kPa. We thus have

$$\frac{1.1\times 10^9}{2.5} = \frac{10^5 \times 4.5}{h} \quad \text{or} \quad h = 0.00102 \text{ m} \quad \text{or} \quad 1.02 \text{ mm}$$

4.2. A vertical axis circular cylindrical wine storage tank, fabricated from stainless steel, has a total height of 8 m, a radius of 150 cm, and is filled to a depth of 6 m with wine. An inert gas occupies the 2-m height H_0 above the gas-wine interface and is pressurized to a value of p_0 of 80 kPa, as shown in Fig. 4-4.

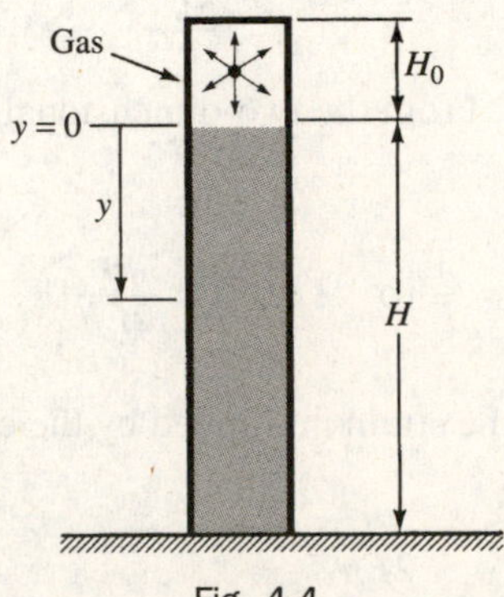

Fig. 4-4

If the working stress in the steel is 185 MPa, determine the required wall thickness. The specific weight of the wine is 9800 N/m^3.

*A pressure of 10^{-6} torr = $10^{-6} \times (100\,000/760) = 0.000132$ Pa.

SOLUTION: If there were no gas pressure above the surface of the wine, the pressure (in any direction) at any depth y below the liquid-free surface is given as $p = \gamma y$, where γ is the specific weight of the wine.

The total pressure at the base ($y = H$) is thus $(p_0 + \gamma H)$ so that from Eq. (4.2), the hoop stress is

$$\sigma_h = \frac{(p_0 + \gamma H)r}{h}$$

where h is the tank wall thickness.

For vertical equilibrium the upward thrust of the gas pressure p_0 must be balanced by axial stresses σ_a distributed uniformly around the tank wall at the tank bottom, as shown in Fig. 4-5. Thus

$$\Sigma F_y = \sigma_a(2\pi r)h - p_0 \pi r^2 = 0 \qquad \therefore \sigma_a = \frac{p_0 r}{2h}$$

The hoop stress σ_h is clearly larger than the axial stress σ_a and thus controls design. We have

$$\frac{(80000 + 9800 \times 6)\text{Pa} \times 1.5 \text{ m}}{h} = 185 \times 10^6 \text{Pa}$$

from which the thickness is found to be $h = 0.0011$ m or 1.1 mm.

Fig. 4-5

4.3. Consider a laminated pressure vessel composed of two thin coaxial cylinders as shown in Fig. 4-6. In the state prior to assembly there is a slight "interference" between these cylinders, i.e., the inner one is too large to slide into the outer one. The outer cylinder is heated, placed on the inner, and allowed to cool, thus providing a "shrink fit." If both the cylinders are of steel and the mean diameter of the assembly is 100 mm, find the tangential stresses in each shell arising from the shrinking if the initial interference (of diameters) is 0.25 mm. The thickness of the inner shell is 2.5 mm, and that of the outer shell 2 mm. Take $E = 200$ GPa.

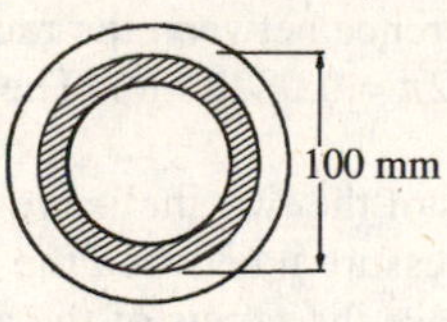

Laminated pressure vessel

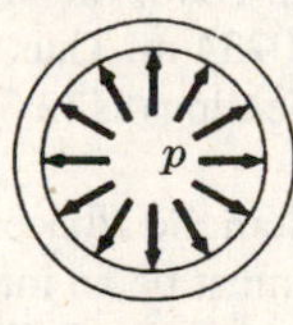

Outer cylinder

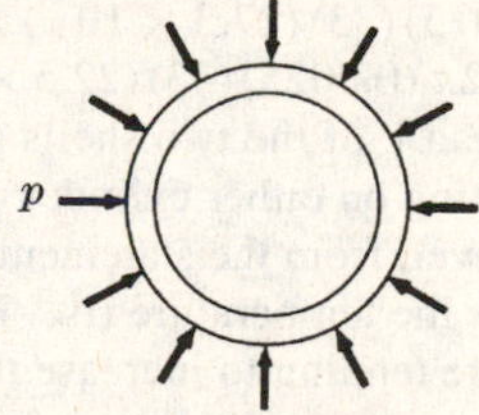

Inner cylinder

Fig. 4-6

SOLUTION: There is evidently an interfacial pressure p acting between the two shells. It is to be noted that there are no external applied loads. The pressure p increases the diameter of the outer shell and decreases the diameter of the inner so that the inner shell just fits inside the outer. The radial expansion of a cylinder due to pressure p was found using Eq. (4.4) to be pr^2/Eh. No axial forces are acting in this problem. The increase in radius of the outer shell due to p, plus the decrease in radius of the inner one due to p, must equal the initial interference between radii, or 0.25/2 mm. Thus we have

$$\frac{p(0.05)^2}{(200\times10^9)(0.0025)}+\frac{p(0.05)^2}{(200\times10^9)(0.002)}=0.000125\text{ m}\qquad\therefore p=11.1\times10^6\text{ Pa}\quad\text{or}\quad p=11.1\text{ MPa}$$

This pressure, illustrated in the above figures, acts between the cylinders after the outer one has been shrunk onto the inner one. In the inner cylinder this pressure p gives rise to a stress

$$\sigma_h=\frac{pr}{h}=\frac{(11.1\times10^6)(0.05)}{(0.0025)}=-222\times10^6\text{ Pa}\qquad\text{or}\qquad -222\text{ MPa}$$

In the outer cylinder the circumferential stress due to the pressure p is

$$\sigma_h=\frac{pr}{h}=\frac{(11.1\times10^6)(0.05)}{(0.002)}=277\times10^6\text{ Pa}\qquad\text{or}\qquad 277\text{ MPa}$$

If, for example, the laminated shell is subject to a uniform internal pressure, these shrink-fit stresses would merely be added to the stresses found by the use of the formula given by Eq. (4.2).

4.4. The thin steel cylinder just fits over the inner aluminum cylinder as shown in Fig. 4-7. Find the tangential stresses in each shell due to a temperature rise of 33°C. Do not consider the effects introduced by the accompanying axial expansion. (This arrangement is sometimes used for storing corrosive fluids.) Use

$$E_{st}=200\text{ GPa}\qquad \alpha_{st}=17.3\times10^{-6}/°\text{C}$$
$$E_{al}=72\text{ GPa}\qquad \alpha_{al}=22.5\times10^{-6}/°\text{C}$$

SOLUTION: The simplest approach is to first consider the two shells to be separated from one another so that they are no longer in contact.

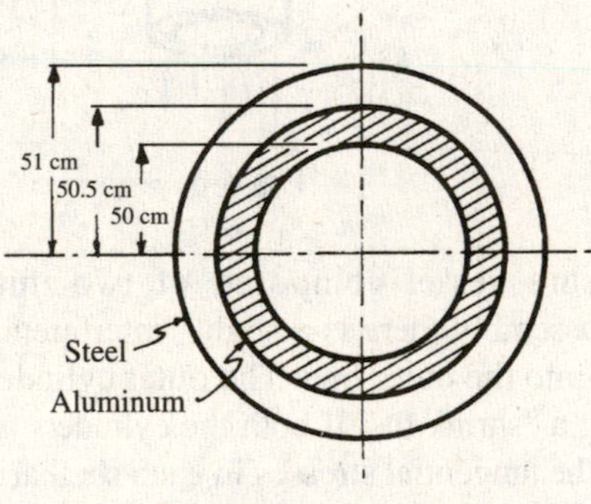

Fig. 4-7

Due to the temperature rise of 33°C, the circumference of the steel shell increases by an amount 2π (0.5075) (33) (17.3 × 10^{-6}) = 0.00182 m. Also, the circumference of the aluminum shell increases an amount 2π (0.5025) (33) (22.5 × 10^{-6}) = 0.00234 m. Thus the interference between the radii, i.e., the difference in radii, of the two shells (due to the heating) is (2.34 – 1.82)/2π = 0.0828 mm. There are no external loads acting on either cylinder.

However, from the statement of the problem the adjacent surfaces of the two shells are obviously in contact after the temperature rise. Hence there must be an interfacial pressure p between the two surfaces, i.e., a pressure tending to increase the radius of the steel shell and decrease the radius of the aluminum shell so that the aluminum shell may fit inside the steel one. Such a pressure is shown in the free-body diagrams of Fig. 4-8.

Using Eq. (4.4), the change of radius of a cylinder due to a uniform radial pressure p (with no axial forces acting) was found to be pr^2/Eh. Consequently, the increase of radius of the steel shell due to p, added to the decrease of radius of the aluminum one due to p, must equal the interference; thus

$$\frac{p(0.5075)^2}{(200\times10^9)(0.005)}+\frac{p(0.5025)^2}{(72\times10^9)(0.005)}=0.0828\times10^{-3}\qquad\text{or}\qquad p=86\,300\text{ Pa}$$

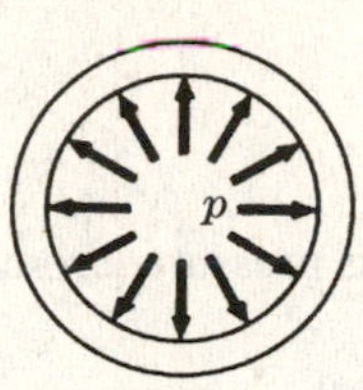

Steel cylinder

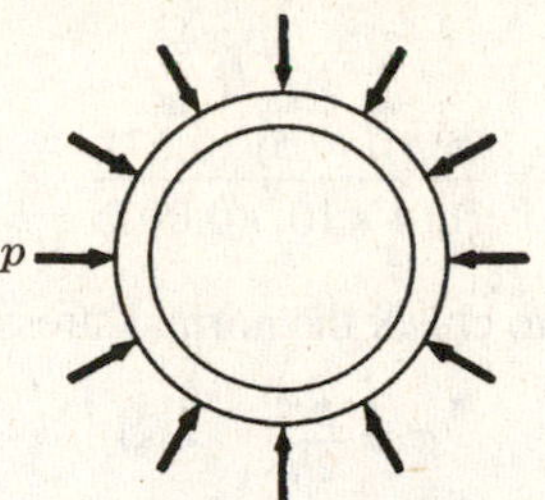

Aluminum cylinder

Fig. 4-8

This interfacial pressure creates the required continuity at the common surface of the two shells when they are in contact. Using the formula for the tangential stress, $\sigma_h = pr/h$, we find the tangential stresses in the steel and aluminum shells to be, respectively,

$$\sigma_{st} = \frac{86\,300(0.5075)}{0.005} = 8.76 \times 10^6 \text{ Pa} \qquad \text{and} \qquad \sigma_{al} = \frac{86\,300(0.5025)}{0.005} = -8.67 \times 10^6 \text{ Pa}$$

4.5. A 20-m-diameter spherical tank is to be used to store gas. The shell plating is 10 mm thick and the working stress of the material is 125 MPa. What is the maximum permissible gas pressure p?

SOLUTION: From Eq. (4.6), the tensile stress in all directions is uniform and given by $\sigma_a = pr/2h$. Substituting:

$$p = \frac{2h\sigma_a}{r}$$

$$= \frac{2 \times 0.01 \times 125 \times 10^6}{10} = 2.5 \times 10^5 \text{ Pa} \qquad \text{or} \qquad 250 \text{ kPa}$$

4.6. The undersea research vehicle *Alvin* has a spherical pressure hull 1 m in radius and shell thickness of 30 mm. The pressure hull is steel having a yield point of 700 MPa. Determine the depth of submergence that would set up the yield point stress in the spherical shell. Consider sea water to have a constant specific weight of 10 kN/m^3. Use $p = \gamma y$ where y is the depth.

SOLUTION: From Eq. (4.6), the compressive stress due to the external hydrostatic pressure is given by $\sigma_a = pr/2y$. The hydrostatic pressure corresponding to yield is thus

$$p = \frac{2y\sigma_a}{r}$$

$$= \frac{2 \times 0.03 \times 700 \times 10^6}{1.0} = 42 \times 10^6 \text{ Pa} \qquad \therefore p = 42$$

Since $p = \gamma y$, where γ is the specific weight of the sea water, we have

$$42 \times 10^6 = 10\,000 \text{ h} \qquad \therefore y = 4200 \text{ m}$$

It should be noted that this neglects the possibility of buckling of the sphere due to hydrostatic pressure as well as effects of entrance ports on its strength. These factors, beyond the scope of this treatment, result in a true operating depth of 1650 m.

4.7. A thin-walled titanium alloy spherical shell has a 1-m inside diameter and is 7 mm thick. It is completely filled with an unpressurized, incompressible liquid. Through a small hole an additional 1000 cm^3 of the same liquid is pumped into the shell, thus increasing the shell radius. Find the pressure after the additional liquid has been introduced and the hole closed. For this titanium allow $E = 114$ GPa and the tensile yield point of the material to be 830 MPa.

SOLUTION: The initial volume of the spherical shell is

$$V = \frac{4}{3}\pi r^3 = \frac{4}{3}\pi(0.5)^3 = 0.5236 \text{ m}^3$$

The volume of liquid pumped in is $1000 \times 10^{-6} = 0.001$ m^3, so that the final volume of the incompressible liquid is 0.5236 m^3 + 0.001 m^3 = 0.5246 m^3, which is equal to the volume of the expanded shell. The relation between pressure and volume change was found in Eq. (4.12) to be

$$\Delta V = \frac{2\pi p r^4}{Eh}(1 - \nu)$$

Substituting,

$$0.001 = \frac{(2\pi)p(0.5)^4(0.67)}{(114\times 10^9)(0.007)} \qquad \therefore\ p = 3.03\times 10^6 \text{ Pa} \qquad \text{or} \qquad 3.03 \text{ MPa}$$

It is well to check the normal stress in the titanium shell due to this pressure. From Eq. (4.6) we have

$$\sigma = \frac{pr}{2h}$$

$$= \frac{(3.03\times 10^6)(0.05)}{2(0.007)} = 109\times 10^6 \text{ Pa} \qquad \text{or} \qquad 109 \text{ MPa}$$

which is well below the yield point of the material.

SUPPLEMENTARY PROBLEMS

4.8. One proposed design for an energy-efficient automobile involves an on-board tank storing hydrogen which would be released to a fuel cell. The tank is to be cylindrical, 0.4 m in diameter, made of type 302 stainless steel having a working stress in tension of 290 MPa, and closed by hemispherical end caps. The hydrogen would be pressurized to 15 MPa, when the tank is initially filled. Determine the required wall thickness of the tank. *Ans.* $h = 5.2$ mm

4.9. A vertical axis circular cylindrical water storage tank of cross-sectional area A is filled to a depth of 15 m. The tank is 4 m in radius and is made of steel having a yield point of 240 MPa. If a safety factor of 2 is applied, determine the required tank wall thickness. The pressure in the water is γy where y is measured from the free surface. Use $\gamma_{water} = 9800$ N/m^3. *Ans.* $h = 4.9$ mm

4.10. A vertical cylindrical gasoline storage tank is 30 m in diameter and is filled to a depth of 15 m with gasoline whose specific weight is 7260 N/m^3. If the yield point of the shell plating is 250 MPa and a safety factor of 2.5 is adequate, calculate the required wall thickness at the bottom of the tank. *Ans.* $h = 16.7$ mm

4.11. The research deep submersible *Aluminaut* has a cylindrical pressure hull of outside diameter 2.6 m and a wall thickness of 20 cm. It is constructed of 7079-T6 aluminum alloy, having an yield point of 400 MPa. Determine the hoop stress in the cylindrical portion of the pressure hull when the vehicle is at its operating depth of 5000 m below the surface of the sea. Use the mean diameter of the shell in calculations, and consider sea water to weigh 10 kN/m^3. *Ans.* 300 MPa

4.12. Derive an expression for the increase of volume per unit volume of a thin-walled circular cylinder subjected to a uniform internal pressure p. The ends of the cylinder are closed by circular plates. Assume that the radial expansion is constant along the length.

Ans. $\dfrac{\Delta V}{V} = \dfrac{pr}{Eh}\left(\dfrac{5}{2} - 2\nu\right)$

4.13. Calculate the increase in volume per unit volume of a thin-walled steel circular cylinder closed at both ends and subjected to a uniform internal pressure of 0.5 MPa. The wall thickness is 1.5 mm, the radius 350 mm, and $\nu = \frac{1}{3}$. Consider $E = 200$ GPa. *Ans.* $\Delta V/V = 10^{-3}$

4.14. Consider a laminated cylinder consisting of a thin steel shell "shrunk" on an aluminum one. The thickness of each is 2.5 mm and the mean diameter of the assembly is 10 cm. The initial "interference" of the shells prior to assembly is 0.1 mm measured on a diameter. Find the tangential stress in each shell caused by this shrink fit. For aluminum $E = 70$ GPa and for steel $E = 200$ GPa. *Ans.* $\sigma_{st} = 51.8$ MPa, $\sigma_{al} = -51.8$ MPa

4.15. A spherical tank for storing gas under pressure is 25 m in diameter and is made of structural steel 15 mm thick. The yield point of the material is 250 MPa and a safety factor of 2.5 is adequate. Determine the maximum permissible internal pressure, assuming the welded seams between the various plates are as strong as the solid metal. *Ans.* $p = 0.24$ MPa

4.16. A thin-walled spherical shell is subject to a temperature rise ΔT which is constant at all points in the shell as well as through the shell thickness. Find the increase in volume per unit volume of the shell. Let α denote the coefficient of thermal expansion of the material. *Ans.* $3\alpha(\Delta T)$

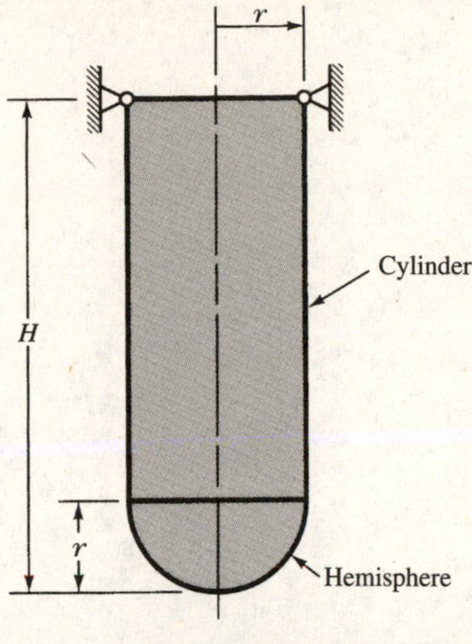

Fig. 4-9

4.17. A liquid storage tank consists of a vertical axis circular cylindrical shell closed at its lower end by a hemispherical shell as shown in Fig. 4-9. The weight of the system is carried by a ring-like support at the top and the lower extremity is unsupported. A liquid of specific weight γ entirely fills the open container. Determine the peak circumferential and meridional stress in the cylindrical region of the assembly, as well as the peak stresses in the hemispherical region. Assume $p = \gamma y$.

Ans. Cylinder: $\sigma_h = \dfrac{\gamma r}{h}(H - r) \qquad \sigma_a = \dfrac{\gamma r}{2h}\left(H - \dfrac{r}{3}\right)$

Hemisphere: $\sigma_a = \dfrac{\gamma H r}{2h}$

Torsion

5.1 Introduction

Consider a bar rigidly clamped at one end and twisted at the other end by a torque (twisting moment) $T = Fd$ applied in a plane perpendicular to the axis of the bar, as shown in Fig. 5-1. Such a bar is in *torsion*. An alternative representation of the torque is the curved arrow shown in the figure.

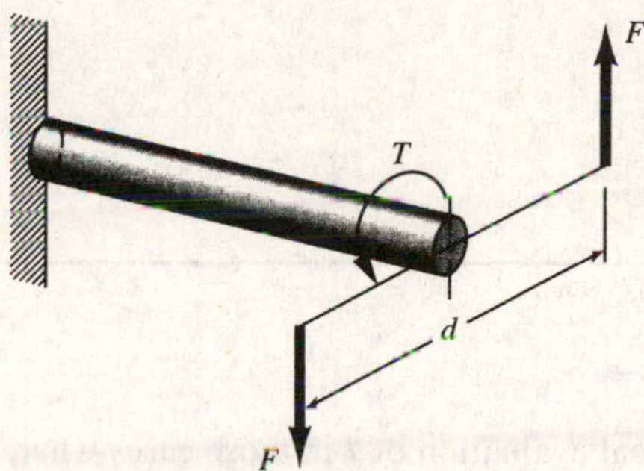

Fig. 5-1 Torque applied to a circular shaft.

Occasionally a number of couples act along the length of a shaft. In that case it is convenient to introduce a new quantity, the *twisting moment*, which for any section along the bar is defined to be the algebraic sum of the moments of the applied couples that lie to one side of the section in question. The choice of side is arbitrary.

Polar Moment of Inertia

A mathematical property of the geometry of the cross section which occurs in the study of the stresses set up in a circular shaft subject to torsion is the *polar moment of inertia J*, defined in a statics course. It is included for quick reference in Table 5.1 on the next page, for several common shapes. For a hollow circular shaft of outer diameter D_o with a concentric circular hole of diameter D_i the polar moment of inertia of the cross-sectional area is given by

$$J = \frac{\pi}{32}(D_o^4 - D_i^4) \tag{5.1}$$

The polar moment of inertia for a solid shaft is obtained by setting $D_i = 0$.

Occasionally it is convenient to rewrite the above equation in the form

$$\begin{aligned} J &= \frac{\pi}{32}(D_o^2 + D_i^2)(D_o^2 - D_i^2) \\ &= \frac{\pi}{32}(D_o^2 + D_i^2)(D_o + D_i)(D_o - D_i) \end{aligned} \tag{5.2}$$

Table 5.1. Properties of Selected Areas

Shape	Centroid	Moment of Inertia
h, C, b, x	$x_C = b/2$	$I_C = bh^3/12$ $I_x = bh^3/3$
h, C, b, x	$y_C = h/3$	$I_C = bh^3/36$ $I_x = bh^3/12$
a, C, x	$x_C = 0$	$I_C = \pi a^4/4$ $J = \pi a^4/2$
C, a, x	$y_C = 4a/3\pi$	$I_x = \pi a^4/8$

This last form is useful in numerical evaluation of J in those cases where the difference $(D_o - D_i)$ is small. See Problem 5.3.

Let us derive an expression relating the applied twisting moment acting on a shaft of circular cross section and the shearing stress at any point in the shaft. In Fig. 5-2, the shaft is shown loaded by the two torques T in static equilibrium. One fundamental assumption is that a plane section of the shaft normal to its axis before loads are applied remains plane and normal to the axis after loading. This may be verified experimentally for circular shafts, but this assumption is not valid for shafts of noncircular cross section.

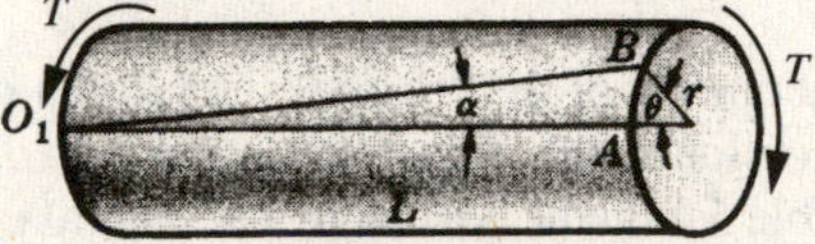

Fig. 5-2 Torque acting on a section of a shaft.

A generator on the surface of the shaft, denoted by O_1A in Fig. 5-2, deforms into the configuration O_1B after torsion has occurred. The angle between these configurations is denoted by α. By definition, the shearing strain γ on the surface of the shaft is

$$\gamma = \tan \alpha \approx \alpha$$

where the angle α is assumed to be small. From the geometry of the figure,

$$\alpha = \frac{AB}{L} = \frac{r\theta}{L}$$

Hence

$$\gamma = \frac{r\theta}{L} \tag{5.3}$$

But since a diameter of the shaft prior to loading is assumed to remain a diameter after torsion has been applied, the shearing strain at a general distance ρ from the center of the shaft may likewise be written $\gamma_\rho = \rho\theta/L$. Consequently the shearing strains of the longitudinal fibers vary linearly as the distances from the center of the shaft.

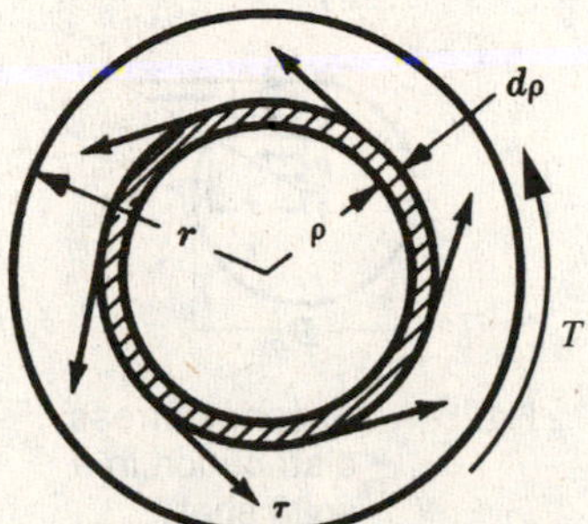

Fig. 5-3 Shearing stress acting on a differential area element.

Since we are concerned only with the linear range where the shearing stress is proportional to shearing strain, it is evident that the shearing stresses of the longitudinal fibers vary linearly as the distances from the center of the shaft. Obviously the distribution of shearing stresses is symmetric around the geometric axis of the shaft. They have the appearance shown in Fig. 5-3. For equilibrium, the sum of the moments of these distributed shearing forces over the entire circular cross section is equal to the torque T. Thus we have

$$T = \int_0^r \tau\rho \, dA \tag{5.4}$$

where dA represents the area of the shaded element shown in Fig. 5-3. However, the shearing stress varies as the distance from the axis; hence

$$\frac{\tau_\rho}{\rho} = \frac{\tau_r}{r} = \text{constant}$$

where the subscripts on the shearing stress denote the distances of the element from the axis of the shaft.

Consequently we may write

$$T = \int_0^r \frac{\tau_\rho}{\rho}(\rho^2) \, dA = \frac{\tau_\rho}{\rho} \int_0^r \rho^2 \, dA$$

since the ratio τ_ρ/ρ is a constant. However, the expression $\int_0^r \rho^2 dA$ is by definition the polar moment of inertia of the cross-sectional area. Hence the desired relationship is

$$T = \frac{\tau_\rho J}{\rho} \quad \text{or} \quad \tau_\rho = \frac{T\rho}{J} \tag{5.5}$$

It is to be emphasized that this expression holds *only* if no points of the bar are stressed beyond the proportional limit of the material.

5.2 Torsional Shearing Stress

So, for either a solid or a hollow circular shaft subject to a twisting moment T the *torsional shearing stress* τ at a distance ρ from the center of the shaft is written as

$$\tau = \frac{T\rho}{J} \tag{5.6}$$

For applications see Problems 5.1, 5.2, 5.5, 5.6, and 5.10. This stress distribution varies from zero at the center of the shaft (if it is solid) to a maximum at the outer fibers, as shown in Fig. 5-4. It is to be emphasized that no points of the bar are stressed beyond the proportional limit.

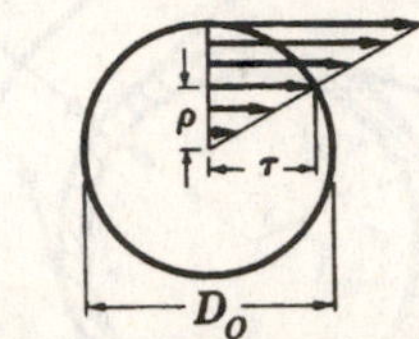

Fig. 5-4 Shearing stress distribution in a solid shaft.

5.3 Shearing Strain

The amount of twist of a shaft is often of interest. Let us determine the angle of twist of a shaft subjected to a torque T. The ratio of the shear stress τ to the shear strain γ is called the *shear modulus* and, in Eq. (2.2) is given by

$$G = \frac{\tau}{\gamma} \tag{5.7}$$

Again the units of G are the same as those of shear stress, since the shear strain is dimensionless.

Using Eq. (5.6) for τ and Eq. (5.3) for γ along with Eq. (5.7), we find the expression for θ to be

$$\theta = \frac{TL}{JG} \tag{5.8}$$

where at the outermost fiber, $\rho = r$.

5.4 Combined Torsion and Axial Loading

In Chapters 3 and 4, equations were formulated for determining normal and shearing stresses on particular planes in shafts being twisted by a torque, in axially loaded bars, and in pressure vessels. The shearing stress is $\tau = T\rho/J$ in a torqued shaft, the normal stress is $\sigma = P/A$ in an axially loaded bar, and the hoop stress is pr/h in a thin-walled pressure vessel. Any of these members can be simultaneously subjected to both torque and axial loads. Providing the strains are small so that a linear relationship exists between loads and strains, the stresses can be superimposed on the same element at any point of interest. Mohr's circle, or the stress transformation equations, can then be used to determine the stress on any plane passing through the point of interest.

Consider a circular shaft subjected to both torsion and an axial load, as sketched in Fig. 5-5. An element on the outermost fibers is shown in Fig. 5-5(*a*) and enlarged in Fig. 5-5(*b*) with $\sigma_x = P/A$ and $\tau_{xy} = T\rho/J$. Mohr's circle, as sketched in Fig. 5-5(*c*), can then be used to find the maximum normal stress and the maximum shearing stress. It is very important to calculate the maximum stresses since they are used in the design of structural and machine members. Interest is focused on the outermost fibers since they must resist the largest normal and shearing stresses. On the centerline of the shaft the shearing stress is zero and only the normal stress would act on an element. The maximum normal stress in Fig. 5-5(c) is obviously larger than σ_x, so the largest normal stress exists on the outermost fibers, but at an angle with the x-axis. Problem 5.12 illustrates the application of *combined loading*.

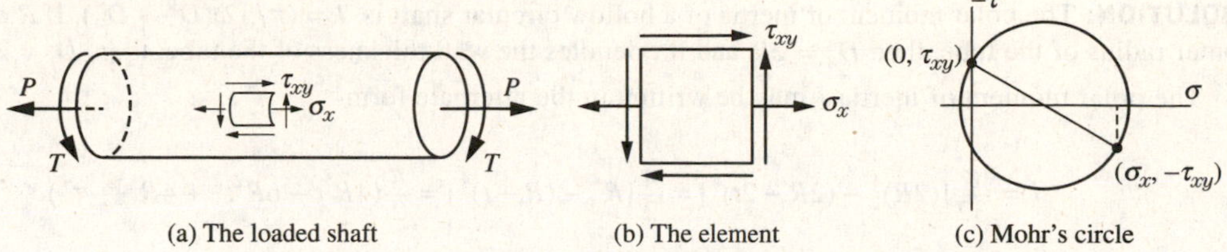

(a) The loaded shaft (b) The element (c) Mohr's circle

Fig. 5-5 A circular shaft subjected to torsion and an axial load.

SOLVED PROBLEMS

5.1. If a twisting moment of 1100 N · m is impressed upon a 4.4-cm-diameter shaft, what is the maximum shearing stress developed? Also, what is the angle of twist in a 150-cm length of the shaft? The material is steel for which $G = 85$ GPa.

SOLUTION: The polar moment of inertia of the cross-sectional area is

$$J = \frac{\pi}{32}(D_o)^4 = \frac{\pi}{32} \times 0.044^4 = 3.68 \times 10^{-7} \text{ m}^4$$

The torsional shearing stress τ at any distance ρ from the center of the shaft is given by Eq. (5.6). The maximum shear stress is developed at the outer fibers where $\rho = 0.022$ m:

$$\tau_{\max} = \frac{1100 \times 0.022}{3.68 \times 10^{-7}} = 65.8 \times 10^6 \text{ Pa} \quad \text{or} \quad 65.8 \text{ MPa}$$

The shear stress varies linearly from zero at the center of the shaft to 65.8 MPa at the outer fibers, as shown in Fig. 5-4.

The angle of twist θ in a 3-m length of the shaft is

$$\theta = \frac{TL}{GJ} = \frac{1100 \times 1.5}{85 \times 10^9 (3.68 \times 10^{-7})} = 0.0527 \text{ rad}$$

5.2. A hollow 3-m-long steel shaft must transmit a torque of 25 kN · m. The total angle of twist in this length is not to exceed 2.5° and the allowable shearing stress is 90 MPa. Determine the inside and outside diameters of the shaft if $G = 85$ GPa.

SOLUTION: Let D_o and D_i designate the outside and inside diameters of the shaft, respectively. The angle of twist is $\theta = TL/GJ$. Thus, in the 3-m length we have

$$2.5° \left(\frac{\pi \text{ rad}}{180°} \right) = \frac{(25000)(3)}{(85 \times 10^9)\,(\pi/32)(D_o^4 - D_i^4)} \qquad \therefore D_o^4 - D_i^4 = (206 \times 10^{-6}) \text{ m}^4 \qquad (1)$$

The maximum shearing stress occurs at the outer fibers where $\rho = D_o/2$. At these points from Eq. (5.6), we have

$$90 \times 10^6 = \frac{(25000)(D_o/2)}{(\pi/32)(D_o^4 - D_i^4)} \qquad \therefore D_o^4 - D_i^4 = (1415 d_o)(10^{-6}) \text{ m}^4 \qquad (2)$$

Comparison of the right-hand sides of equations (1) and (2) indicates that

$$206 \times 10^{-6} = 1415 D_o (10^{-6})$$

and thus $D_o = 0.146$ m or 146 mm. Substitution of this value into either of the equations then gives $D_i = 0.126$ m or 126 mm.

5.3. Let us consider a thin-walled tube subject to torsion. Derive an approximate expression for the applied torque if the working stress in shear is a given constant τ_w. Also, derive an approximate expression for the strength–weight ratio T/W of such a tube.

SOLUTION: The polar moment of inertia of a hollow circular shaft is $J = (\pi/32)(D_o^4 - D_i^4)$. If R denotes the outer radius of the tube, then $D_o = 2R$, and if t denotes the wall thickness of the tube, then $D_i = 2R - 2t$.

The polar moment of inertia J may be written in the alternate form

$$J = \frac{\pi}{32}[(2R)^4 - (2R-2t)^4] = \frac{\pi}{2}[R^4 - (R-t)^4] = \frac{\pi}{2}(4R^3t - 6R^2t^2 + 4Rt^3 - t^4)$$

$$= \frac{\pi}{2}R^4\left[4\left(\frac{t}{R}\right) - 6\left(\frac{t}{R}\right)^2 + 4\left(\frac{t}{R}\right)^3 - \left(\frac{t}{R}\right)^4\right]$$

Neglecting squares and higher powers of the ratio t/R, since we are considering a thin-walled tube, this becomes, approximately, $J = 2\pi R^3 t$.

The ordinary torsion formula [Eq. (5.6)] is $T = \tau_w J/R$. For a thin-walled tube, the expression for the applied torque is

$$T = 2\pi R^2 t \tau_w$$

The weight W of the tube is $W = \gamma LA$, where γ is the specific weight of the material, L the length of the tube, and A the cross-sectional area of the tube. The area is given by

$$A = \pi[R^2 - (R-t)^2] = \pi(2Rt - t^2) = \pi R^2\left[\frac{2t}{R} - \left(\frac{t}{R}\right)^2\right]$$

Again neglecting the square of the ratio t/R for a thin tube, this becomes $A \approx 2\pi Rt$. The strength–weight ratio T/W is given by

$$\frac{T}{W} = \frac{2\pi R^2 t\tau_w}{2\pi RtL\gamma} = \frac{R\tau_w}{L\gamma}$$

The ratio is of considerable importance in aircraft design.

5.4. Consider two solid circular shafts connected by 5-cm- and 25-cm-pitch-diameter gears as in Fig. 5-6(a). Find the angular rotation of D, the right end of one shaft, with respect to A, the left end of the other, caused by the torque of 280 N · m applied at D. The left shaft is steel for which $G = 80$ GPa and the right is brass for which $G = 33$ GPa.

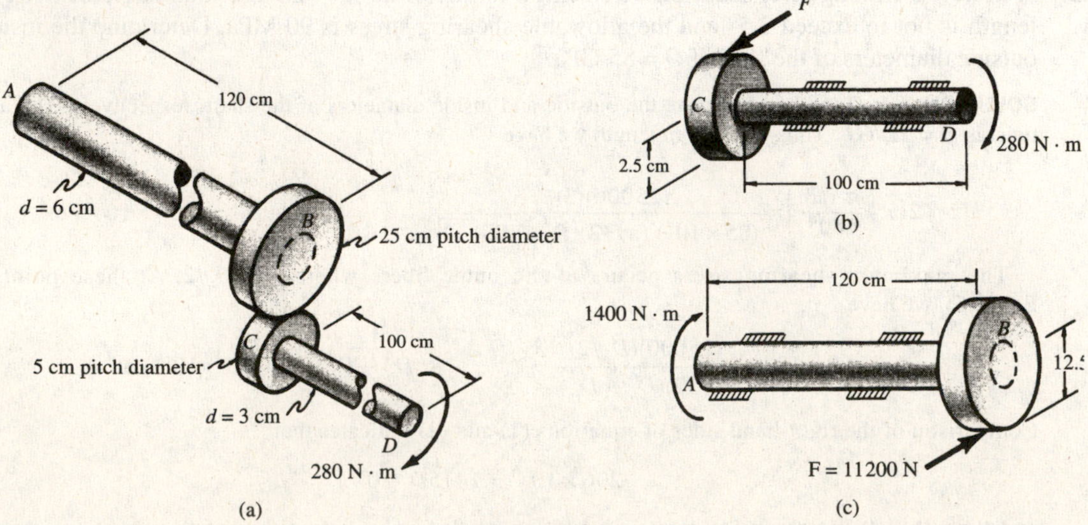

Fig. 5-6

SOLUTION: A free-body diagram of the right shaft *CD* [Fig. 5-6(*b*)] reveals that a tangential force *F* must act on the smaller gear. For equilibrium,

$$Fr = T \qquad 0.025\,F = 280 \qquad \therefore F = 11\,200\ \text{N}$$

The angle of twist of the right shaft is

$$\theta_1 = \frac{TL}{GJ} = \frac{280 \times 1.00}{(33 \times 10^9)\pi \times 0.03^4/32} = 0.1067\ \text{rad}$$

A free-body diagram of the left shaft *AB* is shown in Fig. 5-6(*c*). The force *F* is equal and opposite to that acting on the small gear *C*. This force *F* acts 12.5 cm from the center line of the left shaft; hence it imparts a torque of $0.125(11\,200) = 1400\ \text{N}\cdot\text{m}$ to the shaft *AB*. Because of this torque there is a rotation of end *B* with respect to end *A* given by the angle θ_2, where

$$\theta_2 = \frac{1400 \times 1.20}{(80 \times 10^9)\pi \times 0.06^4/32} = 0.0165\ \text{rad}$$

This angle of rotation θ_2 induces a *rigid-body* rotation of the entire shaft *CD* because of the gears. In fact, the rotation of *CD* will be in the same ratio to that of *AB* as the ratio of the pitch diameters, or 5:1. Thus a rigid-body rotation of 5(0.0165) rad is imparted to shaft *CD*. Superposed on this rigid body movement of *CD* is the angular displacement of *D* with respect to *C*, previously denoted by θ_1.

Hence the resultant angle of twist of *D* with respect to *A* is

$$\theta = 5\theta_2 + \theta_1$$

$$= 5 \times 0.0165 + 0.1067 = 0.189\ \text{rad} \qquad \text{or} \qquad 10.8^\circ$$

5.5. A solid circular shaft is required to transmit 200 kW while turning at 90 rpm (rev/min). The allowable shearing stress is 42 MPa. Find the required shaft diameter.

SOLUTION: The time rate of work (power) is expressed in N · m/s. By definition, 1 N · m/s is 1 W. Power is thus given by $P = T\omega$, where *T* is torque and ω is shaft angular velocity in rad/s. Or, alternatively, $P = 2\pi f T$, where *f* is revolutions per second or hertz. Thus we have

$$200\,000 = \frac{90 \times 2\pi}{60} T$$

$$T = 21\,220\ \text{N}\cdot\text{m}$$

The outer fiber shearing stresses are maximum and given by

$$\tau = \frac{T(d/2)}{\pi d^4/32} = \frac{16T}{\pi d^3}$$

Thus,

$$42 \times 10^6 = \frac{16(21\,220)}{\pi d^3} \qquad \therefore d = 0.137\ \text{m}$$

5.6. A solid circular shaft has a uniform diameter of 5 cm and is 4 m long. At its midpoint 65 hp is delivered to the shaft by means of a belt passing over a pulley. This power is used to drive two machines, one at the left end of the shaft consuming 25 hp and one at the right end consuming the remaining 40 hp. Determine the maximum shearing stress in the shaft and also the relative angle of twist between the two extreme ends of the shaft. The shaft turns at 200 rpm and the material is steel for which $G = 80$ GPa.

SOLUTION: In the left half of the shaft we have 25 hp which corresponds to a torque T_1 given by

$$P = T\omega \qquad 25 \times 746 = T_1 \frac{200 \times 2\pi}{60} \qquad \therefore T_1 = 890\ \text{N}\cdot\text{m}$$

where we have used 1 hp = 746 W. Similarly, in the right half we have 40 hp corresponding to a torque T_2 given by

$$40 \times 746 = T_2 \frac{200 \times 2\pi}{60} \qquad \therefore T_2 = 1425\ \text{N}\cdot\text{m}$$

The maximum shearing stress consequently occurs in the outer fibers in the right half and is given by

$$\tau_\rho = \frac{T\rho}{J} \qquad \text{or} \qquad \tau = \frac{1425 \times 0.025}{\pi \times 0.05^4/32} = 58.1 \times 10^3 \text{ Pa}$$

The angles of twist of the left and right ends relative to the center are, respectively, using $\theta = TL/GJ$,

$$\theta_1 = \frac{890 \times 2}{80 \times 10^9(\pi \times 0.05^4/32)} = 0.0363 \text{ rad} \qquad \text{and} \qquad \theta_2 = \frac{1425 \times 2}{80 \times 10^9(\pi \times 0.05^4/32)} = 0.0581 \text{ rad}$$

Since θ_1 and θ_2 are in the same direction, the relative angle of twist between the two ends of the shaft is $\theta = \theta_2 - \theta_1 = 0.0218$ rad.

5.7. A circular shaft is clamped at one end, free at the other, and loaded by a uniformly distributed twisting moment of magnitude t per unit length along its length [see Fig. 5-7(*a*)]. The *torsional rigidity* of the bar is GJ. Find the angle of twist of the free end of the shaft.

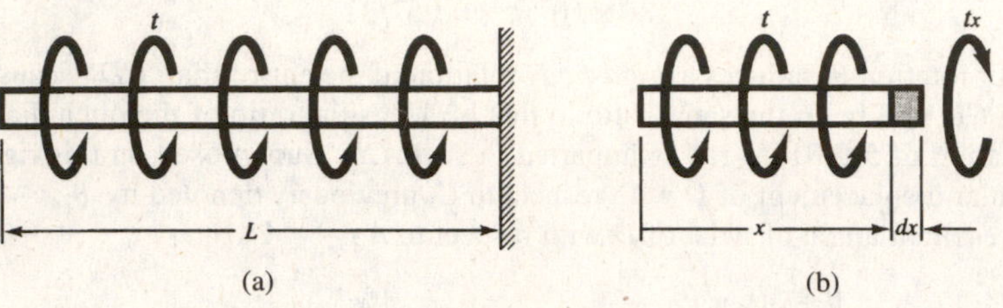

Fig. 5-7

SOLUTION: A free-body diagram of the portion of the bar between the left end and the section x is shown in Fig. 5-7(*b*). An element of length dx is shown in that figure and we wish to determine the angular rotation of the cylindrical element of length dx. For equilibrium of moments about the axis of the bar, a twisting moment tx must act at the right of the section shown. This twisting moment tx imparts to the element of length dx an angular rotation [from Eq. (5.8)]

$$d\theta = \frac{(tx)\,dx}{GJ}$$

The total rotation of the left end with respect to the right end is found by the integration of all such elemental angles of twist to be

$$\theta = \int_0^L \frac{(tx)\,dx}{GJ} = \frac{tL^2}{2GJ}$$

5.8. A steel shaft ABC, of constant circular cross section and of diameter 80 mm, is clamped at the left end A, loaded by a twisting moment of 6000 N · m at its midpoint B, and elastically restrained against twisting at the right end C (see Fig. 5-8). At end C the bar ABC is attached to vertical steel bars each of 16-mm diameter. The upper bar MN is attached to the end N of a horizontal diameter of the 80-mm bar ABC and the lower bar PQ is attached to the other end Q of this same horizontal diameter, as shown in Fig. 5-8(*a*). For all materials $E = 200$ GPa and $G = 80$ GPa. Determine the peak shearing stress in bar ABC as well as the tensile stress in bar MN.

SOLUTION: Let us consider that bars MN and PQ are temporarily disconnected from the bar ABC. Then, from Eq. (5.8) the angle of twist at B relative to A is

$$\theta = \frac{TL}{GJ} = \frac{(6000)(0.75)}{(80 \times 10^9)\,(\pi/32)\,(0.08)^4} = 0.01399 \text{ rad}$$

Since no additional twisting moments act between B and C, this same angle of twist due to the 6000-N · m loading exists at C, called θ_C.

From Fig. 5-8(*b*) the horizontal diameter NQ of bar ABC must rotate to some final position, indicated by the dotted line. This is due to extension Δ of each of the vertical bars, which is accompanied by an axial force P in each bar. For a small angle of rotation θ, we have $\Delta = (0.040 \text{ m})\theta_C$. The axial forces P constitute a couple of magnitude $P(0.08 \text{ m}) = T_C$ which must act at the end C of bar ABC when the vertical bars are once again considered to be attached to the horizontal bar ABC. This couple must act in a sense opposite to the 6000-N · m load since the elastic vertical bars tend to restrain angular rotation.

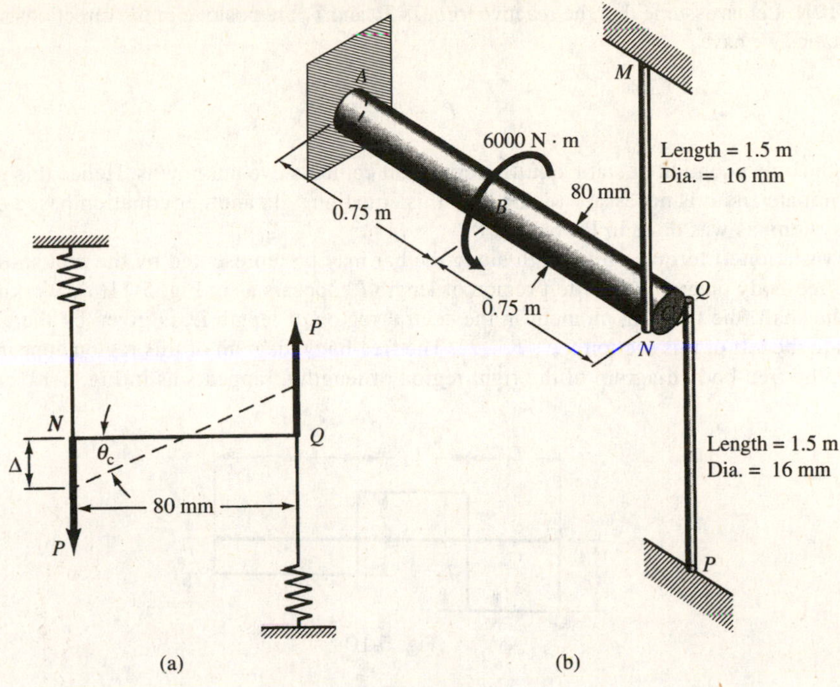

Fig. 5-8

The elongation of each vertical bar may be found to be

$$\Delta = \frac{PL}{AE} = \frac{P(1.5)}{(\pi/4)(0.016)^2 E} = \frac{(T_C/0.08)(1.5)}{(\pi/4)(0.016)^2(200\times 10^9)} = 4.66\times 10^{-7}T_C$$

The angular rotation of end C of bar ABC may now be determined by (a) considering the effect of the twisting moments of 6000 N · m and the end load T_C, and by (b) considering the angular rotation caused by the axial force P in the vertical bars. Thus, for the same rotation of end C we have

$$0.01399 - \frac{T_C\times 1.5}{80\times 10^9(\pi\times 0.08^4/32)} = \frac{4.66\times 10^{-7}T_C}{0.04}$$

Solving, $T_C = 857$ N · m and $P = T_C/0.08 = 10\ 720$ N.

The twisting moment between B and C is 857 N · m and between A and B it is 6000 – 857 = 5143 N · m. Thus, the peak torsional shearing stress occurs at the outer fibers at all points between A and B and is

$$\tau_{max} = \frac{T\rho}{J} = \frac{5143\times 0.04}{\pi\times 0.08^4/32} = 51.2\times 10^6 \text{ Pa} \quad \text{or} \quad 51.2 \text{ MPa}$$

The axial stress in each of the vertical bars is

$$\sigma = \frac{P}{A} = \frac{10\ 720}{\pi(0.008)^2} = 53.3\times 10^6 \text{ Pa} \quad \text{or} \quad 53.3 \text{ MPa}$$

5.9. Determine the reactive torques at the fixed ends of the circular shaft loaded by the couples shown in Fig. 5-9(a). The cross section of the bar is constant along the length. Assume elastic action.

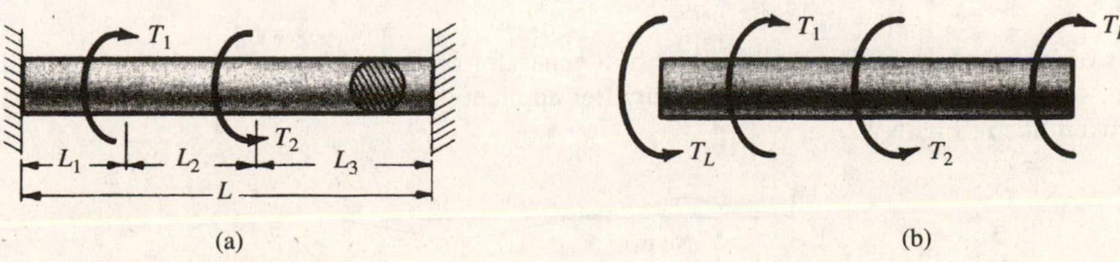

Fig. 5-9

SOLUTION: Let us assume that the reactive torques T_L and T_R are positive in the directions shown in Fig. 5-9(*b*). From statics we have

$$T_L - T_1 + T_2 - T_R = 0 \tag{1}$$

This is the only equation of static equilibrium and it contains two unknowns. Hence this problem is statically indeterminate and it is necessary to augment this equation with another equation based on the deformations of the system, as was done in Problem 5.8.

The variation of torque with length along the bar may be represented by the plot shown in Fig. 5-10.

The free-body diagram of the left region of length L_1 appears as in Fig. 5-11(*a*). Working from left to right along the shaft, the twisting moment in the central region of length L_2 is given by the algebraic sum of the torques to the left of this section, i.e., $T_1 - T_L$. The free-body diagram of this region appears as in Fig. 5-11(*b*). Finally, the free-body diagram of the right region of length L_3 appears as in Fig. 5-11(*c*).

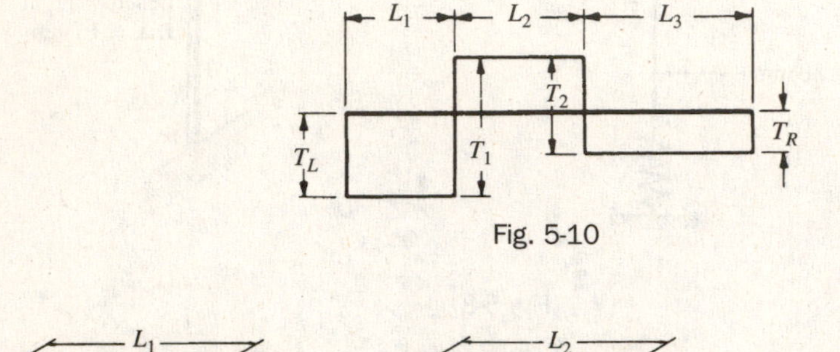

Fig. 5-10

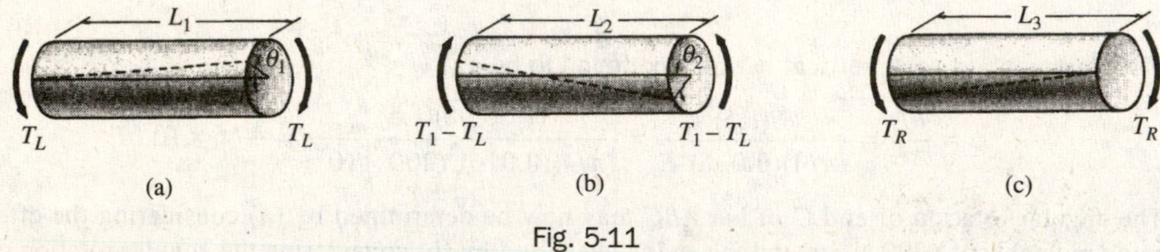

Fig. 5-11

Let θ_1 denote the angle of twist at the point of application of T_1, and θ_2 the angle at T_2. Then from a consideration of the regions of lengths L_1 and L_3 we immediately have

$$\theta_1 = \frac{T_L L_1}{GJ} \quad \text{and} \quad \theta_2 = \frac{T_R L_3}{GJ} \tag{2}$$

The original position of a generator on the surface of the shaft is shown by a solid line in Fig. 5-11, and the deformed position by a dashed line. Consideration of the central region of length L_2 reveals that the angle of twist of its right end with respect to its left end is $\theta_1 + \theta_2$. Hence, since the torque causing this deformation is $T_1 - T_L$, we have

$$\theta_1 + \theta_2 = \frac{(T_1 - T_L)L_2}{GJ} \tag{3}$$

Solving (1) through (3) simultaneously, we find

$$T_L = T_1 \frac{L_2 + L_3}{L} - T_2 \frac{L_3}{L} \quad \text{and} \quad T_R = -T_1 \frac{L_1}{L} + T_2 \frac{L_1 + L_2}{L}$$

It is of interest to examine the behavior of a generator on the surface of the shaft. Originally it was, of course, straight over the entire length L, but after application of T_1 and T_2 it has the appearance shown by the broken line in Fig. 5-12.

Fig. 5-12

5.10. Consider a composite shaft fabricated from a 6-cm-diameter solid aluminum alloy, $G = 28$ GPa, surrounded by a hollow steel circular shaft of outside diameter 7 cm and inside diameter 6 cm, with $G = 84$ GPa. The two metals are rigidly connected at their juncture. If the composite shaft is loaded by a twisting moment of 154 kN · m, calculate the maximum shearing stress in the steel and also in the aluminum.

SOLUTION: Let T_1 = torque carried by the aluminum shaft and T_2 = torque carried by the steel. For static equilibrium of moments about the geometric axis we have

$$T_1 + T_2 = T = 154\,000 \text{ N} \cdot \text{m}$$

where T = external applied twisting moment. This is the only equation from statics available in this problem. Since it contains two unknowns, T_1 and T_2, it is necessary to supplement it with an additional equation coming from the deformations of the shaft. The structure is thus statically indeterminate.

Such an equation is easily found, since the two materials are rigidly joined; hence their angles of twist must be equal. In a length L of the shaft we have, using the formula $\theta = TL/GJ$,

$$\frac{T_1 L}{28\times10^9(\pi\times0.06^4/32)} = \frac{T_2 L}{84\times10^9[\pi(0.07^4-0.06^4)/32]} \qquad \text{or} \qquad T_1 = 0.391\,T_2$$

This equation, together with the statics equation, may be solved simultaneously to yield

$$T_1 = 43\,000 \text{ N} \cdot \text{m (carried by aluminum)} \qquad \text{and} \qquad T_2 = 111\,000 \text{ N} \cdot \text{m (carried by steel)}$$

The maximum shearing stresses in the steel and the aluminum are, respectively,

$$\tau_2 = \frac{111\,000\times0.035}{\pi(0.07^4-0.06^4)/32} = 3.58\times10^9 \text{ Pa} \qquad \text{and} \qquad \tau_1 = \frac{43\,000\times0.03}{\pi\times0.06^4/32} = 1.01\times10^9 \text{ Pa}$$

5.11. A stepped shaft has the appearance shown in Fig. 5-13. The region AB is aluminum, having $G = 28$ GPa, and the region BC is steel, having $G = 84$ GPa. The aluminum portion is of solid circular cross section 45 mm in diameter, and the steel region is circular with 60-mm outside diameter and 30-mm inside diameter. Determine the maximum shearing stress in each material as well as the angle of twist at B where a torsional load of 4000 N · m is applied. Ends A and C are rigidly clamped.

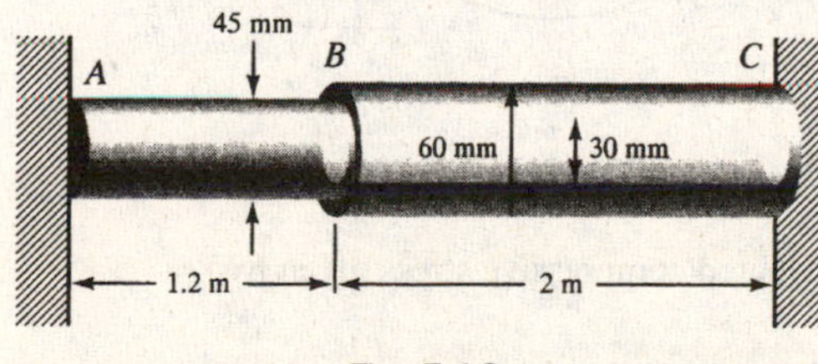

Fig. 5-13

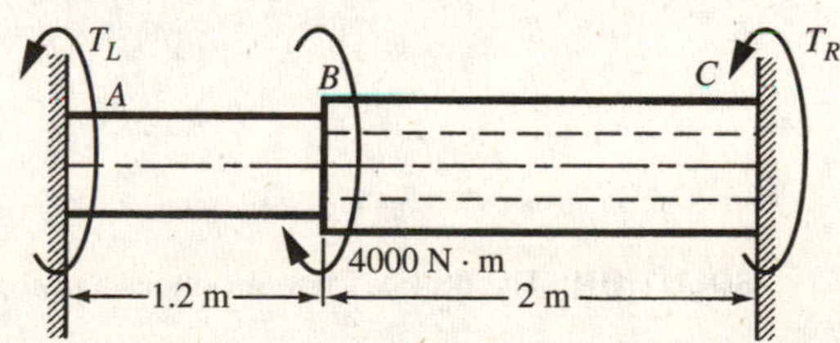

Fig. 5-14

SOLUTION: The free-body diagram of the system is shown in Fig. 5-14. The applied load of 4000 N · m as well as the unknown end reactive torques are as indicated. The only equation of static equilibrium is

$$\Sigma M_x = T_L + T_R - 4000 = 0$$

Since there are two unknowns T_L and T_R, another equation (based upon deformations) is required. This is set up by realizing that the angular rotation at B is the same if we determine it at the right end of AB or the left end of BC. Using Eq. (5.8), we thus have

$$\frac{T_L\times1.2}{(28\times10^9)\pi\times0.045^4/32} = \frac{T_R\times2.0}{(84\times10^9)\pi(0.06^4-0.03^4)/32} \qquad \text{or} \qquad T_L = 0.1875\,T_R$$

Solving for T_L and T_R, we find

$$T_L = 632 \text{ N}\cdot\text{m} \qquad \text{and} \qquad T_R = 3368 \text{ N}\cdot\text{m}$$

The maximum shearing stress in *AB* is given by

$$\tau_{AB} = \frac{T\rho}{J} = \frac{(632)(0.0225)}{\pi(0.045)^4/32} = 35.6 \text{ MPa}$$

and in *BC* by

$$\tau_{BC} = \frac{T\rho}{J} = \frac{(3370)(0.030)}{\pi(0.06^4 - 0.03^4)/32} = 85.0 \text{ MPa}$$

The angle of twist at *B*, using parameters of the region *AB*, is

$$\theta_B = \frac{TL}{GJ} = \frac{(632)(1.2)}{(28\times 10^9)(\pi \times 0.045^4/32)} = 0.0673 \text{ rad} \qquad \text{or} \qquad 3.86°$$

5.12. A hollow shaft of outer radius 140 mm and inner radius 125 mm is subjected to an axial force and a torque, as shown in Fig. 5-15. Calculate the maximum normal and shearing stresses in the shaft.

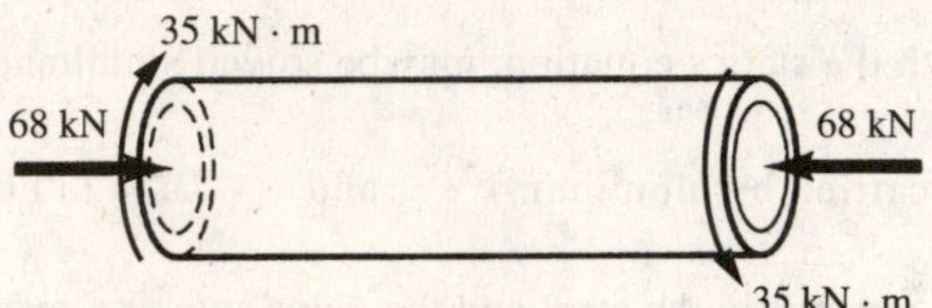

Fig. 5-15

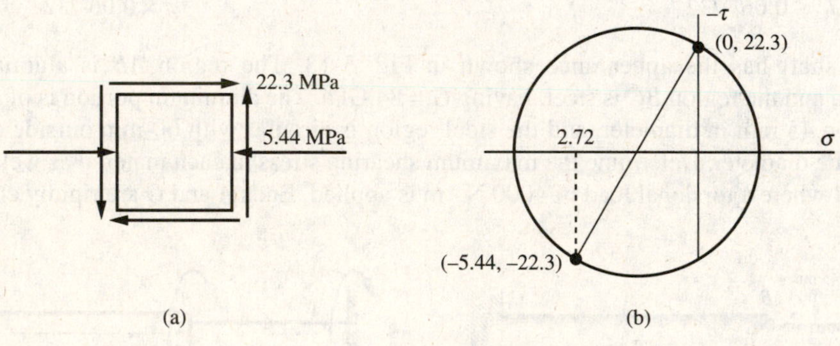

Fig. 5-16

SOLUTION: The 68-kN force produces a uniformly distributed compressive stress given by

$$\sigma_1 = \frac{68\,000}{\pi[(0.140)^2 - (0.125)^2]} = 5.44 \text{ MPa}$$

as shown in Fig. 5-16(*a*). The torsional shearing stresses due to the 35 kN · m torque at the outer fibers of the shell are given by

$$\tau = \frac{T\rho}{J} = \frac{(35\,000)(0.140)}{\pi(0.28^4 - 0.25^4)/32} = 22.3 \times 10^6 \text{ Pa} \qquad \text{or} \qquad 22.3 \text{ MPa}$$

as shown in Fig. 5-16(*a*).

From Mohr's circle the principal stresses are found to be

$$\sigma_{max} = -2.72 + \sqrt{2.72^2 + 22.3^2} = 19.75 \text{ MPa}$$

$$\sigma_{min} = -2.72 - \sqrt{2.72^2 + 22.3^2} = -25.2 \text{ MPa}$$

and the peak shearing stress is 22.47 MPa.

SUPPLEMENTARY PROBLEMS

5.13. If a solid circular shaft of 3.2-cm diameter is subject to a torque T of 300 N · m resulting in an angle of twist of 4.5° in a 2-m length, determine the shear modulus of the material. *Ans.* $G = 74$ GPa

5.14. Determine the maximum shearing stress in a 10-cm-diameter solid shaft carrying a torque of 25 kN · m. What is the angle of twist per unit length if the material is steel for which $G = 80$ GPa? *Ans.* 127 MPa, 0.0318 rad/m

5.15. A propeller shaft in a ship is 350 mm in diameter. The allowable working stress in shear is 50 MPa and the allowable angle of twist is 1° in 15 diameters of length. If $G = 85$ GPa, determine the maximum torque the shaft can transmit. *Ans.* 416 kN · m

5.16. Consider the same shaft described in Problem 5.15 but with a 175-mm axial hole bored throughout its length. The conditions on working stress and angle of twist remain as before. By what percentage is the torsional load-carrying capacity reduced? By what percentage is the weight of the shaft reduced? *Ans.* 6 percent, 25 percent

5.17. A compound shaft is composed of a 70-cm length of solid copper 10 cm in diameter, joined to 90-cm length of solid steel 12 cm in diameter. A torque of 14 kN · m is applied to each end of the shaft. Find the maximum shear stress in each material and the total angle of twist of the entire shaft. For copper $G = 40$ GPa, for steel $G = 80$ GPa. *Ans.* In the copper, 71.3 MPa; in the steel, 41.3 MPa; $\theta = 0.0328$

5.18. In Fig. 5-17, the vertical shaft and the pulleys keyed to it may be considered to be weightless. The shaft rotates with a uniform angular velocity. The known belt pulls are indicated and the three pulleys are rigidly keyed to the shaft. If the working stress in shear is 50 MPa, determine the necessary diameter of a solid circular shaft. Neglect bending of the shaft because of the proximity of the bearings to the pulleys. *Ans.* 29 mm

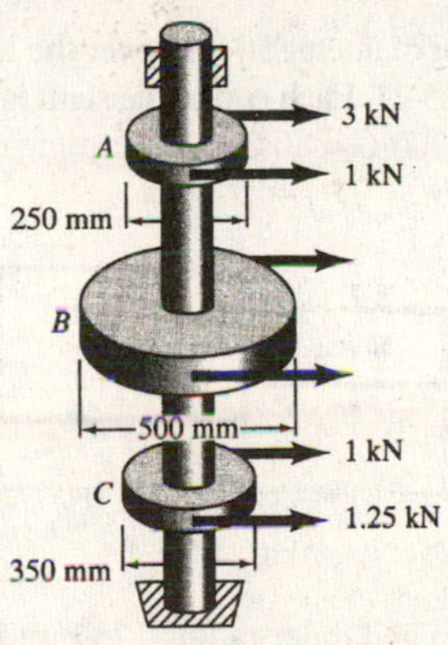

Fig. 5-17

5.19. Determine the reactive torques at the fixed ends of the circular shaft loaded by the three couples shown in Fig. 5-18. The cross section of the bar is constant along the length. *Ans.* $T_L = 3590$ N · m, $T_R = 4790$ N · m

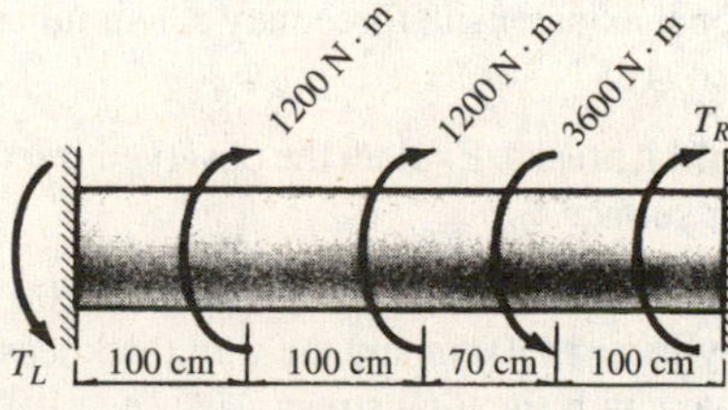

Fig. 5.18

5.20. A bar of circular cross section is clamped at its left end, free at the right, and loaded by a twisting moment t per unit length that is uniformly distributed over the middle third of the bar as shown in Fig. 5-19. Find the angle of twist of the free end of the bar.

Ans. $\frac{2}{9}\frac{tL^2}{GJ}$

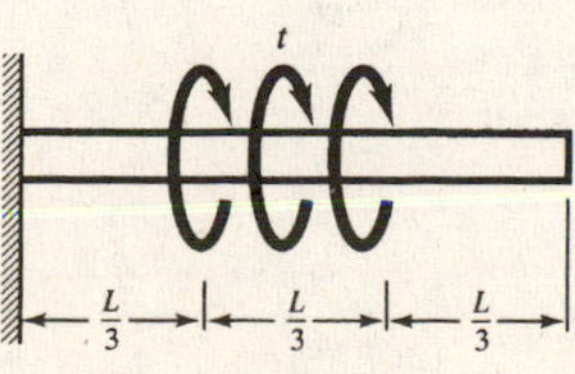

Fig. 5-19

5.21. It is desired to transmit 90 kW by means of a solid circular shaft rotating at 3.5 rad/s. The allowable shearing stress is 45 MPa. Find the required shaft diameter. *Ans.* 77.4 mm

5.22 It is required to transmit 70 hp from a turbine by a solid circular shaft turning at 200 rpm (rev/min). If the allowable shearing stress is 45 MPa, determine the required shaft diameter. *Ans.* 65.6 mm

5.23. A hollow circular shaft whose outside diameter is three times its inner diameter transmits 110 hp at 120 rev/min. If the maximum allowable shearing stress is 50 MPa, find the required outside diameter of the shaft. *Ans.* 89 mm

5.24. A solid circular cross-section shaft is clamped at both ends and loaded by a twisting moment t per unit length over the right section, as shown in Fig. 5-20. Determine the reactive twisting moments at each end of the bar.

Ans. $T_A = \frac{2}{9}tL, T_C = \frac{4}{9}tL$

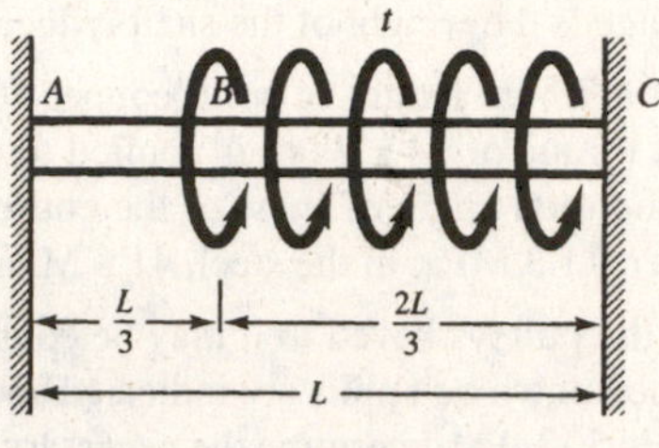

Fig. 5-20

5.25. A circular cross-section steel shaft is of diameter 50 mm over the left 150 mm of length and of diameter 100 mm over the right 150 mm, as shown in Fig. 5-21. Each end of the shaft is loaded by a twisting moment of 1000 N · m (as indicated by the double-headed arrows). If $G = 80$ GPa, determine the angle of twist between the ends of the shaft as well as the peak shearing stress. *Ans.* 1.09°, 40.7 MPa

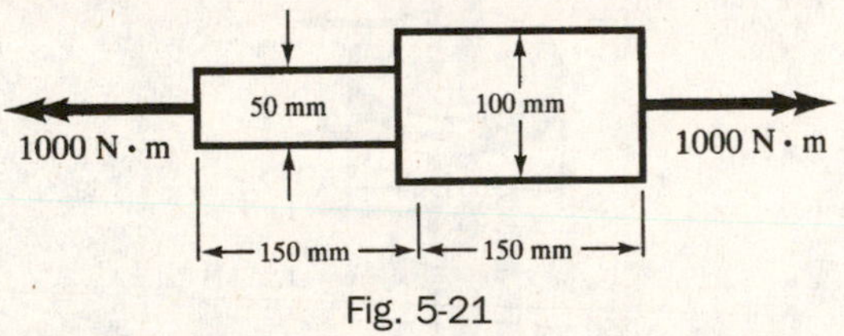

Fig. 5-21

5.26. A hollow shaft of outer radius 140 mm and inner radius 125 mm is subjected to a compressive force of 200 kN and a torque. If the allowable shearing stress is 100 MPa, what is the maximum torque that can be applied? *Ans.* 1570 N · m

5.27. A thin-walled cylinder is 26 cm in diameter and of wall thickness 2.5 mm. The cylinder is subject to an internal pressure of 650 kPa. What additional axial tensile force may act simultaneously without the maximum shearing stress exceeding 40 MPa? *Ans.* 163 N

5.28. Obtain a piece of chalk. Twist it until it breaks. Explain the observed fracture and estimate the angle of the surface of the fracture relative to the outer surface.

5.29. A thin-walled cylindrical shell is subject to an axial compression of 200 kN together with a torsional moment of 3600 N · m. The diameter of the cylinder is 30 cm and the wall thickness 3 mm. Determine the principal stresses in the shell. Also determine the maximum shearing stress. *Ans.* 1.05 MPa, –72.6 MPa, 36.8 MPa

5.30. A shaft 6 cm in diameter is subject to an axial tension of 160 kN together with a twisting moment of 40 N · m. Determine the principal stresses in the shaft. Also determine the maximum shearing stress. *Ans.* 70.6 MPa, –14 MPa, 42.3 MPa

Shearing Force and Bending Moment

6.1 Basics

A bar subject to forces or couples that lie in a plane containing the longitudinal axis of the bar is called a *beam*. The forces are understood to act perpendicular to the longitudinal axis.

Cantilever Beams

If a beam is supported at only one end and in such a manner that the axis of the beam cannot rotate at that point, it is called a *cantilever beam*. This type of beam is illustrated in Fig. 6-1. The left end of the bar is free to deflect but the right end is rigidly clamped. The reaction of the supporting wall upon the beam consists of a vertical force together with a couple acting in the plane of the applied loads shown.

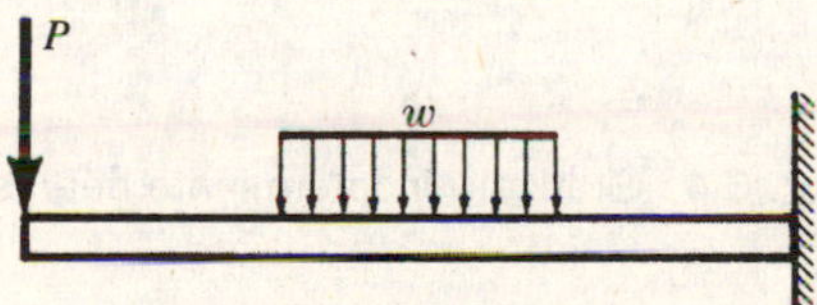

Fig. 6-1 A cantilever beam.

Simple Beams

A beam that is freely supported at both ends is called a *simple beam*. The term "freely supported" implies that the end supports are capable of exerting only forces upon the bar and are not capable of exerting any moments. Thus there is no restraint offered to the angular rotation of the ends of the bar at the supports as the bar deflects under the loads. Two simple beams are sketched in Fig. 6-2.

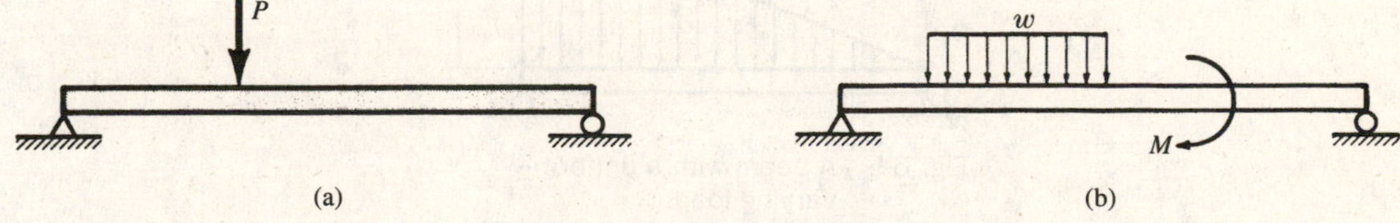

Fig. 6-2 Simple beams.

It is to be observed that at least one of the supports must be capable of undergoing horizontal movement so that no force will exist in the direction of the axis of the beam. Thus, a roller is shown as one of the supports. If neither end were free to move horizontally, then some axial force would arise in the beam as it deforms under the load. Problems of this nature are not considered in this book.

The beam of Fig. 6-2(*a*) is subjected to a concentrated force; that of Fig. 6-2(*b*) is loaded by a uniformly distributed load as well as a couple.

Overhanging Beams

A beam freely supported at two points and having one or both ends extending beyond these supports is termed an *overhanging beam*. Two examples are given in Fig. 6-3.

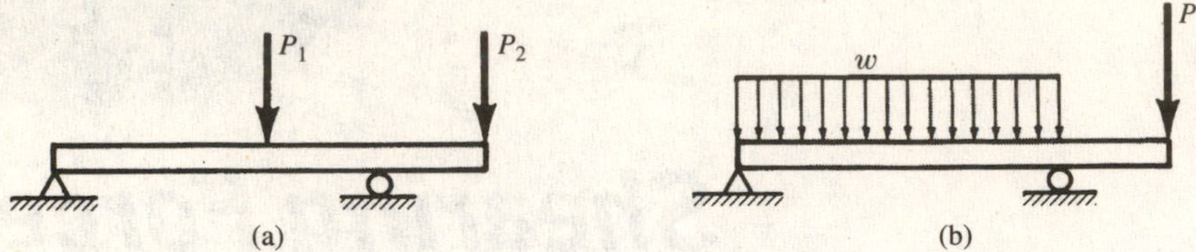

Fig. 6-3 Overhanging beams.

Statically Determinate Beams

All the beams considered above, the cantilevers, simple beams, and overhanging beams, are ones in which the reactions of the supports may be determined by use of the equations of static equilibrium. The values of these reactions are independent of the deformations of the beam. Such beams are said to be *statically determinate*.

Statically Indeterminate Beams

If the number of reactions exerted upon the beam exceeds the number of equations of static equilibrium, then the statics equations must be supplemented by equations based upon the deformations of the beam. In this case the beam is said to be *statically indeterminate*. Examples are shown in Fig. 6-4.

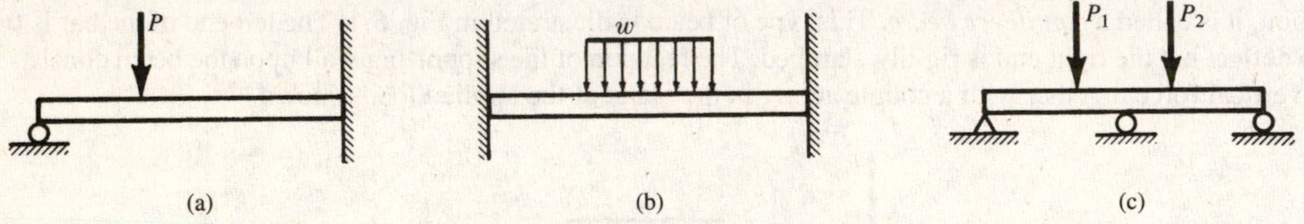

Fig. 6-4 Statically indeterminate beams.

Types of Loading

Loads commonly applied to a beam may consist of concentrated forces (applied at a point), uniformly distributed loads, in which case the magnitude is expressed as a certain number of newtons per meter of length of the beam, or uniformly varying loads. This last type of load is exemplified in Fig. 6-5.

A beam may also be loaded by an applied couple. The magnitude of the couple will be expressed in N · m.

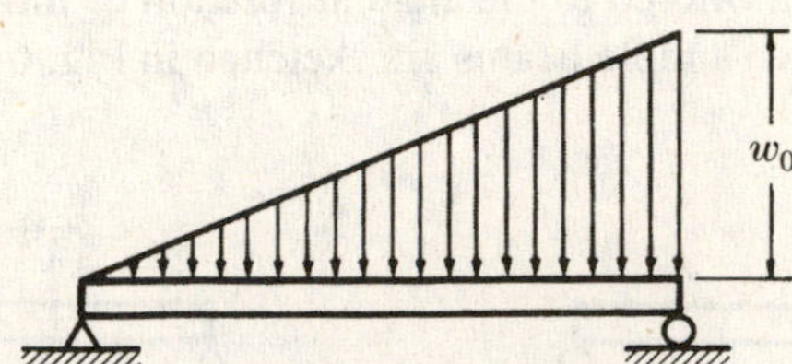

Fig. 6-5 A beam with a uniformly varying load.

6.2 Internal Forces and Moments in Beams

When a beam is loaded by forces and couples, internal stresses arise in the bar. In general, both normal and shearing stresses will occur. In order to determine the magnitude of these stresses at any section of the beam, it is necessary to know the resultant force and moment acting at that section. These may be found by applying the equations of static equilibrium.

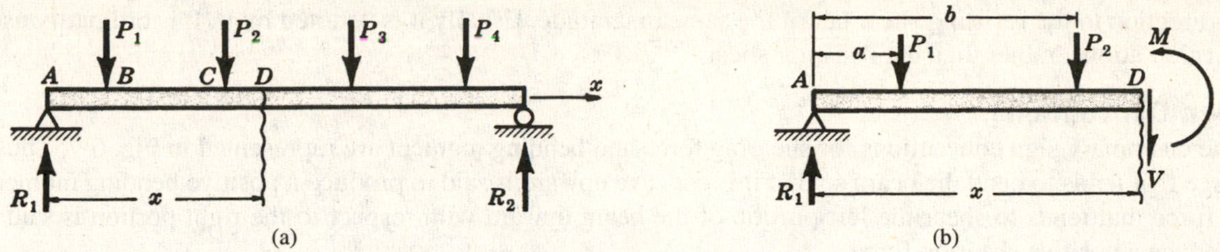

Fig. 6-6 Several forces acting on a simple beam.

Suppose several concentrated forces act on a simple beam as in Fig. 6-6(*a*). It is desired to study the internal stresses across the section at D, located at a distance x from the left end of the beam. To do this let us consider the beam to be cut at D and the portion of the beam to the right of D removed. The portion removed must then be replaced by the effect it exerted upon the portion to the left of D and this effect will consist of a vertical shearing force together with a moment as represented by V and M, respectively, in the free-body diagram of the left portion of the beam shown in Fig. 6-6(*b*).

The force V and the couple M hold the left portion of the bar in equilibrium under the action of the forces R_1, P_1, P_2. The quantities V and M are taken to be positive if they have the senses indicated above.

Resisting Moment

The couple M shown in Fig. 6-6(*b*) at section D is called the *resisting moment*. The magnitude of M may be found by use of a statics equation which states that the sum of the moments of all forces about an axis through D and perpendicular to the plane of the page is zero. Thus

$$\Sigma M_D = M - R_1 x + P_1(x-a) + P_2(x-b) = 0 \quad \text{or} \quad M = R_1 x - P_1(x-a) - P_2(x-b) \qquad (6.1)$$

Thus the resisting moment M is the moment at point D created by the moments of the reaction at A and the applied forces P_1 and P_2. The resisting moment M is due to stresses that are distributed over the vertical section at D. These stresses act in a horizontal direction and are tensile in certain portions of the cross section and compressive in others. Their nature will be discussed in detail in Chapter 7.

Resisting Shear

The vertical force V shown in Fig. 6-6(*b*) is called the *resisting shear* at section D. For equilibrium of forces in the vertical direction,

$$\Sigma F_y = R_1 - P_1 - P_2 - V = 0 \quad \text{or} \quad V = R_1 - P_1 - P_2 \qquad (6.2)$$

This force V is actually the resultant of shearing stresses distributed over the vertical section at D. The nature of these stresses will be studied in Chapter 7.

Bending Moment

The algebraic sum of the moments of the external forces to one side of the section D about an axis through D is called the *bending moment* at D. This is represented by

$$R_1 x - P_1(x-a) - P_2(x-b)$$

for the loading considered above. Thus the bending moment is opposite in direction to the resisting moment but is of the same magnitude. It is usually denoted by M also. Ordinarily the bending moment rather than the resisting moment is used in calculations because it can be represented directly in terms of the external loads.

Shearing Force

The algebraic sum of all the vertical forces to one side, say the left side, of section D is called the *shearing force* at that section. This is represented by $R_1 - P_1 - P_2$ for the above loading. The shearing force is opposite in direction to the resisting shear but of the same magnitude. Usually it is denoted by V. It is ordinarily used in calculations, rather than the resisting shear.

Sign Conventions

The customary sign conventions for shearing force and bending moment are represented in Fig. 6-7. Thus a force that tends to bend the beam so that it is concave upward is said to produce a positive bending moment. A force that tends to shear the left portion of the beam upward with respect to the right portion is said to produce a positive shearing force.

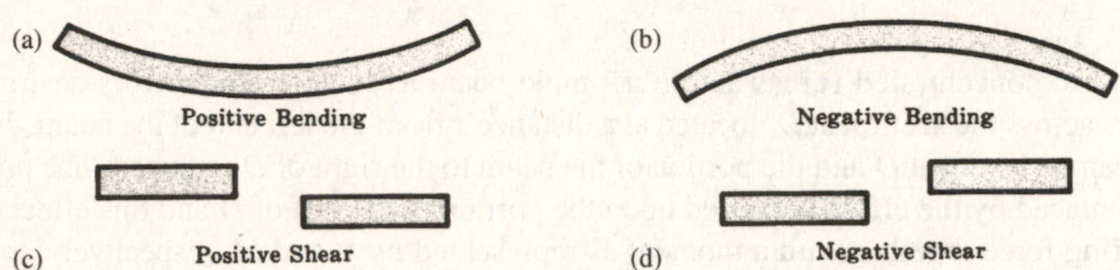

Fig. 6-7 Sign conventions for bending and shear.

An easier method for determining the algebraic sign of the bending moment at any section is to say that upward external forces produce positive bending moments, downward forces yield negative bending moments.

6.3 Shear and Moment Equations with Diagrams

Usually it is convenient to introduce a coordinate system along the beam, with the origin at one end of the beam. It will be desirable to know the shearing force and bending moment at all sections along the beam and for this purpose two equations are written, one specifying the shearing force V as a function of the distance, say x, from one end of the beam, the other giving the bending moment M as a function of x.

The plots of the equations for V and M are known as *shearing force* and *bending moment diagrams*, respectively. In these plots the abscissas (horizontals) indicate the position of the section along the beam and the ordinates (verticals) represent the values of the shearing force and bending moment, respectively. Thus these diagrams represent graphically the variation of shearing force and bending moment at any section along the length of the bar. From these plots it is quite easy to determine the maximum value of each of these quantities.

A simple beam with a varying load indicated by $w(x)$ is sketched in Fig. 6-8. The coordinate system with origin at the left end A is established and distances to various sections in the beam are denoted by the variable x. Let us derive relationships between $w(x)$, $V(x)$, and $M(x)$ at any section of the beam, shown in Fig. 6-8. The beam is subjected to any type of transverse load of the general form shown in Fig. 6-8(*a*). Simple supports are illustrated but the following consideration holds for all types of beams. We will isolate from the beam the element of length dx shown and draw a free-body diagram of it. The shearing force V acts on the left side of the element, and in passing through the distance dx the shearing force V will in general change to $V + dV$. The bending moment M acts on the left side of the element and $M + dM$ on the right side. Since dx is

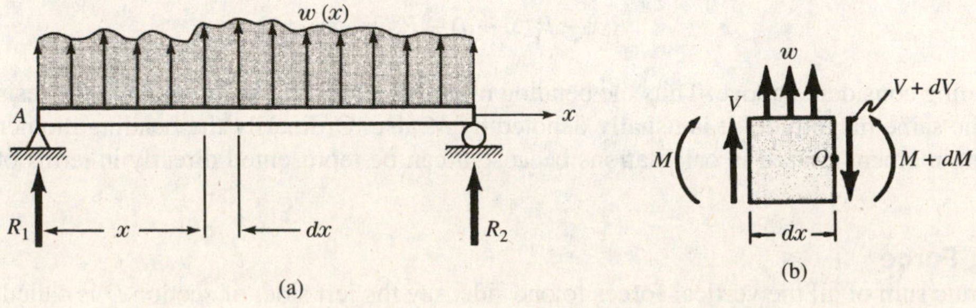

Fig. 6-8 A simple beam with a varying load.

extremely small, the applied load may be taken as uniform over the top of the beam and equal to w. The free-body diagram of this element thus appears as in Fig. 6-8(b). For equilibrium of moments about O, we have

$$\sum M_0 = M - (M + dM) + Vdx + wdx(dx/2) = 0 \qquad \text{or} \qquad dM = Vdx + \frac{1}{2}w(dx)^2 \tag{6.3}$$

Since the last term consists of the product of two differentials, it is negligible compared with the other terms involving only one differential. Hence

$$dM = Vdx \qquad \text{or} \qquad V = \frac{dM}{dx} \tag{6.4}$$

Thus the shearing force is equal to the rate of change of the bending moment with respect to x.

This equation will prove to be of considerable value in drawing shearing force and bending moment diagrams. For example, from this equation it is evident that if the sketching force is positive at a certain section of the beam then the slope of the bending moment diagram is also positive at that point. Also, it demonstrates that an abrupt change in shear, corresponding to a concentrated force, is accompanied by an abrupt change in the slope of the bending moment diagram.

Further, at those points where the shear is zero, the slope of the bending moment diagram is zero. At those points where the tangent to the moment diagram is horizontal, the moment may have a maximum or minimum value. This follows from the usual calculus technique of obtaining maximum or minimum values of a function by equating the first derivative of the function to zero. Thus in Fig. 6-9 if the curves shown represent portions of a bending moment diagram then critical values may occur at points A and B.

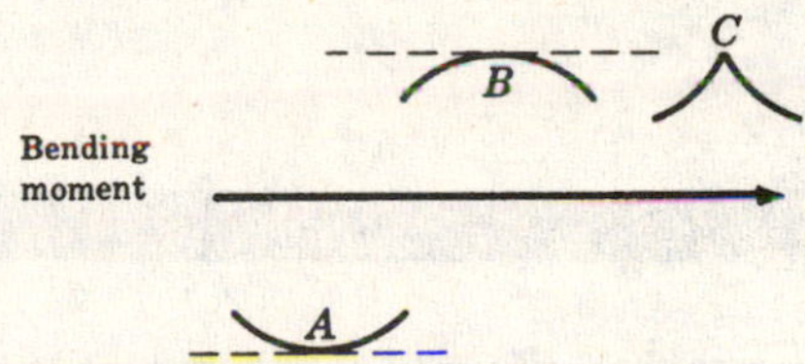

Fig. 6-9 Maximum and minimum values for M(x).

To establish the direction of concavity at a point such as A or B, we may form the second derivative of M with respect to x, that is, d^2M/dx^2. If the value of this second derivative is positive, then the moment diagram is concave upward, as at A, and the moment assumes a minimum value. If the second derivative is negative, the moment diagram is concave downward, as at B, and the moment assumes a maximum value.

However, it is to be carefully noted that the calculus method of obtaining critical values by use of the first derivative does not indicate possible maximum values at a cusp-like point in the moment diagram, if one occurs, such as that shown at C. If such a point is present, the moment there must be determined numerically and then compared to other values that are possibly critical.

Lastly, for vertical equilibrium of the element we have

$$wdx + V - (V + dV) = 0 \qquad \text{or} \qquad w = \frac{dV}{dx} \tag{6.5}$$

This relation will be of value in establishing shearing force diagrams.

6.4 Singularity Functions

The techniques discussed in the preceding Sections 6.2 and 6.3 are adequate if the loadings are continuously varying over the length of the beam. However, if concentrated forces or moments are present, a distinct pair of shearing force and bending moment equations must be written for each interval between such concentrated forces or moments. Although this presents no fundamental difficulties, it usually leads to very cumbersome results. As we shall see in Chapter 8, these results are particularly unwieldy to work with in dealing with deflections of beams.

At least some compactness of representation may be achieved by introduction of *singularity* or *half-range* functions. Let us introduce, by definition, the pointed brackets $\langle x - a \rangle$ and define this quantity to be zero if $(x - a) < 0$, that is, $x < a$, and to be simply $(x - a)$ if $(x - a) > 0$, that is, $x > a$. That is, a half-range function is defined to have a value only when the argument is positive. When the argument is positive, the pointed brackets behave just as ordinary parentheses. The singularity function

$$f_n(x) = \langle x - a \rangle^n \tag{6.6}$$

obeys the integration law

$$\int_0^x \langle y - a \rangle^n dy = \frac{\langle x - a \rangle^{n+1}}{n+1} \qquad \text{for} \qquad n \geq 0$$

The singularity function is very well suited for representation of shearing forces and bending moments in beams subject to loadings of interest. This is clear since, say in Problem 6.8 for shearing force, the effect of the distributed load is not present (explicitly) in the Eq. (1) for V for points along the beam to the left of w, but it immediately appears in the equation for V when one considers values of x to the right of the point where the distributed load begins.

The use of singularity functions for the representations of shearing force and bending moment makes it possible to describe each of these quantities by a single equation along the entire length of the beam, no matter how complex the loading may be. Most important, the singularity function approach leads to simple computer implementation.

SOLVED PROBLEMS

6.1. For the cantilever beam subject to the uniformly distributed load of w N/m of length, as shown below in Fig. 6-10(a), write equations for the shearing force and bending moment at any point along the length of the bar. Also sketch the shearing force and bending moment diagrams.

SOLUTION: It is not necessary to determine the reactions at the supporting wall. We shall choose the axis of the beam as the x-axis of a coordinate system with origin O at the left end of the bar. To determine the shearing force and bending moment at any section of the beam a distance x from the free end, we may replace the portion of the distributed load to the left of this section by its resultant. As shown by the dashed vector in Fig. 6-10(b), the resultant is a downward force of wx N acting midway between O and the section at x. Note that none of the load to the right of the section is included in calculating this resultant. Such a resultant force tends to shear the portion of the bar to the left of the section downward with respect to the portion to the right. By our sign convention this constitutes negative shear.

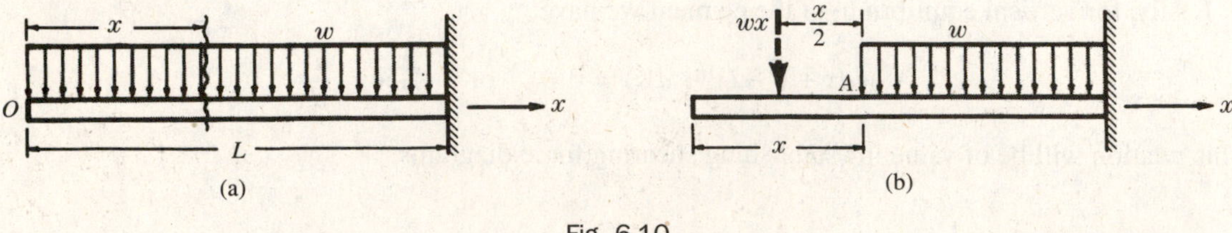

Fig. 6-10

The shearing force at this section x is defined to be the sum of the forces to the left of the section. In this case, the sum is wx acting downward; hence

$$V = -wx$$

This equation indicates that the shear is zero at $x = 0$ and when $x = L$ it is $-wL$. Since V is a first-degree function of x, the shearing force plots as a straight line connecting these values at the ends of the beam. It has the appearance shown in Fig. 6-11(a).

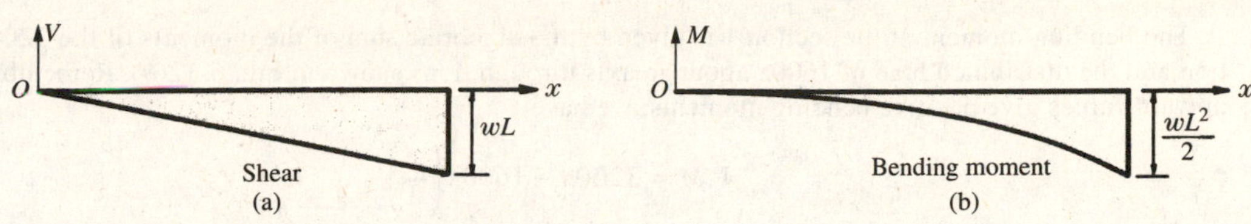

Fig. 6-11

The bending moment at this same section x is defined to be the sum of the moments of the forces to the left of this section about an axis through point A and perpendicular to the plane of the page. This sum of the moments is given by the moment of the resultant about an axis through A, that is,

$$M = -wx\left(\frac{x}{2}\right)$$

The minus sign is necessary because downward loads indicate negative bending moments. By this equation the bending moment is zero at the left end of the bar and $-wL^2/2$ at the clamped end when $x = L$. The variation of bending moment is parabolic along the bar and may be plotted as in Fig. 6-11(b).

It is to be noted that a downward uniform load as considered here leads to a bending moment diagram that is concave downward. This could be established by taking the second derivative of M with respect to x, the derivative in this particular case being $-w$. Since the second derivative is negative, calculus tells us that the curve must be concave downward.

6.2. Consider a simply supported beam 4 m long and subjected to a uniformly distributed vertical load of 1600 N/m, as shown in Fig. 6-12(a). Draw shearing force and bending moment diagrams.

SOLUTION: The total load on the beam is 6400 N, and from symmetry each of the end reactions is 3200 N. We shall now consider any cross section of the beam at a distance x from the left end. The shearing force at this section is given by the algebraic sum of the forces to the left of this section and these forces consist of the 3200 N reaction and the distributed load of 1600 N/m extending over a length x. We may replace the portion of the distributed load to the left of the section at x by its resultant, which is $1600x$ acting downward as shown by the dashed vector in Fig. 6-12(b). The shearing force at x is then given by

$$V = 3200 - 1600x$$

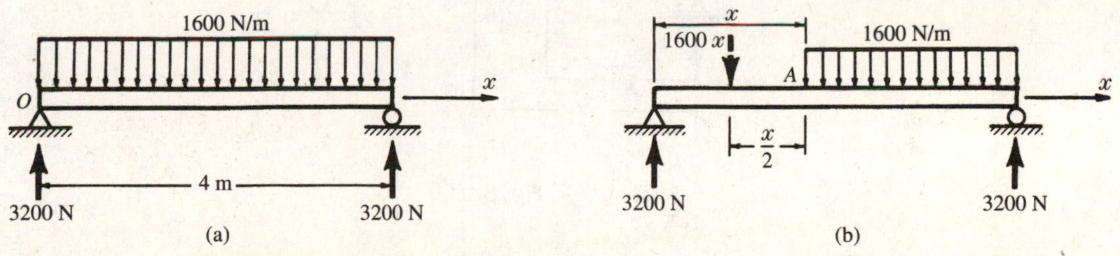

Fig 6-12

Since there are no concentrated loads acting on the beam, this equation is valid at all points along its length. The variation of shearing force along the length of the bar may then be represented by a straight line connecting the two end-point values. The shear diagram is shown in Fig. 6-13(a). The shear is zero at the center of the beam.

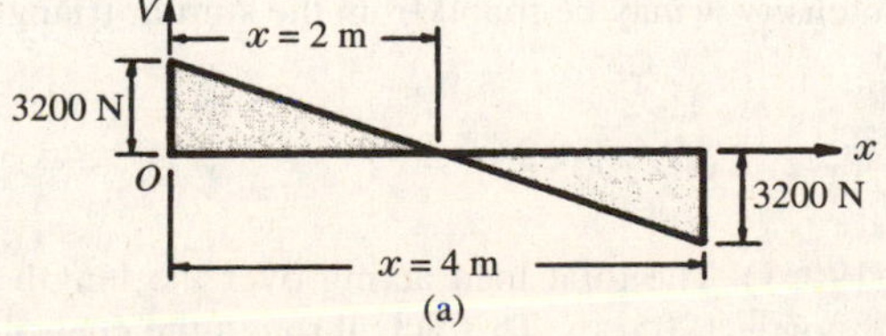

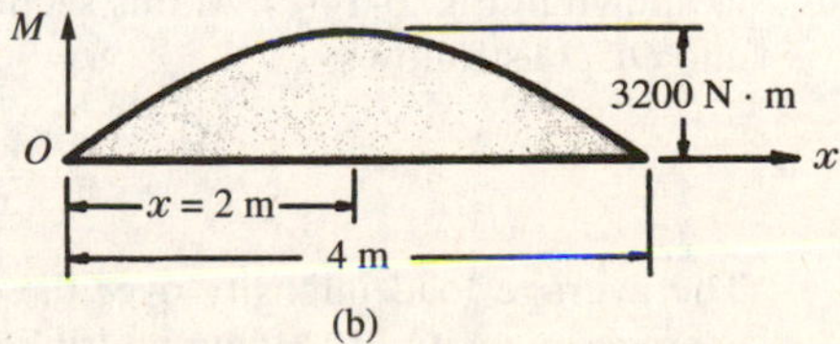

Fig. 6-13

The bending moment at the section x is given by the algebraic sum of the moments of the 3200-N reaction and the distributed load of $1600x$ about an axis through A, as shown in Fig. 6-12(b). Remembering that upward forces give positive bending moments, we have

$$M = 3200x - 1600x\left(\frac{x}{2}\right)$$
$$= 3200x - 800x^2$$

Since the load is uniformly distributed, the resultant indicated by the dashed vector acts at a distance $x/2$ from A, i.e., at the midpoint of the uniform load to the left of the section x where the bending moment is being calculated. From the above equation it is evident that the bending moment is represented by a parabola along the length of the beam as shown in Fig. 6-13(b). Since the bar is simply supported, the moment is zero at either end and, because of the symmetry of loading, the bending moment must be a maximum at the center of the beam where $x = 2$ m. The maximum bending moment is

$$M_{max} = 3200(2) - 800(2)^2 = 3200 \text{ N} \cdot \text{m}$$

6.3. The simply supported beam shown in Fig. 6-14(a) carries a vertical load that increases uniformly from zero at the left end to a maximum value of 8000 N/m at the right end. Draw the shearing force and bending moment diagrams.

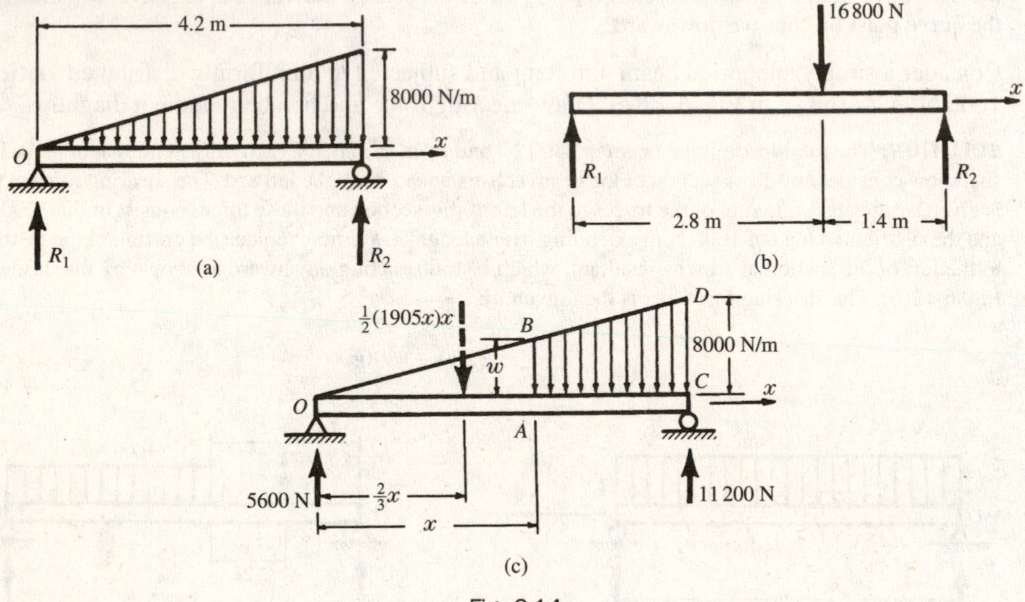

Fig. 6-14

SOLUTION: For the purpose of determining the reactions R_1 and R_2 the entire distributed load may be replaced by its resultant which will act through the centroid of the triangular loading diagram. Since the load varies from 0 at the left end to 8000 N/m at the right end, the average intensity is 4000 N/m acting over a length of 4.2 m. Hence the total load is 16 800 N applied 2.8 m to the right of the left support. The free-body diagram to be used in determining the reactions is shown in Fig. 6-14(b). Applying the equations of static equilibrium to this bar, we find $R_1 = 5600$ N and $R_2 = 11\,200$ N.

However, this resultant cannot be used for the purpose of drawing shear and moment diagrams. We must consider the distributed load and determine the shear and moment at a section, a distance x from the left end as shown in Fig. 6-14(c). At this section x the load intensity w may be found from the similar triangles OAB and OCD as follows:

$$\frac{w}{x} = \frac{8000}{4.2} \qquad \text{or} \qquad w = 1905x$$

The average load intensity over the length x is $\frac{1}{2}(1905x)$. The total load acting over the length x is the average intensity of loading multiplied by the length, or $\frac{1}{2}(1905x)x$. This acts through the centroid of the triangular region OAB shown, i.e., through a point located a distance $\frac{2}{3}x$ from O. The resultant of this portion of the distributed load is indicated by the dashed vector in Fig. 6-14(*c*). No portion of the load to the right of the section x is included in this resultant force.

The shearing force and bending moment at *A* are now readily found to be

$$V = 5600 - \frac{1}{2}(1905x)x = 5600 - 952x^2$$

$$M = \int V dx = 5600x - 317x^2$$

The shearing force thus plots as a parabola, having a value 5600 N when $x = 0$ and $-11\,200$ N when $x = 4.2$ m. The bending moment vanishes at the ends and assumes a maximum value where the shear is zero. This is true because $V = dM/dx$, and hence the point of zero shear must be the point where the tangent to the moment diagram is horizontal. This point of zero shear may be found by setting $V = 0$:

$$0 = 5600 - 952x^2 \qquad \text{or} \qquad x = 2.43 \text{ m}$$

The bending moment at this point is found to be

$$M_{x=2.43} = 5600(2.43) - 317(2.43)^3 = 9060 \text{ N} \cdot \text{m}$$

The plots of the shear and moment equations appear in Fig. 6-15.

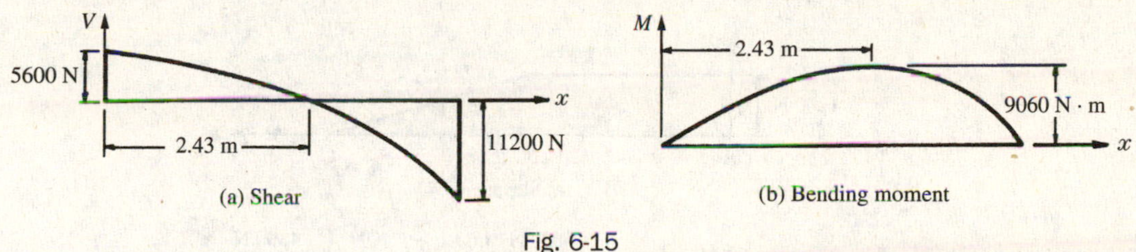

Fig. 6-15

6.4. The cantilever beam *AC* in Fig. 6-16 is loaded by the uniform load of 600 N/m over the length *BC* together with the couple of magnitude 4800 N · m at the tip *C*. Determine the shearing force and bending moment diagrams.

SOLUTION: The reactions at *A* must consist of a vertical shearing force together with a moment. To find these reactions, we write the statics equations

$$\Sigma F_y = R_A - (600)(2) = 0 \tag{1}$$

$$\Sigma M_A = M_A - 4800 - (1200) \cdot (3) = 0 \tag{2}$$

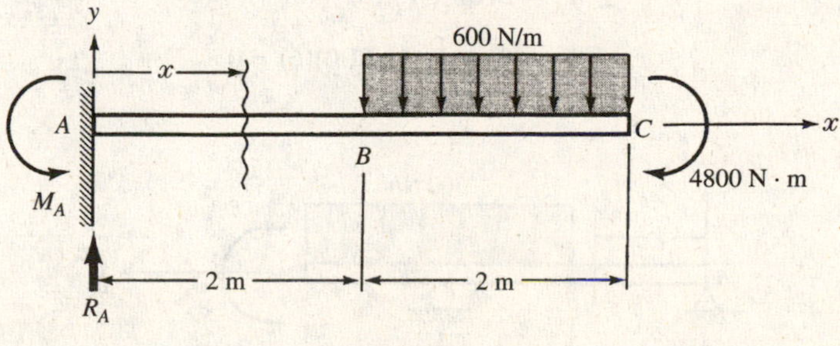

Fig. 6-16

Solving,

$$R_A = 1200 \text{ N} \qquad M_A = 8400 \text{ N} \cdot \text{m}$$

The shearing force at any point, say, a distance x to the right of A is given by the sum of all forces to the left of x. Thus we must write the two equations

$$V = 1200 \text{ N} \qquad 0 < x < 2 \text{ m} \tag{3}$$

$$V = 1200 - 600(x - 2) \qquad 2 < x < 4 \text{ m} \tag{4}$$

Likewise, the bending moment at this point x is given by the sum of the moments of all forces (and couples) to the left of x about point x. This is given by the two equations

$$M = -8400 + 1200x \qquad 0 < x < 2 \text{ m} \tag{5}$$

$$M = 8400 + 1200x - 300(x - 2)^2 \qquad 2 < x < 4 \text{ m} \tag{6}$$

Plots of Eqs. (3) through (6) appear in Figs. 17(a) and (b), respectively. The nature of the concave region of the bending moment in BC is determined by taking the second derivative of the bending moment Eq. (6) in BC:

$$\frac{d^2M}{dx^2} = -600$$

Since this is negative for values of x in BC, the plot in BC of bending moment is concave downward. The bending moment in AB is seen from Eq. (5) to be a linear function of x; hence the bending moment in AB plots as a straight line connecting the end couple of –8400 N · m with the bending moment at B of –6000 N · m, as determined from Eq. (6).

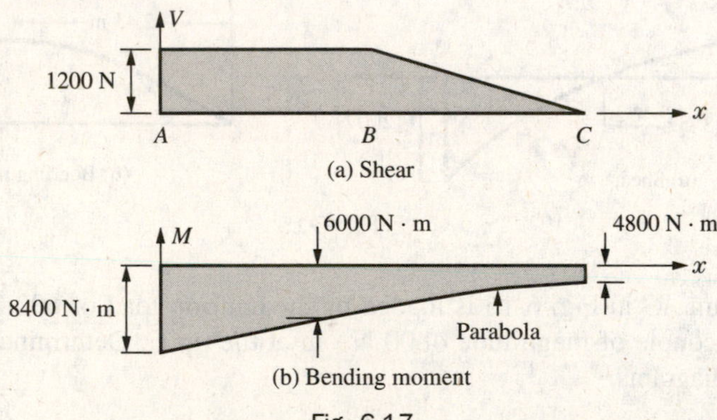

Fig. 6-17

6.5. The beam AC is simply supported at A and C and subject to the uniformly distributed load of 300 N/m plus the couple of magnitude 2700 N · m as shown in Fig. 6-18. Write equations for shearing force and bending moment and make sketches of these equations.

SOLUTION: It is necessary to first determine the reactions from the equilibrium equations

$$\Sigma M_A = 2700 + R_C(6) - (300)(6)(6) = 0 \tag{1}$$

$$\Sigma F_y = R_A + R_C - (300)(6) = 0 \tag{2}$$

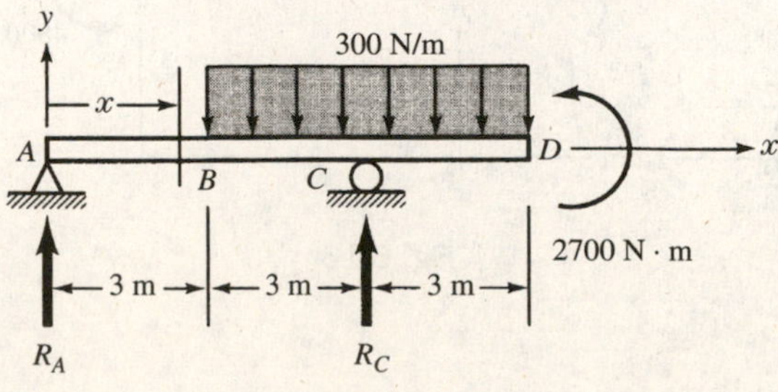

Fig. 6-18

Solving,

$$R_A = 450\,\text{N} \qquad R_C = 1350\,\text{N}$$

For the coordinate x as shown, the shearing force at a distance x from point A is described by the three relations

$$V = 450 \qquad 0 < x < 3\,\text{m} \tag{3}$$

$$V = 450 - 300(x - 3) \qquad 3 < x < 6\,\text{m} \tag{4}$$

$$V = 450 - 300(x - 3) + 1350 \qquad 6 < x < 9\,\text{m} \tag{5}$$

Likewise the bending moment in each of these three regions of the beam is described by

$$M = 450x \qquad 0 < x < 3\,\text{m} \tag{6}$$

$$M = 450x - 300(x - 3)\left(\frac{x-3}{2}\right) \qquad 3 < x < 6\,\text{m} \tag{7}$$

$$M = 450x - 300\frac{(x-3)^2}{2} + 1350(x - 6) \qquad 6 < x < 9\,\text{m} \tag{8}$$

Sketches of these equations appear in Fig. 6-19. In regions BC and CD it is necessary to determine that the second derivative of the bending moment from Eq. (7) and Eq. (8) is negative in each of these regions, and that hence in each case the curvature of the bending moment plot is concave downward.

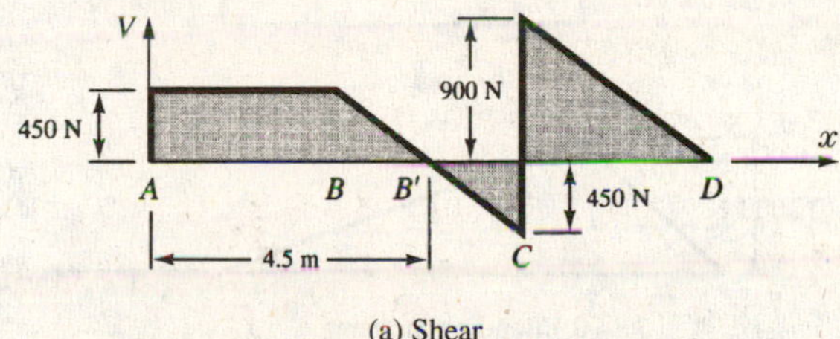

(a) Shear

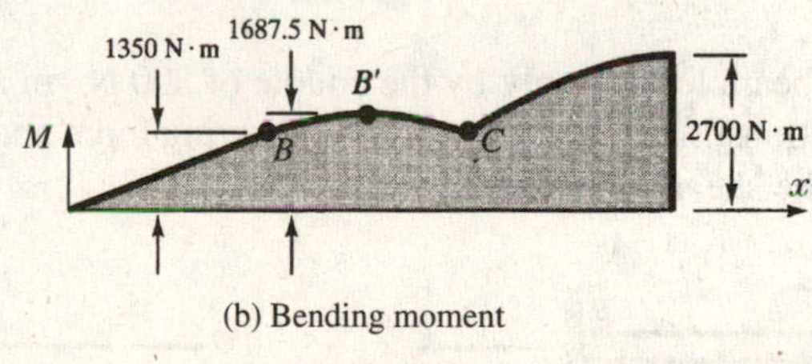

(b) Bending moment

Fig. 6-19

6.6. Use singularity functions to write equations for the shearing force and bending moment at any position in the simply supported beam shown in Fig. 6-20.

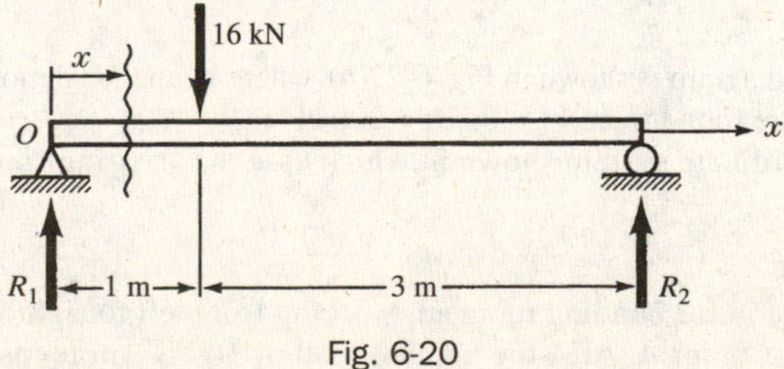

Fig. 6-20

SOLUTION: From statics the reactions are easily found to be

$$R_1 = 12000\,\text{N} \qquad R_2 = 4000\,\text{N}$$

For the coordinate system shown, with origin at O, we may write

$$V = 12000\langle x\rangle^0 - 16000\langle x-1\rangle^0 \tag{1}$$

which indicates that $V = 12\,000\ N$ if $x < 1$ m and $V = 12\,000 - 16\,000 = -4000\ N$ if $x > 1$ m.

Similarly,

$$M = 12000\langle x\rangle^1 - 16000\langle x-1\rangle^1 \tag{2}$$

which tells us that $M = 12\,000x$ if $x < 1$ m and $M = 12\,000x - 16\,000(x-1)$ if $x > 1$ m.

The relations (1) and (2) hold for all values of x provided we apply the definition of singularity functions. Use of these equations leads to the shearing force and bending moment diagrams shown in Fig. 6-21.

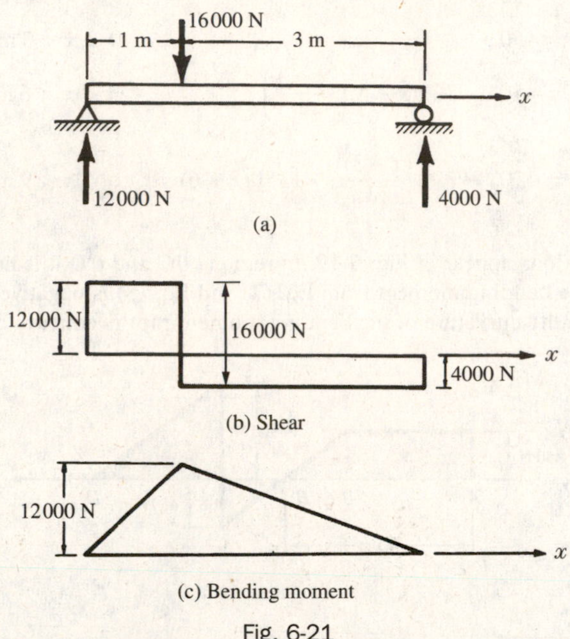

Fig. 6-21

6.7. Consider a cantilever beam loaded only by the couple of 300 N · m applied as shown in Fig. 6-22(*a*). Using singularity functions, write equations for the shearing force and bending moment at any position in the beam and plot the shear and moment diagrams.

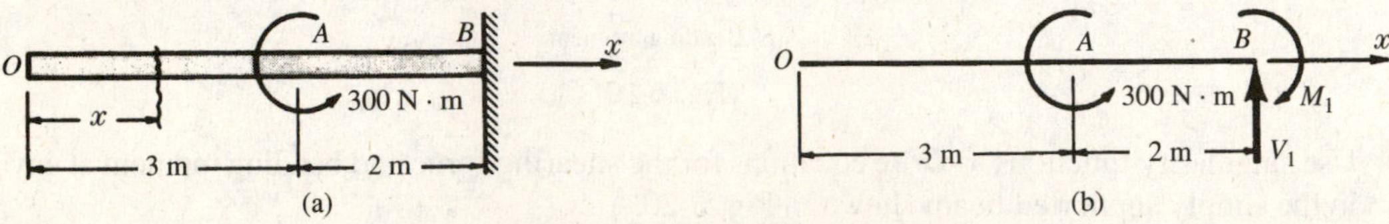

Fig. 6-22

SOLUTION: A free-body diagram is shown in Fig. 6-22(*b*), where V_1 and M_1 denote the reactions of the supporting wall. From statics these are found to be $V_1 = 0$, $M_1 = 300$ N · m.

We introduce the coordinate system shown in which case the shearing force everywhere is

$$V = 0 \tag{1}$$

In writing the expression for bending moment, working from left to right it is clear that there is no bending moment to the left of point A. At A the applied load of 300 N · m tends to bend the portion AB into a curvature that is concave downward, which according to our sign convention is negative bending. Thus the bending moment anywhere in the beam is

$$M = -300\langle x-3\rangle^0 \tag{2}$$

Plots of (1) and (2) appear in Fig. 6-23.

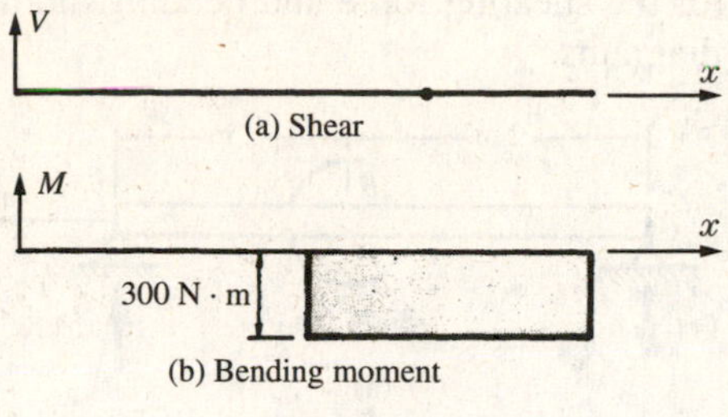

(a) Shear

(b) Bending moment

Fig. 6-23

6.8. Consider a cantilever beam loaded by a concentrated force at the free end together with a uniform load distributed over the right half of the beam [see Fig. 6-24(*a*)]. Using singularity functions, write equations for the shearing force and bending moment at any point in the beam and plot the shear and moment diagrams.

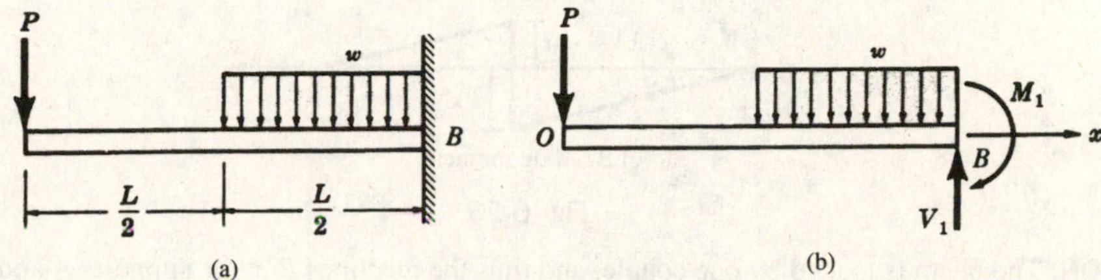

(a) (b)

Fig. 6-24

SOLUTION: A free-body diagram is shown in Fig. 6-27(*b*). From statics, the wall reactions are found to be

$$V_1 = P + \frac{wL}{2} \qquad M_1 = PL + \frac{wL^2}{8}$$

although for the case of a cantilever it is not necessary to find these prior to writing shearing force and bending moment equations, since there are no unknown loads to the left of the wall (P and w are assumed known).

With the coordinate system shown, with origin at O, the effect of the concentrated force P as well as the distributed load is to produce negative shear. Thus we may write

$$V = -P\langle x\rangle^0 - w\left\langle x - \frac{L}{2}\right\rangle^1 \tag{1}$$

which indicates shearing force at any position x if one applies the definition of the bracketed term.

Likewise, the bending moment at any position x is [we use the free-body or integrate (1)]

$$M = -P\langle x\rangle^1 - \frac{w}{2}\left\langle x - \frac{L}{2}\right\rangle^2 \tag{2}$$

The loaded beam together with plots of the shear and moment equations are shown in Fig. 6-25.

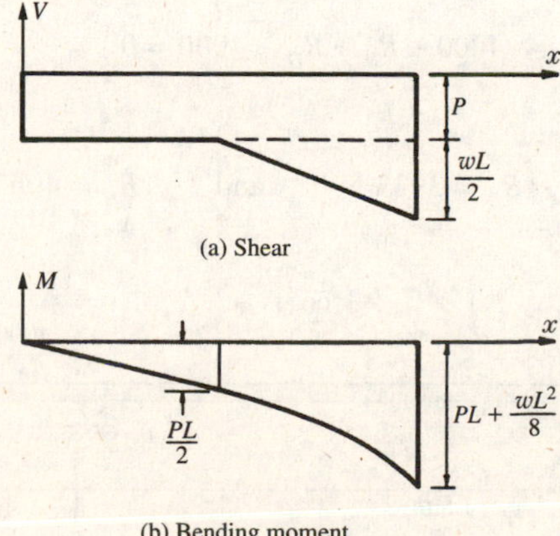

(a) Shear

(b) Bending moment

Fig. 6-25

6.9. In Fig. 6-26(a) a simply supported beam is loaded by the couple of 3 kN · m. Using singularity functions, write equations for the shearing force and bending moment at any point in the beam and plot the shear and moment diagrams.

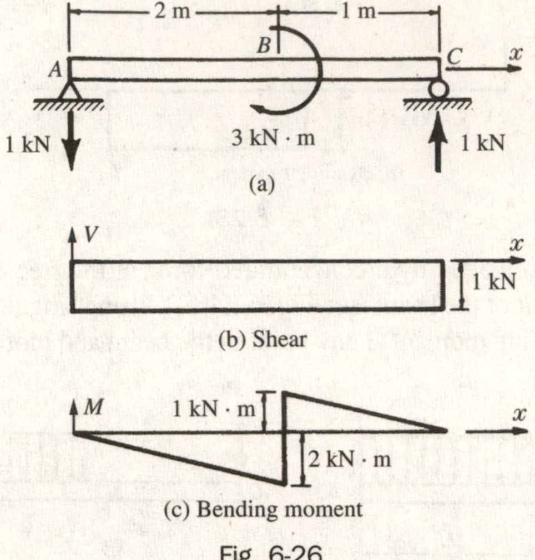

Fig. 6-26

SOLUTION: The beam is loaded by one couple, and thus the reactions R at the supports A and C must constitute another couple. These reactions appear as in Fig. 6-26(b). For equilibrium,

$$\sum M_A = 3R - 3 = 0 \qquad \therefore R = 1\text{ kN}$$

Between A and B the shearing force is negative and also the bending moment is negative. Therefore the expressions for V and M are

$$V = -(1)\langle x\rangle^0$$
$$M = -(1)\langle x\rangle^1 + 3\langle x-2\rangle^0$$

Shear and moment diagrams are plotted in Figs. 6-26(b) and 6-26(c), respectively. The bending moment diagram exhibits an abrupt jump or discontinuity at the point where the couple is applied.

6.10. The overhanging beam AE is subject to uniform normal loadings in the regions AB and DE, together with a couple acting at the midpoint C as shown in Fig. 6-27. Using singularity functions, write equations for the shearing force and bending moment at any point in the beam and plot the shear and moment diagram.

SOLUTION: To first determine the reactions, we have from statics

$$\sum M_B = (4000)(1)(0.5) + 200 + R_D(3) - (4000)(1)(3.5) = 0 \tag{1}$$

$$\sum F_y = -4000 + R_B + R_D - 4000 = 0 \tag{2}$$

Solving,

$$R_D = 3933\text{ N} \qquad \text{and} \qquad R_B = 4067\text{ N} \tag{3}$$

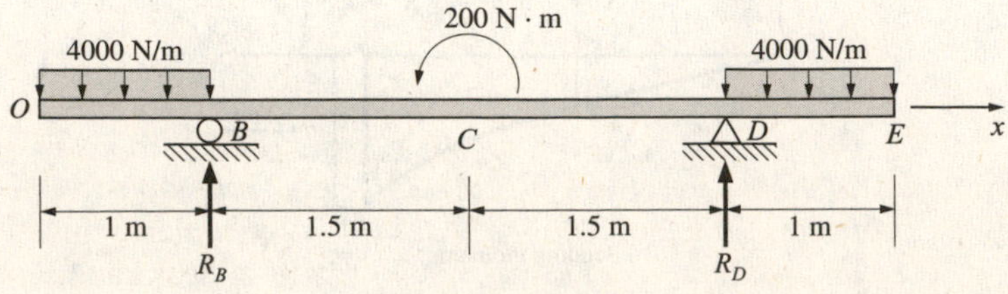

Fig. 6-27

For the coordinate system shown and using the definition of the singularity function, we may write

$$V = \overset{①}{-4000\langle x\rangle^1} + \overset{②}{4000\langle x-1\rangle^1} + \overset{③}{4067\langle x-1\rangle^0} + \overset{④}{3933\langle x-4\rangle^0} - \overset{⑤}{4000\langle x-4\rangle^1} \tag{4}$$

$$M = \overset{⑥}{-4000\frac{\langle x\rangle^2}{2}} + \overset{⑦}{-4000\frac{\langle x-1\rangle^2}{2}} + \overset{⑧}{4067\langle x-1\rangle^1} - \overset{⑨}{200\langle x-2.5\rangle^0}$$
$$+\overset{⑩}{3933\langle x-4\rangle^1} - \overset{⑪}{4000\frac{\langle x-4\rangle^2}{2}} \tag{5}$$

Equations (4) and (5) each contain quantities designated by the numerals circled above the terms. Terms may be interpreted as follows for shearing force V:

I. The shearing force V acting in region AB of Fig. 6-27 is, for any value of the coordinate x in AB, simply the sum of all applied downward normal forces to the left of x, i.e., $4000x$, which is term ①. Such forces tend to produce the type of displacement shown in Fig. 6-7(*d*), hence we must prefix the load $4000\langle x\rangle$ by a negative sign.

II. Continuing, the first term ① in Eq. (4) holds for all values of x ranging from $x = 0$ to $x = 5$ m. That is, the singularity functions are defined as being zero if the quantity in brackets $\langle\ \rangle$ is negative, but there is no way to specify an upper bound on the coordinate x shown in term ①. Consequently, we must annul the downward 4000 N/m load to the right of point B and this may be accomplished by adding an upward (positive) uniform load to the right of B, i.e., for all values of $x > 1$ m, which is term ②. But this upward uniform load has now annulled the actual downward uniform load in region DE. We will return to this shortly.

III. Immediately to the right of B the upward reaction R_B has a shear effect of 4067 N upward so that it tends to produce displacement such as shown in Fig. 6-7(*c*), which we term positive, hence the positive sign in term ③.

IV. The applied couple of 200 N · m has no force effect in any direction, hence does not appear in Eq. (4).

V. Immediately to the right of D the upward reaction R_D has a shear effect of 3933 N upward so that it tends to produce displacement such as shown in Fig. 6-7(*c*), which we term positive, hence the positive sign in term ④.

VI. As mentioned in (II), the true downward uniform load in DE has temporarily been annulled, hence we must introduce the term ⑤ to return it and make the external loading correct.

Equation (4) in terms of singularity functions now correctly specifies the vertical shear at all points on the beam from O to E. A plot of this is given below in Fig. 6-28(*a*).

In a nearly comparable manner, the bending moment from O to E may be written, except that now account must be taken of the applied moment of 200 N · m at C. The moment equation is given in (5) and a plot of it from O to E appears in Fig. 6-28(*b*). The curves at both ends of the moment diagrams are parabolas.

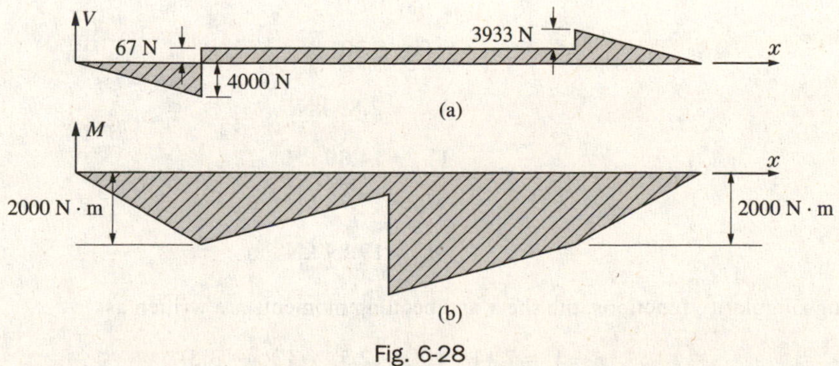

Fig. 6-28

6.11. The simply supported beam AD is subject to a uniform load over the segment BC together with a concentrated force applied at C as shown in Fig. 6-29. Using singularity functions, write equations for the shearing force and bending moment at any point in the beam and plot shear and moment diagrams.

SOLUTION: The vertical reactions at A and D must first be determined from statics:

$$\Sigma M_A = 4.5R_D - 12(3.5) - (20)(3.5) = 0$$

$$R_D = 24.89 \text{ kN}$$

$$\Sigma F_y = R_A + 24.89 - 12 - 20 = 0$$

$$R_A = 7.11 \text{ kN}$$

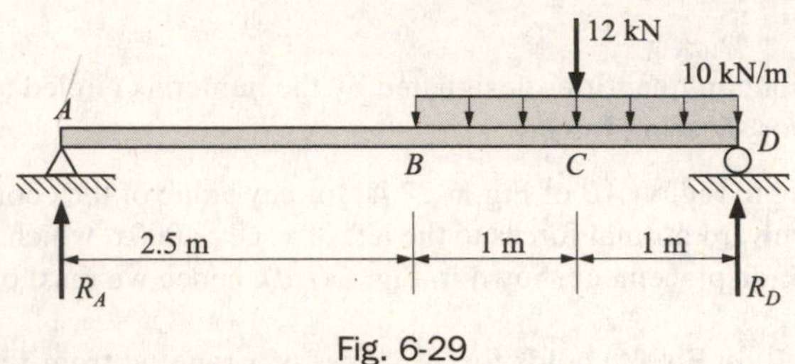

Fig. 6-29

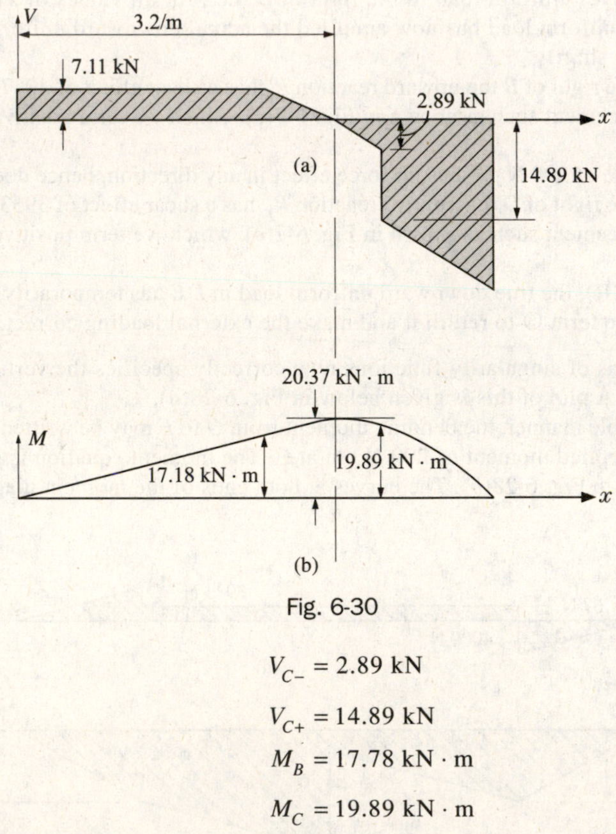

Fig. 6-30

$$V_{C-} = 2.89 \text{ kN}$$

$$V_{C+} = 14.89 \text{ kN}$$

$$M_B = 17.78 \text{ kN} \cdot \text{m}$$

$$M_C = 19.89 \text{ kN} \cdot \text{m}$$

Using singularity functions, the shear and bending moments are written as

$$V = 7.11 - 10\langle x - 2.5\rangle^1 - 12\langle x - 3.5\rangle^0$$

$$M = 7.11\langle x\rangle^1 - 10\frac{\langle x - 2.5\rangle^2}{2} - 12\langle x - 3.5\rangle^1$$

From these equations the shear and moment diagrams may be plotted as shown in Figs. 6-30(a) and (b), respectively.

SUPPLEMENTARY PROBLEMS

6.12. A simply supported beam is subject to a uniform load of 2 kN/m over the region shown in Fig. 6-31. Determine the maximum shearing force and bending moment in the beam. *Ans.* 2000 N, 2000 N · m

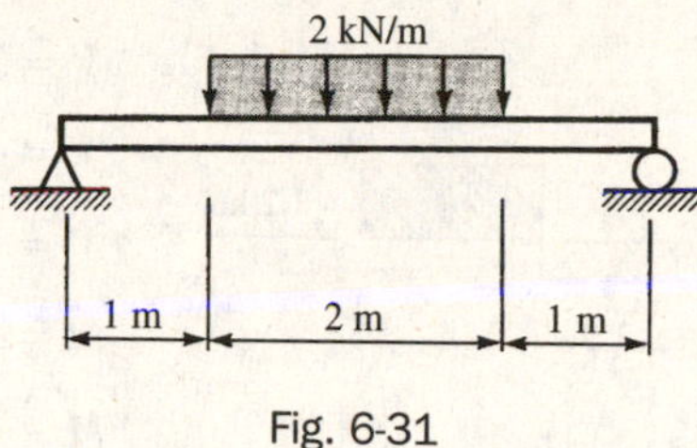

Fig. 6-31

6.13. Determine the maximum shearing force and bending moment in the simply supported beam shown in Fig. 6-32. *Ans.* 15 383 N, 8897 N · m

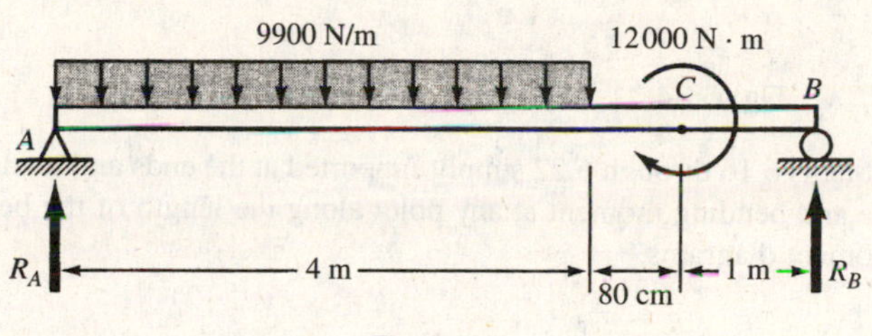

Fig. 6-32

For the cantilever beams loaded as shown in Fig. 6-33 and Fig. 6-34, write equations for the shearing force and bending moment at any point along the length of the beam. Also, draw the shearing force and bending moment diagrams.

6.14.

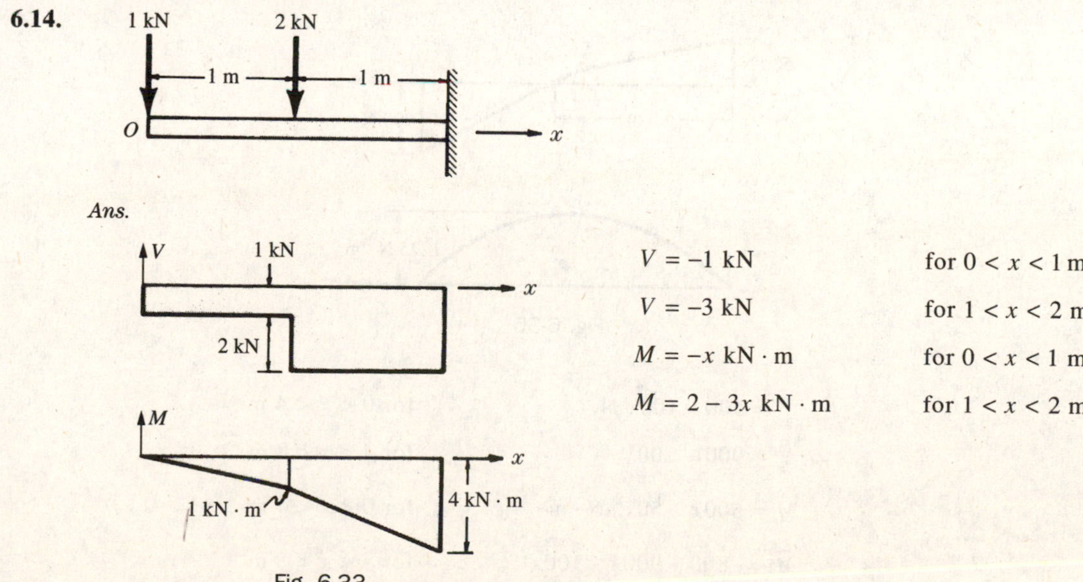

$V = -1$ kN for $0 < x < 1$ m

$V = -3$ kN for $1 < x < 2$ m

$M = -x$ kN · m for $0 < x < 1$ m

$M = 2 - 3x$ kN · m for $1 < x < 2$ m

Fig. 6-33

6.15.

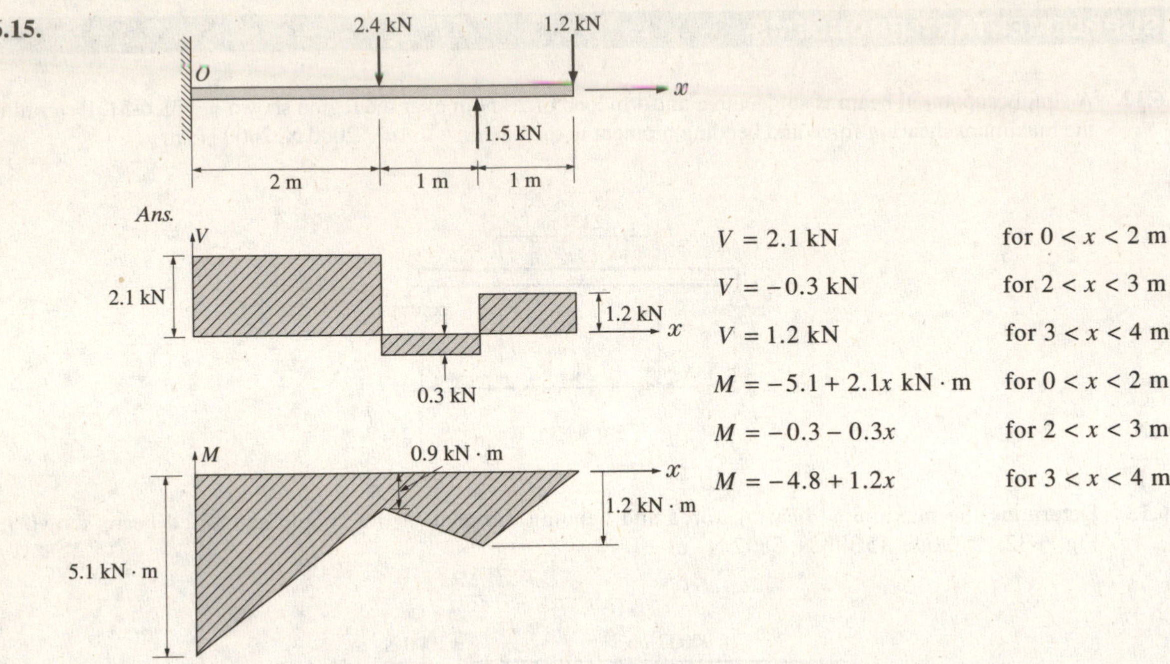

$V = 2.1$ kN for $0 < x < 2$ m

$V = -0.3$ kN for $2 < x < 3$ m

$V = 1.2$ kN for $3 < x < 4$ m

$M = -5.1 + 2.1x$ kN · m for $0 < x < 2$ m

$M = -0.3 - 0.3x$ for $2 < x < 3$ m

$M = -4.8 + 1.2x$ for $3 < x < 4$ m

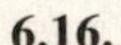

Fig. 6-34

For the beams of Problems 6.16 through 6.22 simply supported at the ends and loaded as shown, write equations for the shearing force and bending moment at any point along the length of the beam. Also, draw the shearing force and bending moment diagrams.

6.16.

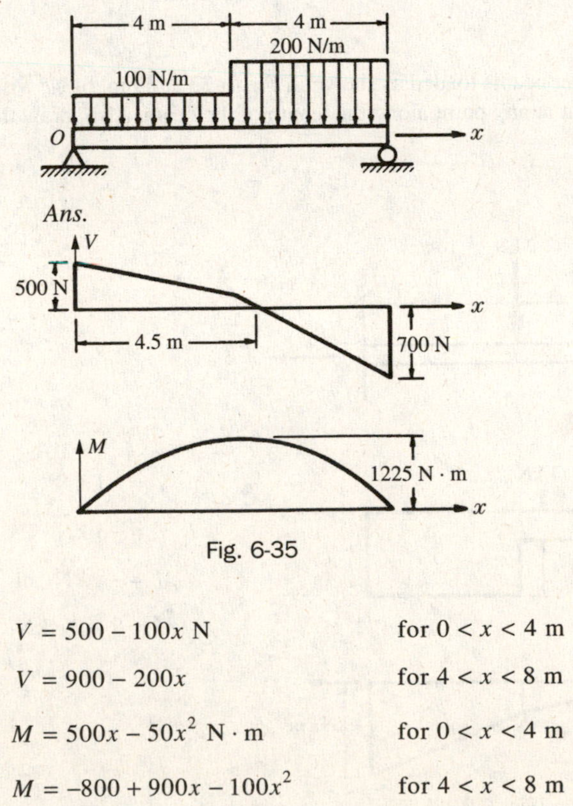

Fig. 6-35

$V = 500 - 100x$ N for $0 < x < 4$ m

$V = 900 - 200x$ for $4 < x < 8$ m

$M = 500x - 50x^2$ N · m for $0 < x < 4$ m

$M = -800 + 900x - 100x^2$ for $4 < x < 8$ m

6.17.

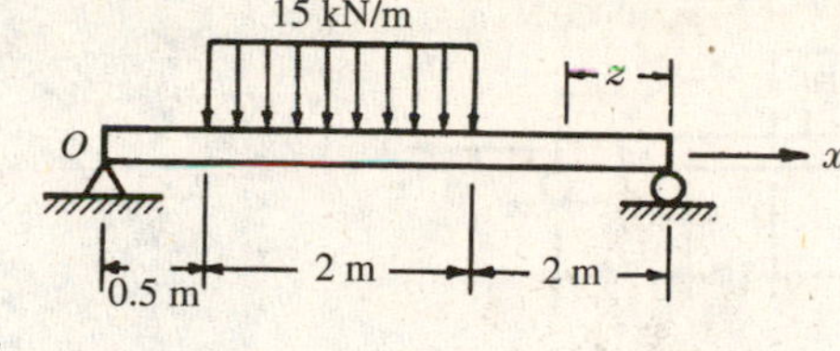

Ans.

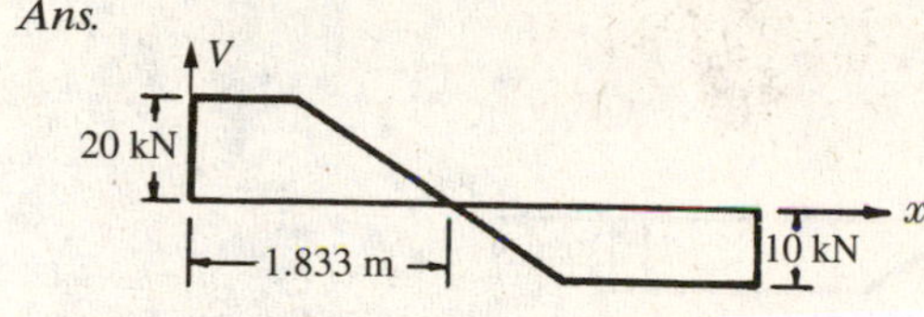

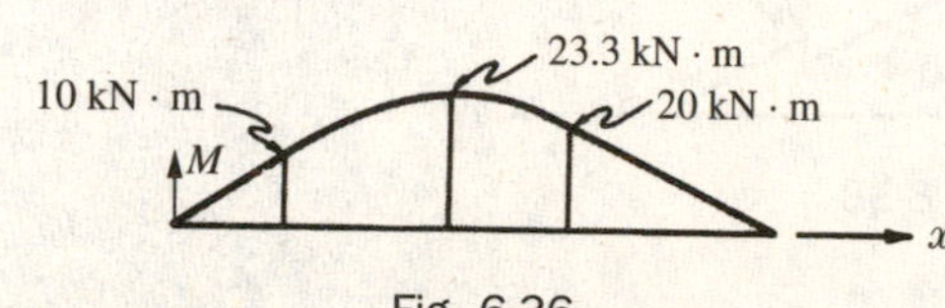

Fig. 6-36

$V = 20$ kN	for $0 < x < 0.5$ m
$V = 27.5 - 15x$	for $0.5 < x < 2.5$ m
$V = -10$ kN	for $2.5 < x < 4.5$ m
$M = 20x$ kN · m	for $0 < x < 0.5$ m
$M = -1.875 + 27.5x - 7.5x^2$	for $0.5 < x < 2.5$ m
$M = 10z$ kN · m	for $0 < z < 2$ m

6.18.

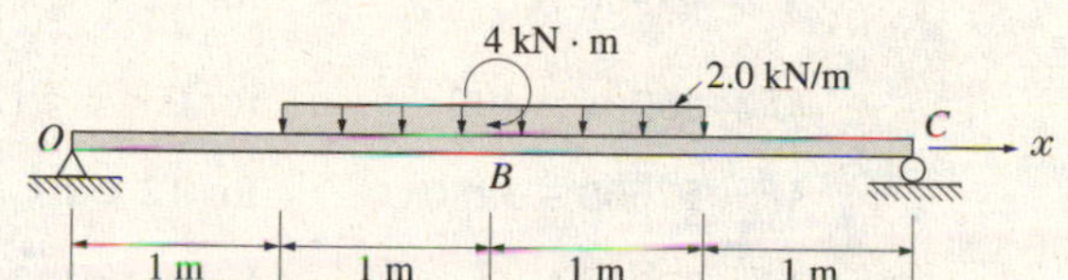

Ans.

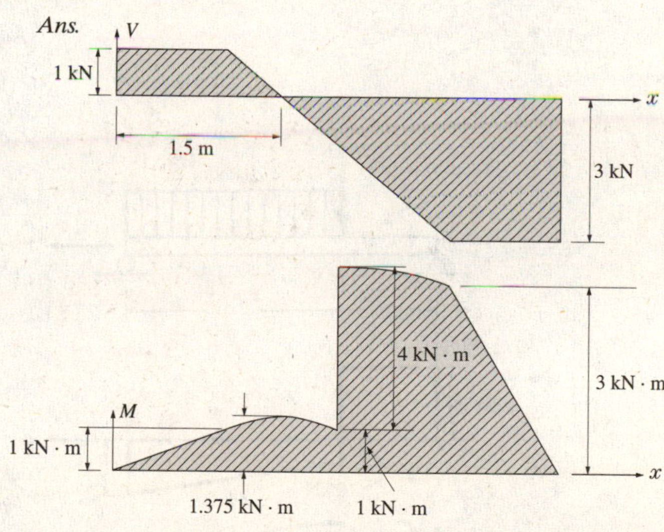

Fig. 6-37

$V = 1$ kN	for $0 < x < 1$ m
$V = 3 - 2x$	for $1 < x < 3$ m
$V = 3$ kN	for $3 < x < 4$ m
$M = x$ kN · m	for $0 < x < 1$ m
$M = -1 + 3x - x^2$	for $1 < x < 2$ m
$M = 3 + 3x - x^2$	for $2 < x < 3$ m
$M = 12 - 3x$	for $3 < x < 4$ m

6.19.

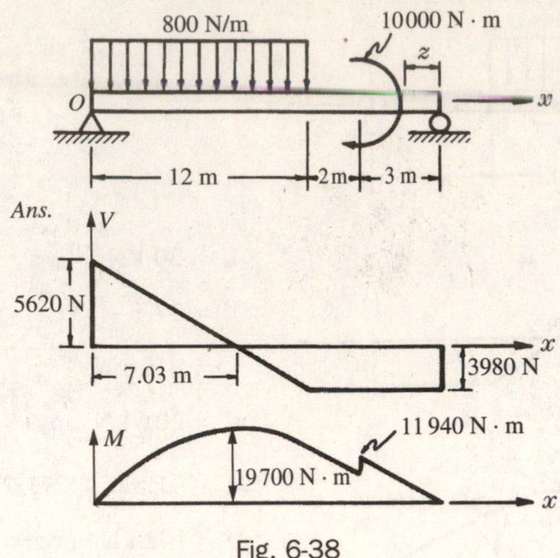

Fig. 6-38

$V = 5620 - 800x$ for $0 < x < 12$ m

$V = -3980$ N for $12 < x < 17$ m

$M = 5620x - 400x^2$ for $0 < x < 12$ m

$M = 57600 - 3980x$ for $12 < x < 14$ m

$M = 3980z$ N · m for $0 < z < 3$ m

6.20.

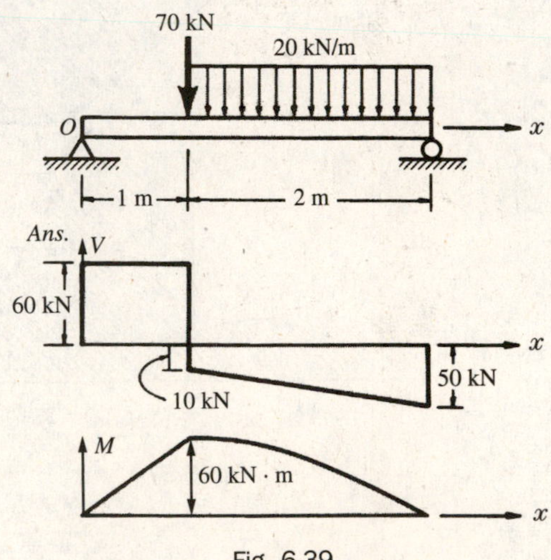

Fig. 6-39

$V = 60$ kN for $0 < x < 1$ m

$V = 10 - 20x$ for $1 < x < 3$ m

$M = 60x$ kN · m for $0 < x < 1$ m

$M = 60 + 10x - 10x^2$ for $1 < x < 3$ m

6.21.

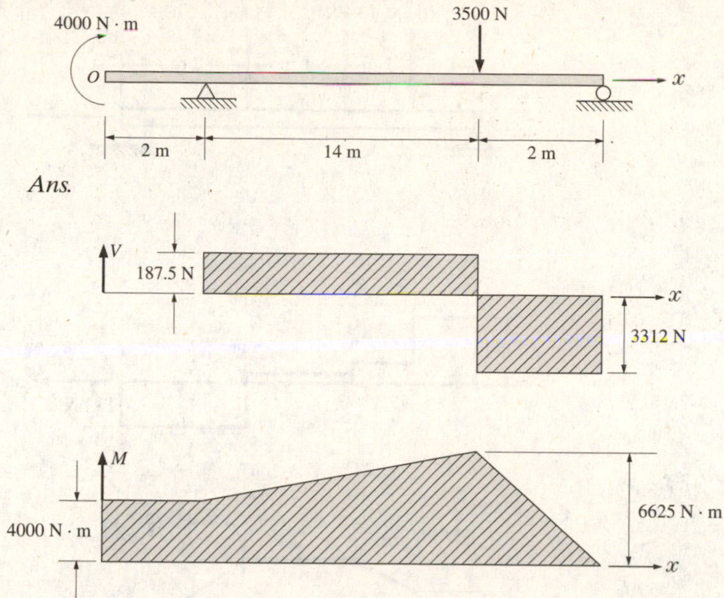

Fig. 6-40

$V = 0$	for $0 < x < 2$ m
$V = 187.5$ N	for $2 < x < 16$ m
$V = -3312.5$ N	for $16 < x < 18$ m
$M = 4000$ N · m	for $0 < x < 2$ m
$M = 3625 + 187.5x$	for $2 < x < 16$ m
$M = 59620 - 3312x$	for $16 < x < 18$ m

6.22.

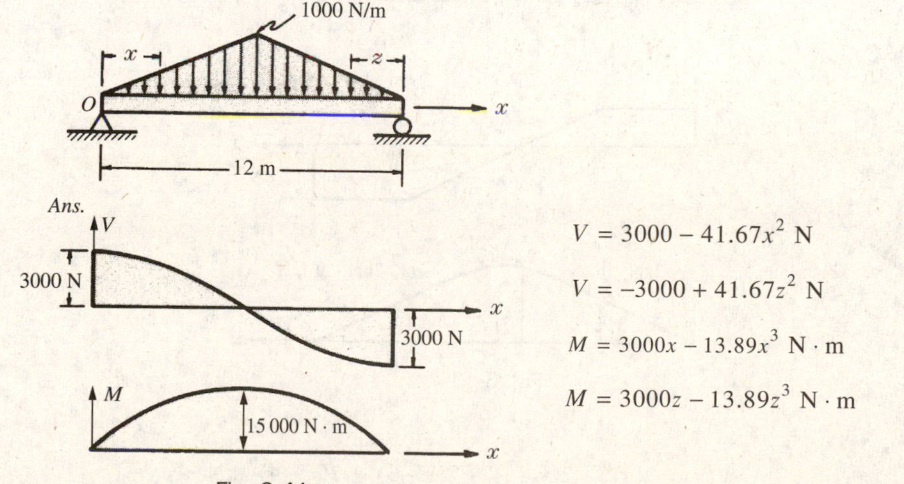

$V = 3000 - 41.67x^2$ N	for $0 < x < 6$ m
$V = -3000 + 41.67z^2$ N	for $0 < z < 6$ m
$M = 3000x - 13.89x^3$ N · m	for $0 < x < 6$ m
$M = 3000z - 13.89z^3$ N · m	for $0 < z < 6$ m

Fig. 6.41

For Problems 6.23 through 6.26 use singularity functions to write the equations for shearing force and bending moment at any point in the beam. Plot the corresponding diagrams.

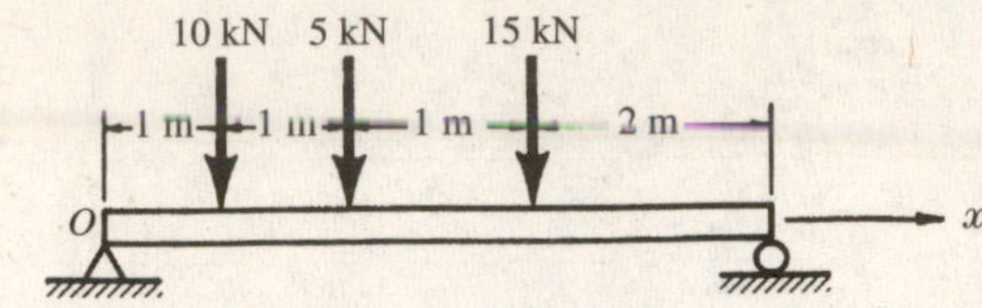

Ans.

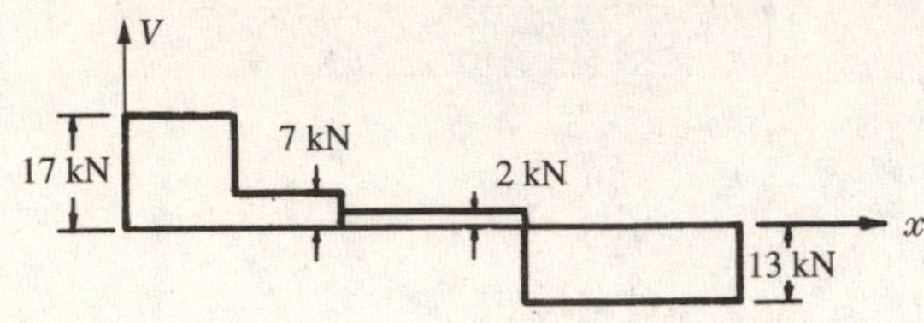

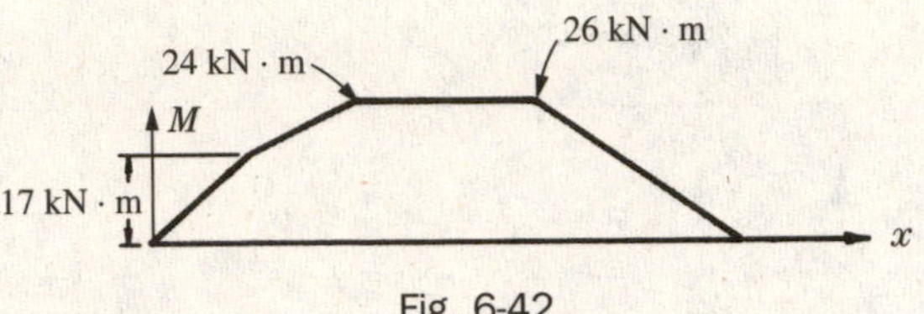

Fig. 6-42

6.23.

$$V(x) = 17\langle x\rangle^0 - 10\langle x-1\rangle^0 - 5\langle x-2\rangle^0 - 15\langle x-3\rangle^0 \text{ kN}$$

$$M(x) = 17\langle x\rangle^1 - 10\langle x-1\rangle^1 - 5\langle x-2\rangle^1 - 15\langle x-3\rangle^1 \text{ kN}\cdot\text{m}$$

6.24.

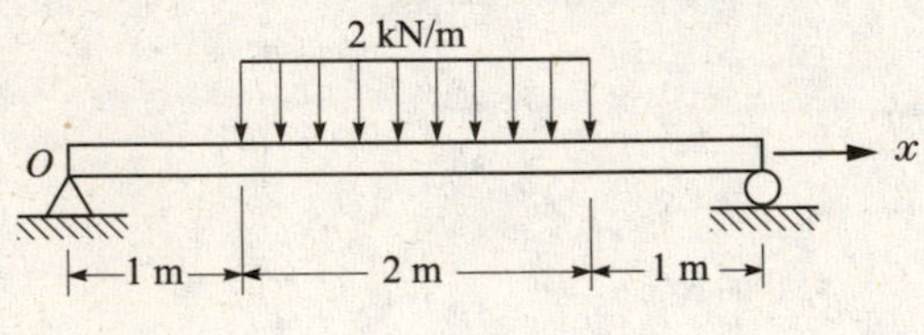

Ans.

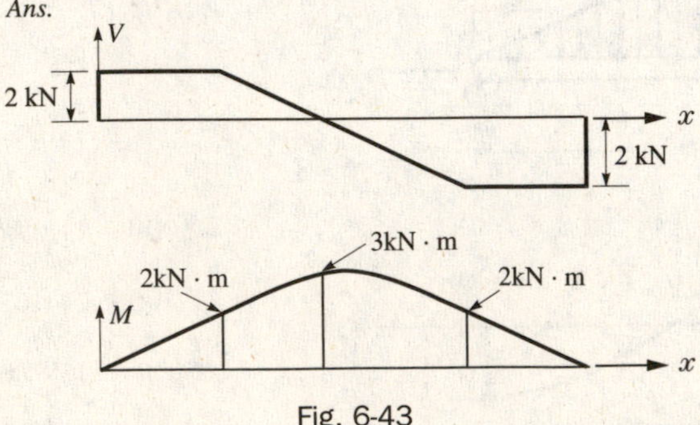

Fig. 6-43

$$V(x) = 2\langle x\rangle^0 - 2\langle x-1\rangle^1 + 2\langle x-3\rangle^1 + 2\langle x-4\rangle^0 \text{ kN}$$

$$M(x) = 2\langle x\rangle^1 - 1\langle x-1\rangle^2 + 1\langle x-3\rangle^2 + 2\langle x-4\rangle^1 \text{ kN}\cdot\text{m}$$

6.25.

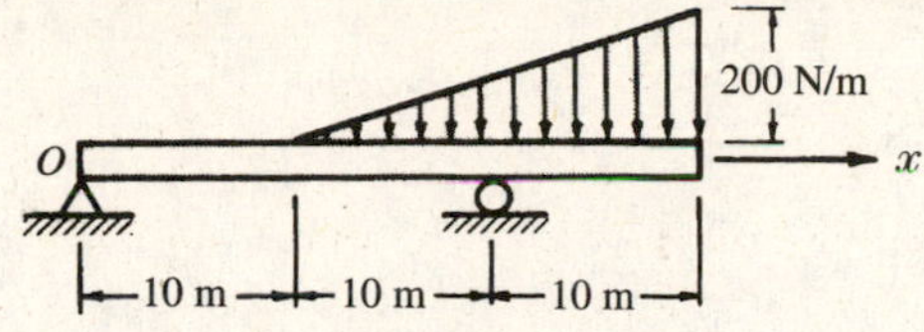

Ans.

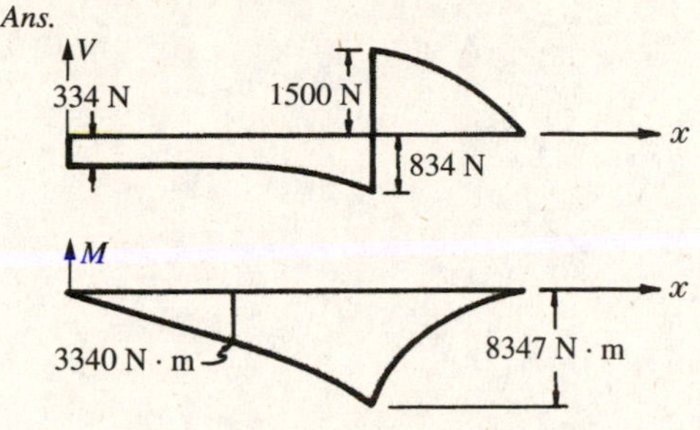

Fig. 6-44

$$V\langle x\rangle = -334\langle x\rangle^0 - 5\langle x-10\rangle^2 + 2334\langle x-20\rangle^0 \text{ N}$$

$$M\langle x\rangle = -334\langle x\rangle^1 - \frac{5}{3}\langle x-10\rangle^3 + 2334\langle x-20\rangle^1 \text{ N}\cdot\text{m}$$

6.26.

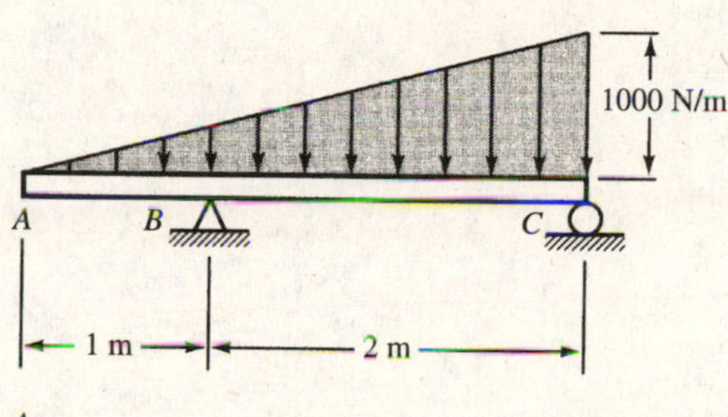

Ans.

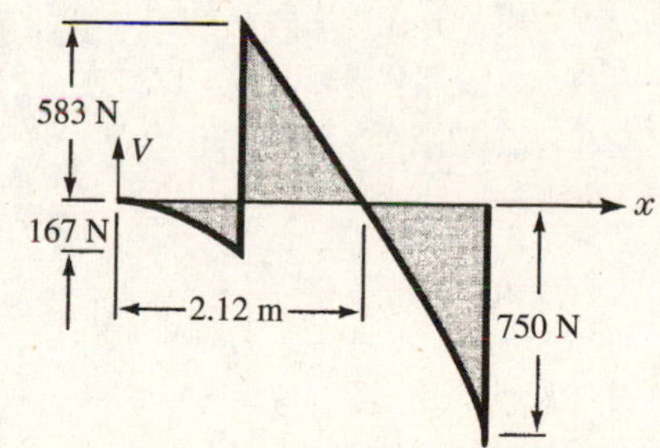

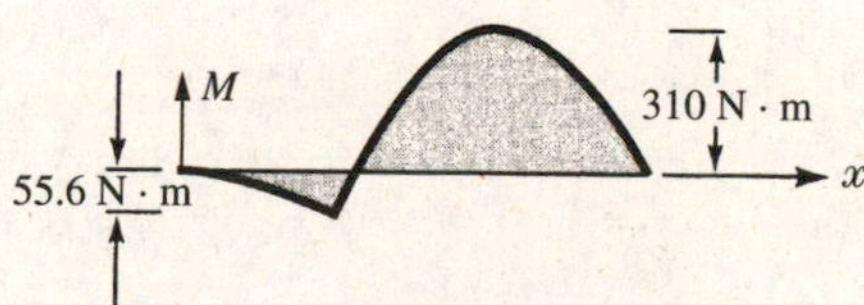

Fig. 6-45

$$V = -166.7\langle x\rangle^2 + 750\langle x-1\rangle^0$$

$$M = -55.6\langle x\rangle^3 + 750\langle x-1\rangle^1$$

Stresses in Beams

7.1 Basics

It is convenient to imagine a beam to be composed of an infinite number of thin longitudinal fibers. Each longitudinal fiber is assumed to act independently of every other fiber. The beam of Fig. 7-1, for example, will deflect downward and the fibers in the lower part of the beam undergo extension, while those in the upper part are shortened. These changes in the lengths of the fibers set up stresses in the fibers. Those that are extended have tensile stresses acting on the fibers in the direction of the longitudinal axis of the beam, while those that are shortened are subject to compressive stresses.

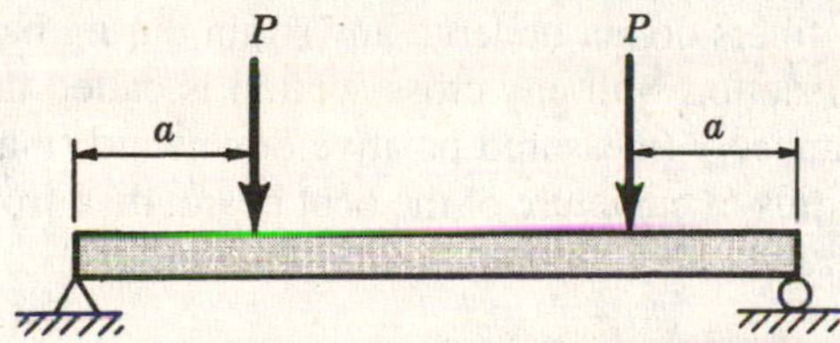

Fig. 7-1 A simple beam.

Neutral Surface

There always exists one surface in the beam containing fibers that do not undergo any extension or compression, and thus are not subject to any tensile or compressive stress. This surface is called the *neutral surface* of the beam.

The intersection of the neutral surface with any cross section of the beam perpendicular to its longitudinal axis is called the *neutral axis*. All fibers on one side of the neutral axis are in a state of tension, while those on the opposite side are in compression.

Bending Moment

The algebraic sum of the moments of the external forces to one side of any cross section of the beam about an axis through that section is called the *bending moment* at that section. This concept was discussed in Chapter 6.

7.2 Normal Stresses in Beams

Let us find a relationship between the bending moment acting at any section in a beam and the bending stress at any point in this same section. Assume Hooke's law holds so that all fibers in the beam are in the elastic range of the material. The beam shown in Fig. 7-2(*a*) is loaded by the two couples M and consequently is in static equilibrium. Since the bending moment has the same value at all points along the bar, the beam is said to be in a condition of *pure bending*. To determine the distribution of bending stress in the beam, let us cut the beam by a plane passing through it in a direction perpendicular to the geometric axis of the bar. In this manner the forces under investigation become external to the new body formed, even though they were internal effects with regard to the original uncut body.

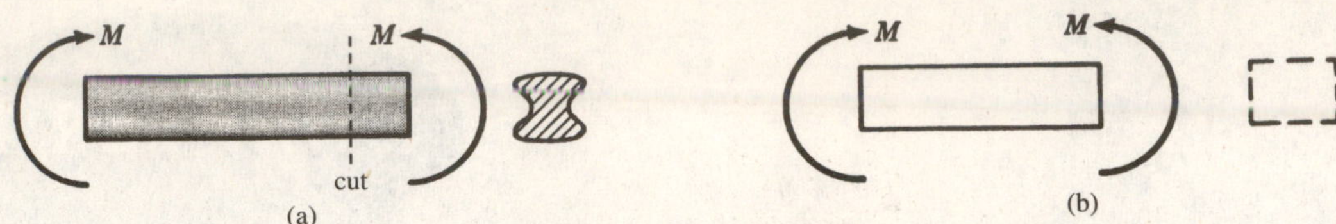

Fig. 7-2 A beam in pure bending.

The free-body diagram of the portion of the beam to the left of this cutting plane now appears as in Fig. 7-2(*b*). A moment *M* must act over the cross section cut by the plane so that the left portion of the beam will be in static equilibrium. The moment *M* acting on the "cut" section represents the effect of the right portion of the beam on the left portion. Since the right portion has been removed, it must be replaced by its effect on the left portion and this effect is represented by the moment *M*.

It is convenient to consider the beam to be composed of an infinite number of thin longitudinal fibers. It is assumed that every longitudinal fiber is subject only to axial tension or compression. Further, it is assumed that a plane section of the beam normal to its axis before loads are applied remains plane and normal to the axis after loading. Finally, it is assumed that the material follows Hooke's law.

Next, let us consider two adjacent cross sections *aa* and *bb* marked on the side of the beam, as shown in Fig. 7-3. Prior to loading, these sections are parallel to each other. After the applied moments have acted on the beam, these sections are still planes but they have rotated with respect to each other to the positions shown, where *O* represents the center of curvature of the beam. Evidently the fibers on the upper surface of the beam are in a state of compression, while those on the lower surface are in tension. The line *cd* is the trace of the surface in which the fibers do not undergo any strain during bending and this surface is called the *neutral surface*, and its intersection with any cross section is called the *neutral axis*. The elongation of the longitudinal fiber at a distance *y* (measured positive downward) may be found by drawing line *de* parallel to *aa*. If ρ denotes the radius of curvature of the bent beam, then from the similar triangles *cOd* and *edf* we find the strain of this fiber to be

$$\epsilon = \frac{\overline{ef}}{\overline{cd}} = \frac{\overline{de}}{\overline{cO}} = \frac{y}{\rho} \tag{7.1}$$

Thus, the strains of the longitudinal fibers are proportional to the distance *y* from the neutral axis.

Since Hooke's law holds, $E = \sigma/\epsilon$, or $\sigma = E\epsilon$, it immediately follows that the stresses existing in the longitudinal fibers are proportional to the distance *y* from the neutral axis, or

$$\sigma = \frac{Ey}{\rho} \tag{7.2}$$

Let us consider a beam of rectangular cross section, although the derivation actually holds for any cross section which has a longitudinal plane of symmetry. In this case, these longitudinal, or bending, stresses appear as in Fig. 7-4.

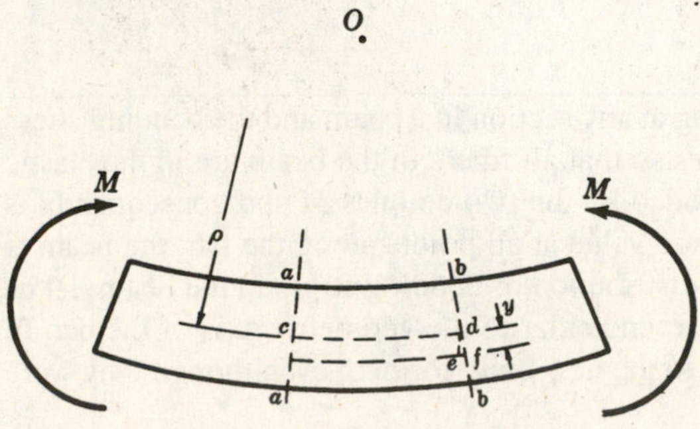

Fig. 7-3 A deformed section.

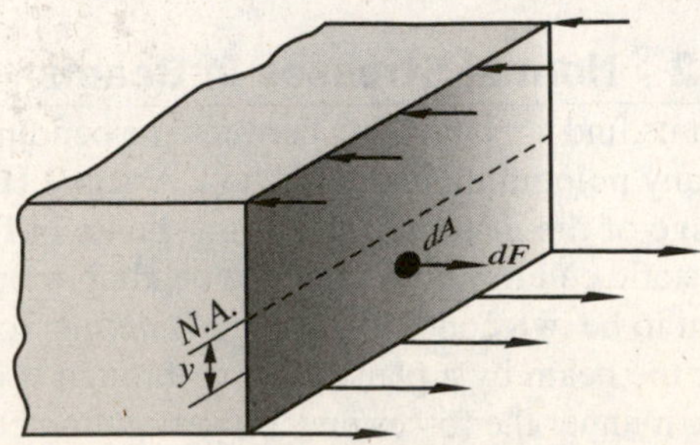

Fig. 7-4 Bending stresses on a beam cross section.

Let dA represent an element of area of the cross section at a distance y from the neutral axis. The stress acting on dA is given by the above expression and consequently the force on this element is the product of the stress and the area dA, that is,

$$dF = \frac{Ey}{\rho} dA \tag{7.3}$$

However, the resultant longitudinal force acting over the cross section is zero (for the case of pure bending) and this condition may be expressed by the summation of all forces dF over the cross section. This is done by integration:

$$\int \frac{Ey}{\rho} dA = \frac{E}{\rho} \int y\, dA = 0 \tag{7.4}$$

Evidently $\int y\, dA = 0$. However, this integral represents the first moment of the area of the cross section with respect to the neutral axis, since y is measured from that axis. But, from statics we may write $\int y\, dA = \bar{y}A$, where $\bar{y}$ is the distance from the neutral axis to the centroid of the cross-sectional area. From this, $\bar{y}A = 0$, and since A is not zero, then $\bar{y} = 0$. Thus the neutral axis always passes through the centroid of the cross section, provided Hooke's law holds.

The moment of the elemental force dF about the neutral axis is given by

$$dM = y\, dF = y\left(\frac{Ey}{\rho} dA\right) \tag{7.5}$$

The resultant of the moments of all such elemental forces summed over the entire cross section must be equal to the bending moment M acting at that section and thus we may write

$$M = \int \frac{Ey^2}{\rho} dA \tag{7.6}$$

But $I = \int y^2\, dA$ and thus we have

$$M = \frac{EI}{\rho} \tag{7.7}$$

It is to be carefully noted that this moment of inertia of the cross-sectional area is computed with respect to the axis through the centroid of the cross section. But previously we had

$$\sigma = \frac{Ey}{\rho} \tag{7.8}$$

Eliminating ρ from these last two equations, we obtain

$$\sigma = \frac{My}{I} \tag{7.9}$$

This formula gives the so-called *bending* or *flexural stresses* in the beam. In it, M is the bending moment at any section, I the moment of inertia of the cross-sectional area about an axis through the centroid of the cross section, and y the distance from the neutral axis (which passes through the centroid) to the fiber on which the stress σ acts.

The value of y at the outer fibers of the beam is frequently denoted by c. At these outer fibers the bending stresses are maximum and there we may write

$$\sigma = \frac{Mc}{I} \tag{7.10}$$

Section Modulus

Equation (7.10) can be written as

$$\sigma = \frac{M}{I/c} = \frac{M}{S} \tag{7.11}$$

where the ratio $I/c = S$ is called the *section modulus*, with units m^3.
This form is convenient because values of S are available in handbooks for a wide range of standard structural steel shapes (see Tables 7-1 and 7-2 at the end of this chapter).

7.3 Shearing Stresses in Beams

In a beam loaded by transverse forces acting perpendicular to the axis of the beam, not only are bending stresses produced parallel to the axis of the bar but shearing stresses also act over cross sections of the beam perpendicular to the axis of the bar. Let us express the intensity of these shearing stresses in terms of the shearing force at the section and the properties of the cross section. The theory to be developed applies only to a cross section of rectangular shape. However, the results of this analysis are commonly used to give approximate values of the shearing stress in other cross sections having a plane of symmetry.

Let us consider an element of length dx cut from a beam as shown in Fig. 7-5. We shall denote the bending moment at the left side of the element by M and that at the right side by $M + dM$, since in general the bending moment changes slightly as we move from one section to an adjacent section of the beam. If y is measured upward from the neutral axis, then the bending stress at the left section a-a is given by

$$\sigma = \frac{My}{I} \tag{7.12}$$

where I denotes the moment of inertia of the entire cross section about the neutral axis. This stress distribution is illustrated above in Eq. (7.12). Similarly, the bending stress at the right section b-b is

$$\sigma' = \frac{(M + dM)y}{I} \tag{7.13}$$

Let us now consider the equilibrium of the shaded element $acdb$. The force acting on an area dA of the face ac is merely the product of the intensity of the force and the area; thus

$$\sigma dA = \frac{My}{I} dA \tag{7.14}$$

The sum of all such forces over the left face ac is found by integration to be

$$\int_y^c \frac{My}{I} dA \tag{7.15}$$

Likewise, the sum of all normal forces over the right face bd is given by

$$\int_y^c \frac{(M + dM)y}{I} dA \tag{7.16}$$

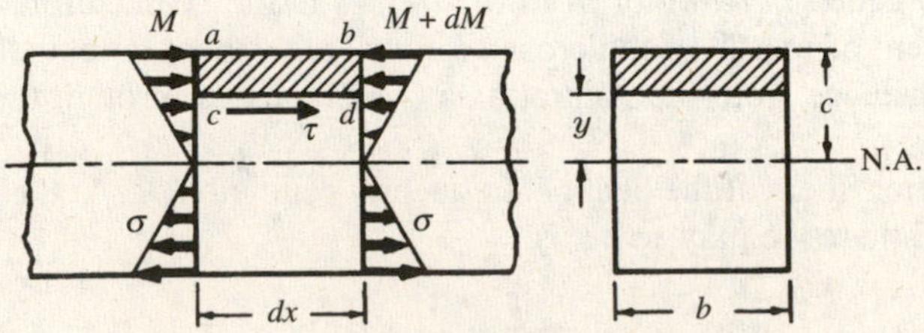

Fig. 7-5 A section of a rectangular beam.

Evidently, since these two integrals are unequal, some additional horizontal force must act on the shaded element to maintain equilibrium. Since the top face *ab* is assumed to be free of any externally applied horizontal forces, then the only remaining possibility is that there exists a horizontal shearing force along the lower face *cd*. This represents the action of the lower portion of the beam on the shaded element. Let us denote the shearing stress along this face by τ as shown. Also, let b denote the width of the beam at the position where τ acts. Then the horizontal shearing force along the face *cd* is $\tau b\,dx$. For equilibrium of the element *acdb* we have

$$\Sigma F_x = \int_y^c \frac{My}{I}\,dA - \int_y^c \frac{(M+dM)y}{I}\,dA + \tau b\,dx = 0 \tag{7.17}$$

Solving,

$$\tau = \frac{1}{Ib}\frac{dM}{dx}\int_y^c y\,dA \tag{7.18}$$

But from Eq. (6.8) we have $V = dM/dx$, where V represents the shearing force (newtons) at the section. Substituting,

$$\tau = \frac{V}{Ib}\int_y^c y\,dA \tag{7.19}$$

The integral in this last equation, Eq. (7.19), represents the first moment of the shaded cross-sectional area about the neutral axis of the beam. This area is always the portion of the cross section that is above the level at which the desired shear acts. This first moment of area is denoted by Q in which case the above formula becomes

$$\tau = \frac{VQ}{Ib} \tag{7.20}$$

The units of Q are m^3.

The shearing stress τ just determined acts horizontally as shown in Fig. 7-5. However, let us consider the equilibrium of a differential element of thickness t taken from any beam and subject to a shearing stress τ_1 on its lower face, as shown in Fig. 7-6. The total horizontal force on the lower face is $\tau_1 t dx$. For equilibrium of forces in the horizontal direction, an equal force but acting in the opposite direction must act on the upper face, hence the shear stress intensity there too is τ_1. These two forces give rise to a couple of magnitude $\tau_1 t\,dx\,dy$. The only way in which equilibrium of the element can be maintained is for another couple to act over the vertical faces. Let the shear stress intensity on these faces be denoted by τ_2. The total force on either vertical face is $\tau_2 t\,dy$. For equilibrium of the moments about the center of the element we have

$$\Sigma M_c = \tau_1 t\,dx\,dy - \tau_2 t\,dy\,dx = 0 \quad \text{or} \quad \tau_1 = \tau_2 \tag{7.21}$$

Thus we have the interesting conclusion that the shearing stresses on any two perpendicular planes through a point on a body are equal. Consequently, not only are there shearing stresses τ acting horizontally at any point in the beam, but shearing stresses of an equal intensity also act vertically at that same point.

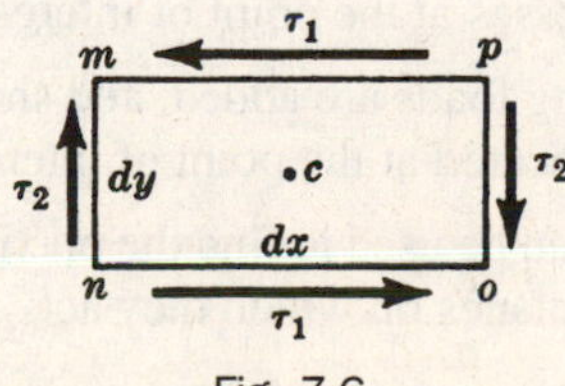

Fig. 7-6

In summary, when a beam is loaded by transverse forces, both horizontal and vertical shearing stresses arise in the beam. The vertical shearing stresses are of such magnitudes that their resultant at any cross section is exactly equal to the shearing force V at that same section.

From Eq. (7.20) it is evident that the maximum vertical shearing stress occurs at the neutral axis of the beam, whereas the vertical shearing stress at the outer fibers is always zero. In contrast, the normal stress varies from zero at the neutral axis to a maximum at the outer fibers.

In a beam of rectangular cross section, as shown in Fig. 7-7, the above equation for shearing stress becomes

$$\tau = \frac{V}{2I}(c^2 - y^2) \tag{7.22}$$

where τ denotes the shearing stress on a fiber at a distance y from the neutral axis. The distribution of vertical shearing stress over the rectangular cross section is thus parabolic, varying from zero at the outer fibers where $y = c$ to a maximum at the neutral axis where $y = 0$.

Both the above equations, Eqs. (7.21) and (7.22), for shearing stress give the vertical as well as the horizontal shearing stresses at a point (see Fig. 7-6), since the intensities of shearing stresses in these two directions are always equal.

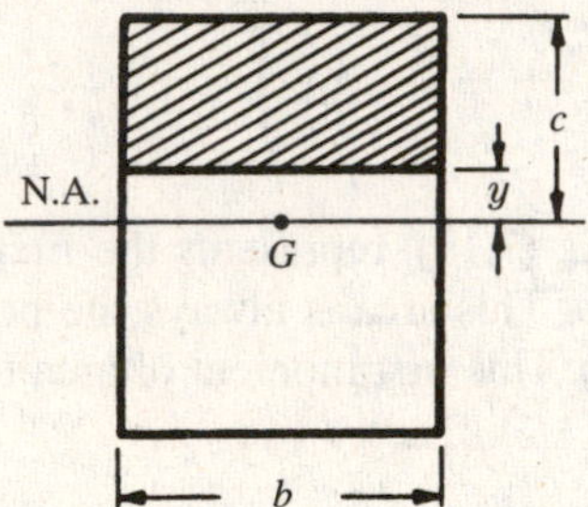

Fig. 7-7 The crosssection of a rectangular beam.

7.4 Combined Loading

The stresses caused by axial forces, torques, and bending loads have now been analyzed. There are situations, however, where all three types of loading are present. An efficient method to account for more than one load is to calculate the normal and shearing stresses acting at a point of interest on planes normal and tangential to the applied loads, and then use Mohr's circle to calculate the maximum and minimum stresses acting at that point on planes oriented at some angle. Such maximum stresses lead to failure in machine and structural components.

Solved Problems will illustrate the procedure and Supplementary Problems will provide the reader the opportunity to gain experience at solving such problems. The procedure follows several general steps providing the proportional limit of the material is not exceeded so that superposition is allowed (the material must remain in the elastic range so that there is a linear relationship between stress and deformation). The steps are

1. Draw a free-body diagram of the member of interest.
2. Pass a section through the point of interest where the maximum stresses are anticipated.
3. Use the equations of equilibrium to find the forces and moments acting on the section containing the point of interest.
4. Calculate the normal and shearing stresses at the point of interest using the formulas already developed.
5. Normal stresses from axial and bending loads are added, and shearing stresses from torsion and vertical shear loads are added on an element located at the point of interest.
6. Mohr's circle (or the equations) can then be used to find the maximum and minimum normal and shearing stresses at the selected point, and the planes on which they act.

Problems 7.15, 7.16, and 7.17 illustrate the procedure.

SOLVED PROBLEMS

7.1. A beam is loaded by a couple of 1400 N·m at each of its ends, as shown in Fig. 7-8. The beam is of steel and has a rectangular cross section, as shown. Determine the maximum bending stress in the beam and indicate the variation of bending stress over the depth of the beam.

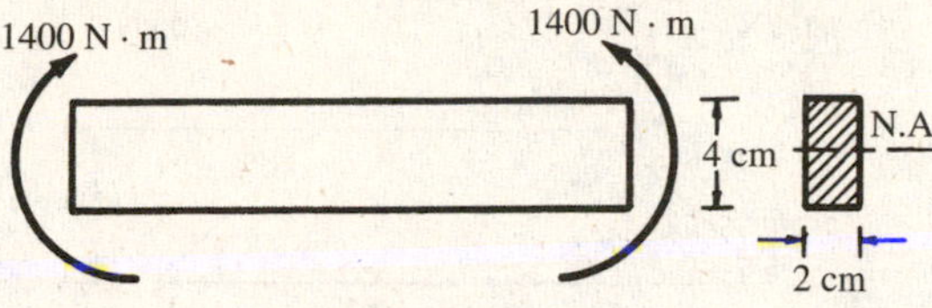

Fig. 7-8

SOLUTION: Bending takes place about the horizontal neutral axis denoted by N.A, which passes through the centroid of the cross section.

The bending stress at a distance y from the neutral axis is given by $\sigma = My/I$, where y is measured from the neutral axis. Since M and I are constant along the length of the bar, the maximum bending stress occurs on those fibers where y takes on its maximum value. These are the fibers along the upper and lower surfaces of the beam. From inspection it is obvious that for the direction of loading shown in Fig. 7-8, the upper fibers are in compression and the lower fibers in tension. For the lower fibers, $y = 2$ cm and the maximum bending stress is

$$\sigma = \frac{Mc}{I} = \frac{1400 \times 0.02}{0.02 \times 0.04^3/12} = 262 \times 10^6 \text{ Pa} \qquad \text{or} \qquad 262 \text{ MPa (ten)}$$

For the fibers along the upper surface, y may be considered to be negative and we have

$$\sigma = \frac{Mc}{I} = \frac{1400 \times 0.02}{0.02 \times 0.04^3/12} = 262 \times 10^6 \text{ Pa} \qquad \text{or} \qquad 262 \text{ MPa (comp)}$$

Thus the peak stresses are 262 MPa in tension for all fibers along the lower surface of the beam and 262 MPa in compression for all fibers along the upper surface. According to the formula $\sigma = My/I$, the bending stress varies linearly from zero at the neutral axis to a maximum at the outer fibers and hence the variation over the depth of the beam may be sketched as in Fig. 7-9.

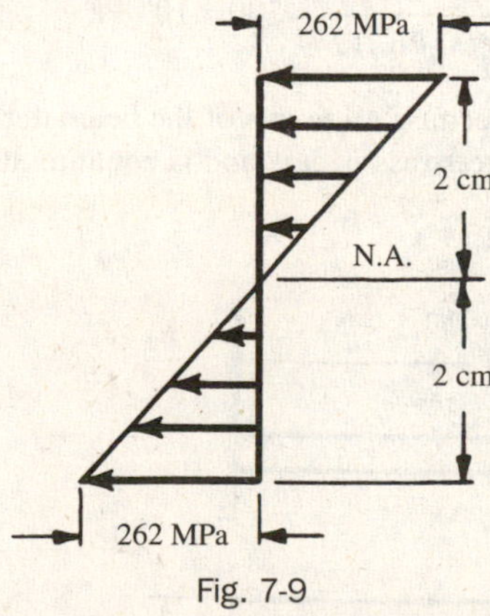

Fig. 7-9

7.2. A beam of circular cross section is 18 cm in diameter. It is simply supported at each end and loaded by two concentrated loads of 80 kN each, applied 90 cm from the ends of the beam as shown in Fig. 7-10(*a*). Determine the maximum bending stress in the beam.

SOLUTION: Here the moment is not constant along the length of the beam, as it was in Problem 7.1. The bending moment diagram of Fig. 7-10(*b*) was obtained by the methods of Chapter 6. It is to be noted that the portion of

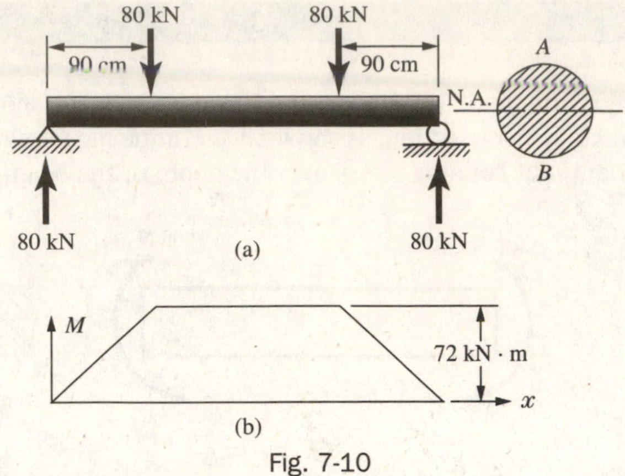

Fig. 7-10

the beam between the two downward loads is in *pure bending* and everywhere in that region the bending moment is equal to $80 \times 0.9 = 72$ kN · m.

The bending stress at a distance y from the horizontal neutral axis shown is $\sigma = My/I$. The maximum bending stresses occur where y is maximum at A and B. This maximum stress is the same at all such points between the applied loads. At point B, $y = 9$ cm and the stress becomes

$$\sigma = \frac{My}{I} = \frac{72 \times 0.09}{\pi \times 0.18^4/64} = 126\,000 \text{ kPa} \qquad \text{or} \qquad 126 \text{ MPa (ten)}$$

At point A the stress is 126 MPa (comp).

7.3. A steel cantilever beam 6 m in length is subjected to a concentrated load of 1200 N acting at the free end of the bar. The beam is of rectangular cross section, 4 cm wide by 6 cm deep. Determine the magnitude and location of the maximum tensile and compressive bending stresses in the beam. As usual, neglect the weight of the beam.

SOLUTION: The bending moment diagram is triangular with a maximum at the supporting wall, as shown below in Fig. 7-11(a). The maximum bending moment is merely the moment of the 1200-N force, $1200 \times 6 = 7200$ N · m.

The bending stress at a distance y from the neutral axis is $\sigma = My/I$. At the supporting wall, where the bending moment is maximum, the peak tensile stress is

$$\sigma = \frac{My}{I} = \frac{7200 \times 0.03}{0.04 \times 0.06^3/12} = 300 \times 10^6 \text{ Pa} \qquad \text{or} \qquad 300 \text{ MPa}$$

This stress must be in tension because all points of the beam deflect downward. At the lower fibers adjacent to the wall the peak compressive stress occurs and is equal to 300 MPa.

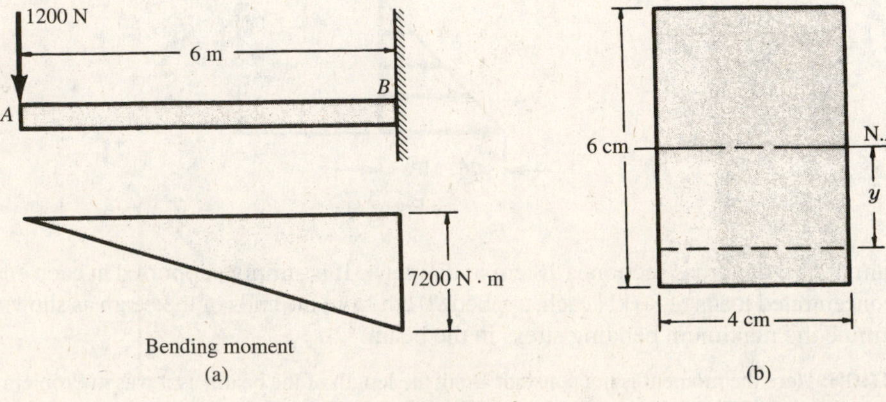

Fig. 7-11

7.4. Let us reconsider Problem 7.3 for the case where the rectangular beam is replaced by a beam with cross section shown in Fig. 7-12. Determine the maximum tensile and compressive bending stresses.

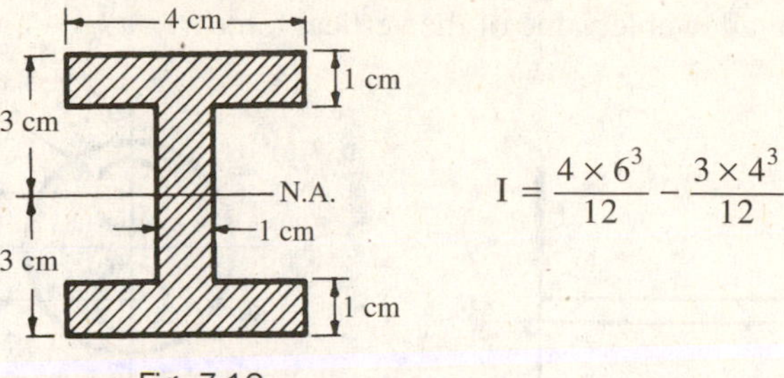

$$I = \frac{4 \times 6^3}{12} - \frac{3 \times 4^3}{12} = 56 \text{ cm}^4$$

Fig. 7-12

SOLUTION: The bending stress at a distance y from the neutral axis is given by $\sigma = My/I$. At the outer fibers, $y = c$ and at the wall where the bending moment is maximum,

$$\sigma = \frac{Mc}{I} = \frac{7200 \times 0.03}{56 \times 10^{-8}} = 386 \times 10^6 \text{ Pa} \qquad \text{or} \qquad 386 \text{ MPa}$$

Again, since the fibers along the top of the beam are stretching, the stress there will be tension. Along the lower face of the beam the fibers are shortening and there the stress is compressive and equal to 386 MPa.

7.5. A cantilever beam 3-m long is subjected to a uniformly distributed load of 30 kN per meter of length. The allowable working stress in either tension or compression is 150 MPa. If the cross section is to be rectangular, determine the dimensions if the height is to be twice as great as the width.

SOLUTION: The bending moment diagram for a uniform load acting over a cantilever beam was determined in Problem 6.1. It was found to be parabolic, varying from zero at the free end of the beam to a maximum at the supporting wall. The loaded beam and the accompanying bending moment diagram are shown in Fig. 7-13. The maximum moment at the wall is given by

$$M_{x=3} = -30(3)(1.5) = -135 \text{ kN} \cdot \text{m}$$

It is to be noted that this problem involves the design of a beam. The only cross section that needs to be considered for design purposes is the one where the bending moment is maximum, i.e., at the supporting wall. Thus we wish to design a rectangular beam to resist a bending moment of 135 kN · m with a maximum bending stress of 150 MPa. The moment of inertia about the neutral axis is given by

$$I = \frac{1}{12}bh^3 = \frac{1}{12}b(2b)^3 = \frac{2}{3}b^4$$

At the cross section of the beam adjacent to the supporting wall the bending stress in the beam is given by $\sigma = My/I$. The maximum bending stress in tension occurs along the upper surface of the beam, and at this surface $y = b$ and $\sigma = 150$ MPa. Then,

$$\sigma = \frac{My}{I} \qquad \text{or} \qquad 150 \times 10^6 = \frac{135\,000\, b}{2b^4/3} \qquad \therefore b = 0.111 \text{ m}$$

from which $b = 110$ mm and $h = 2b = 220$ mm.

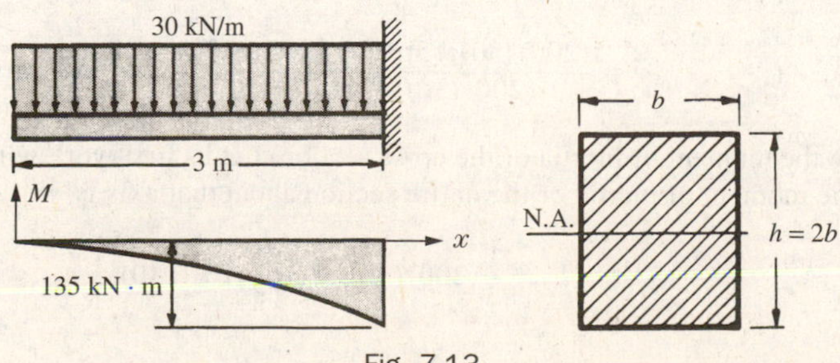

Fig. 7-13

7.6. A cantilever beam is of length 1.5 m, loaded by a concentrated force P at its tip as shown in Fig. 7-14, and is of circular cross section (R = 100 mm), having two symmetrically placed longitudinal holes as indicated. The material is titanium alloy, having an allowable working stress in bending of 600 MPa. Determine the maximum allowable value of the vertical force P.

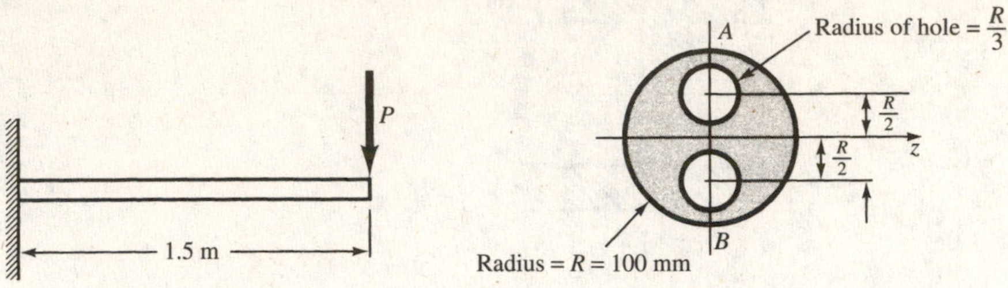

Fig. 7-14

SOLUTION: First, it is necessary to determine the moment of inertia. The moment of inertia of a solid circular cross section about a diametral axis z is $\pi R^4/4$. Using this value for the solid section and subtracting the moments of inertia of each of the holes (from the parallel-axis theorem), we have

$$I = \frac{\pi R^4}{4} - 2\left\{\frac{\pi}{4}\left(\frac{R}{3}\right)^4 + \pi\left(\frac{R}{3}\right)^2\left(\frac{R}{2}\right)^2\right\} = 0.592R^4$$

The bending stresses in the uppermost and lowermost fibers are denoted by points A and B, respectively, are, using R = 0.1 m,

$$\sigma_{\max} = \frac{Mc}{I}$$

$$600 \times 10^6 = \frac{P(1.5) \times 0.1}{0.592 \times 0.1^4} \qquad \therefore P = 237\,000 \text{ N}$$

7.7. The extruded beam shown in Fig. 7-15 is made of aluminum alloy having an allowable working stress in either tension or compression of 90 MPa. The beam is a cantilever, subject to a uniform vertical load. Determine the allowable intensity of uniform loading.

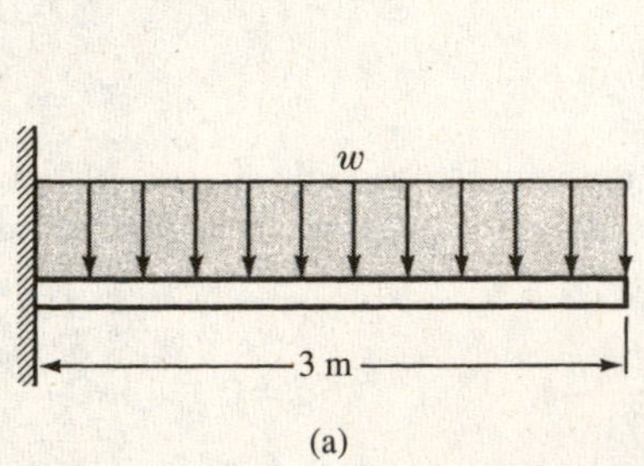

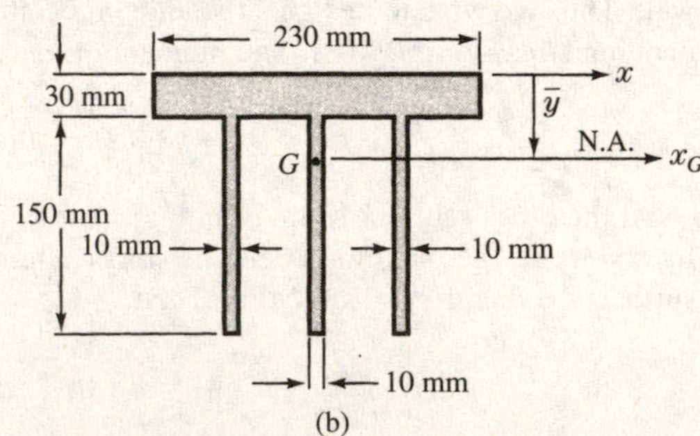

Fig. 7-15

SOLUTION: First, locate the centroid of the cross section:

$$\bar{y} = \frac{(200)(30)(15) + 3(180)(10)(90)}{(200)(30) + 3(180)(10)} = 50.5 \text{ mm}$$

Next, determine the moment of inertia of the cross section. Let us first work with the x-axis through the top of the beam. The moment of inertia of the entire section about that axis is

$$I_x = \frac{1}{3}(200)(30)^3 + 3\left\{\frac{1}{3}(10)(180)^3\right\}$$

$$= 60.12 \times 10^6 \text{ mm}^4$$

and from the parallel axis theorem we may now transfer to the x_G axis through the centroid of the cross section to find

$$I_{XG} = 60.12 \times 10^6 - (11\,400)(50.5)^2$$
$$= 31.05 \times 10^6 \text{ mm}^4$$

The peak bending moment occurs at the supporting wall and is

$$M_{\max} = \frac{wL^2}{2}$$

Next, applying $\sigma = Mc/I$ to the lowermost fibers of the beam, since those are the most distant from the neutral axis through G, we have

$$90 \times 10^6 = \frac{(w \times 3^2/2)(0.1295)}{(31.05 \times 10^6) \times 10^{-12}} \qquad \therefore w = 4800 \text{ N/m}$$

7.8. The simply supported beam AD is loaded by a concentrated force of 80 kN together with a couple of magnitude 30 kN · m, as shown in Fig. 7-16. Use the cross-section of Problem 7.4 but with unknown width b. Determine b if the peak allowable working stress in tension as well as compression is 600 MPa.

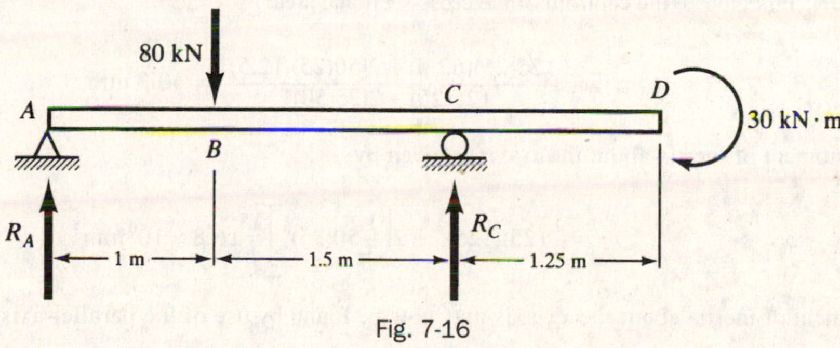

Fig. 7-16

SOLUTION: It is first necessary to determine the reactions at A and C. We have

$$\Sigma M_A: 2.5\, R_c = 30 + 80 \times 1 \qquad \therefore R_c = 44 \text{ kN}$$
$$\Sigma F_y: R_A = 80 - 44 = 36 \text{ kN}$$

From the methods of Chap. 6, we can now construct the moment diagram which appears as in Fig. 7-17. The moment of inertia is

$$I = \frac{b \times 6^3}{12} - \frac{(b-1) \times 4^3}{12} = 12.67b + 5.33 \text{ cm}^4$$

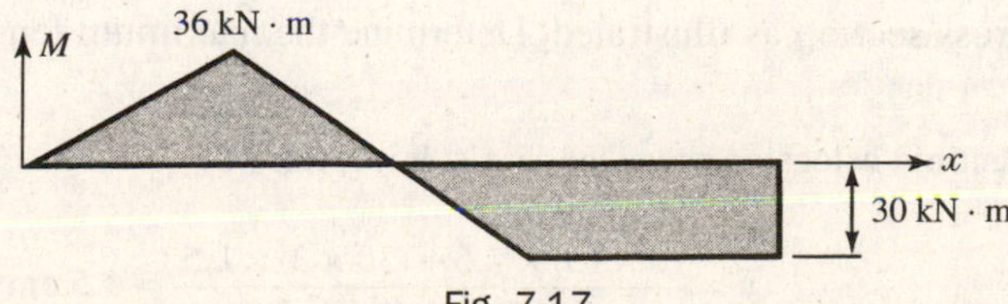

Fig. 7-17

Using the maximum moment of 36 kN · m in $\sigma = Mc/I$, we have

$$600 \times 10^6 = \frac{36\,000 \times 0.03}{(12.67b + 5.33) \times 10^{-8}} \qquad \therefore b = 13.8 \text{ cm}$$

It would be more reasonable to select an I-beam with greater depth.

7.9. A beam is loaded by one couple at each of its ends, the magnitude of each couple being 5 kN · m. The beam is of steel and has a T-type cross section with the dimensions indicated in Fig. 7-18. Determine the maximum tensile stress in the beam and its location, and the maximum compressive stress and its location.

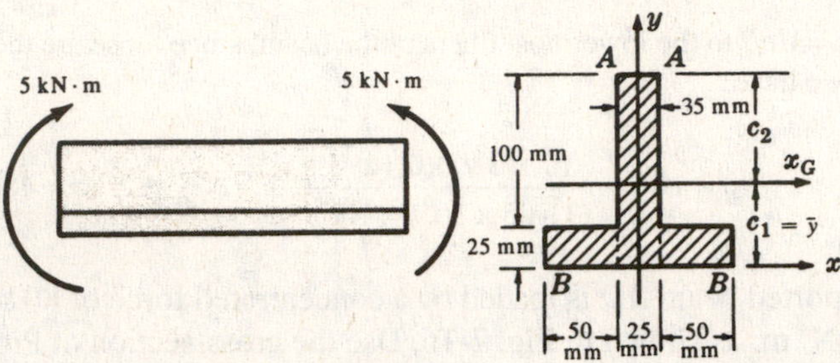

Fig. 7-18

SOLUTION: First, locate the centroid of the cross-sectional area:

$$\bar{y} = \frac{125(25)(62.5) + 2[50(25)(12.5)]}{125(25) + 2[25(50)]} = 40.3 \text{ mm}$$

The moment of inertia about the x-axis is given by

$$I_x = \frac{1}{3}(25)(125)^3 + 2\left[\frac{1}{3}50(25)^3\right] = 16.8 \times 10^6 \text{mm}^4$$

The moment of inertia about the x_G-axis may now be found by use of the parallel-axis theorem. Thus,

$$I_x = I_G + A(\bar{y})^2 \qquad 16.8 \times 10^6 = I_G + 5625(40.3)^2 \qquad \text{and} \qquad I_G = 7.7 \times 10^6 \text{ mm}^4$$

For the loading shown, the fibers below the x_G-axis are in tension. Let c_1 and c_2 denote the distances of the extreme fibers from the neutral axis (x_G) as shown. Obviously $c_1 = 40.3$ mm and $c_2 = 84.7$ mm. The maximum tensile stress occurs in those fibers along B-B and is given by

$$\sigma = \frac{Mc_1}{I} = \frac{5000 \times 0.0403}{(7.7 \times 10^6) \times 10^{-12}} = 26.2 \times 10^6 \text{Pa} \qquad \text{or} \qquad 26.2 \text{ MPa}$$

The maximum compressive stress occurs in those fibers along A-A and is given by

$$\sigma = \frac{Mc_2}{I} = \frac{5000 \times 0.0847}{(7.7 \times 10^6) \times 10^{-12}} = 55 \times 10^6 \text{Pa} \qquad \text{or} \qquad 55 \text{ MPa}$$

7.10. A simply supported beam is loaded by the couple of 1500 N · m as shown in Fig. 7-19. The beam has a channel-type cross section as illustrated. Determine the maximum tensile and compressive stresses in the beam.

SOLUTION: The centroid is located above the x-axis. It is located by

$$\bar{y} = \frac{(6 \times 15) \times 7.5 + (30 \times 3) \times 1.5}{6 \times 15 + 30 \times 3} = 4.5 \text{ cm}$$

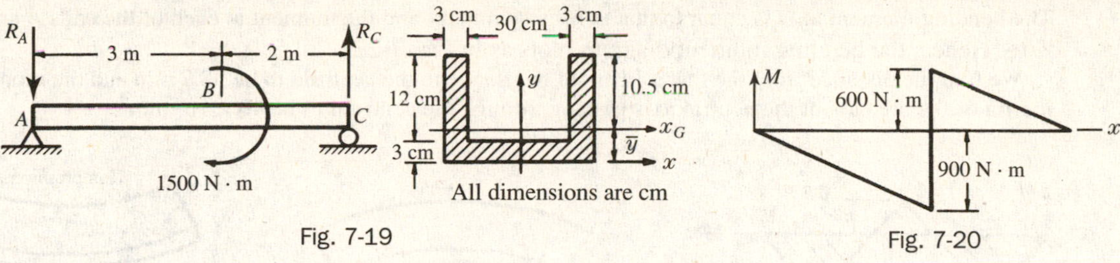

Fig. 7-19

Fig. 7-20

The moment of inertia is

$$I_G = \left(\frac{6 \times 15^3}{3} + \frac{30 \times 3^3}{3}\right) - 180 \times 4.5^2 = 2835 \text{ cm}^4$$

We considered this particular loading in Problem 6.9. The couple formed by the forces at A and C is found from $5R_A = 1500$. Thus $R_A = 300$ N ↓ and $R_C = 300$ N ↑. This provides the tension in the top fibers between A and B and compression in the top fibers between B and C. The bending moment diagram is shown in Fig. 7-20. Just to the left of point B the normal stresses are

$$\sigma_{top} = \frac{900 \times 0.105}{2835 \times 10^{-8}} = 3.33 \times 10^6 \text{ Pa (ten)} \qquad \sigma_{bottom} = \frac{900 \times 0.045}{2835 \times 10^{-8}} = 1.43 \times 10^6 \text{ Pa (comp)}$$

Just to the right of B the normal stresses are

$$\sigma_{top} = \frac{600 \times 0.105}{2835 \times 10^{-8}} = 2.22 \times 10^6 \text{ Pa (comp)} \qquad \sigma_{bottom} = \frac{600 \times 0.045}{2835 \times 10^{-8}} = 0.952 \times 10^6 \text{ Pa (ten)}$$

The maximum tensile and compressive stresses must now be selected from the above four values. The maximum tension is 3.33 MPa occurring in the upper fibers just to the left of point B; the maximum compression is 2.22 MPa occurring in the upper fibers also but just to the right of point B.

7.11. Consider the beam with overhanging ends loaded by the three concentrated forces shown in Fig. 7-21. The beam is simply supported and is of T-type cross section as shown. The material is gray cast iron having an allowable working stress in tension of 35 MPa and in compression of 150 MPa. Determine the maximum allowable value of P.

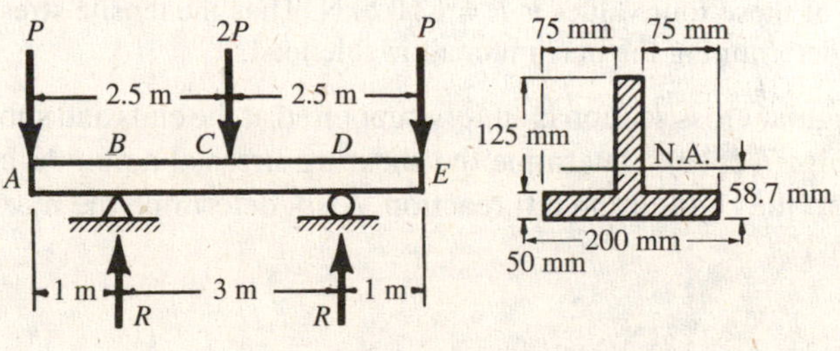

Fig. 7-21

SOLUTION: From symmetry each of the reactions denoted by R is equal to $2P$. The bending moment diagram consists of a series of straight lines connecting the ordinates representing bending moments at the points A, B, C, D, and E. At B the bending moment is given by the moment of the force P acting at A about an axis through B. Thus,

$$M_B = -P \times 1 = -P \text{ N} \cdot \text{m}$$

At C the bending moment is given by the sum of the moments of the forces P and $R = 2P$ about an axis through C. Thus,

$$M_C = -P \times 2.5 + 2P \times 1.5 = 0.5P \text{ N} \cdot \text{m}$$

The bending moment at D is equal to that at B by symmetry and the moment at each of the ends A and E is zero. Hence, the bending moment diagram plots as in Fig. 7-22.

We find the distance from the lower fibers of the flange to the centroid to be 58.7 mm and the moment of inertia of the area about the neutral axis passing through the centroid to be 40×10^6 mm^4.

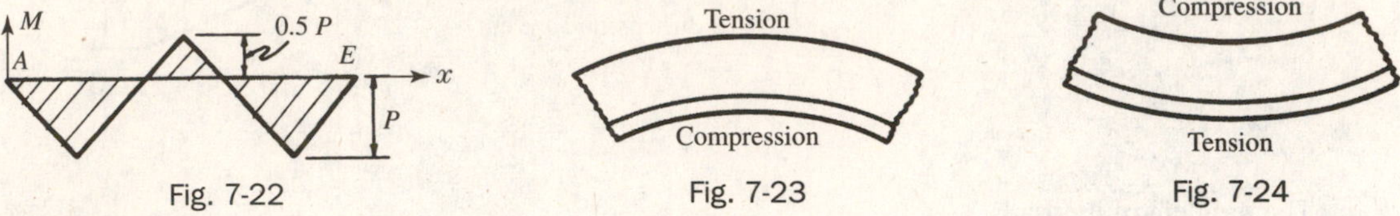

Fig. 7-22 Fig. 7-23 Fig. 7-24

The upper fibers at B and D are in tension and the lower fibers are subject to compression, as shown in Fig. 7-23. We shall first calculate a value of P, assuming that the allowable tensile stress of 35 MPa is realized in the upper fibers. Applying $\sigma = My/I$ to these upper fibers, we find

$$35 \times 10^6 = \frac{(P \times 0.116)}{40 \times 10^6 (10^{-12})} \qquad \text{or} \qquad P = 12\,070 \text{ N}$$

Next, we shall calculate a value of P, assuming that the allowable compressive stress of 150 MPa is set up in the lower fibers at B. Again applying the flexure formula, we find

$$150 \times 10^6 = \frac{(P \times 0.0587)}{40 \times 10^6 (10^{-12})} \qquad \text{or} \qquad P = 102\,200 \text{ N}$$

We shall now examine point C. Since the bending moment is opposite in sign from that at B, the upper fibers are in compression and the lower fibers are subject to tension, as shown in Fig. 7-24. First we will calculate a value of P, assuming that the allowable tension of 35 MPa is set up in the lower fibers. From the flexure formula we find

$$35 \times 10^6 = \frac{0.5P \times 0.0587}{40 \times 10^6 (10^{-12})} \qquad \text{or} \qquad P = 47\,700 \text{ N}$$

We shall now assume that the allowable compression of 150 MPa is set up in the upper fibers. Applying the flexure formula, we have

$$150 \times 10^6 = \frac{(0.5P \times 0.116)}{40 \times 10^6 (10^{-12})} \qquad \text{or} \qquad P = 103\,000 \text{ N}$$

The minimum of these four values is $P = 12.07$ kN. Thus the tensile stress at points B and D is the controlling factor in determining the maximum allowable load.

7-12. A beam of rectangular cross section is simply supported at the ends and subject to the single concentrated force shown in Fig. 7-25(a). Determine the shearing stress at a point 3 cm below the top of the beam at a section 1 m to the right of the left reaction. Also, determine the maximum* shearing stress due to the vertical shear V.

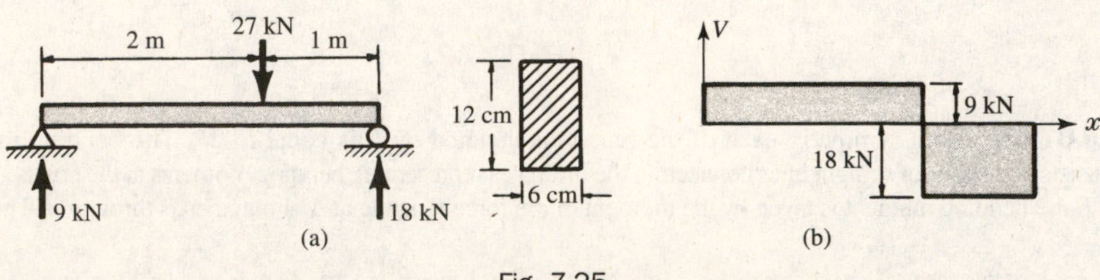

Fig. 7-25

*The shearing stress is undoubtedly maximum at the location where the normal stress is maximum in an outermost fiber, as determined using Mohr's circle.

SOLUTION: The reactions are readily found from statics to be 9 kN and 18 kN as shown. The shearing force diagram for this type of loading appears in Fig. 7-25(*b*).

From the shear diagram, the shearing force acting at a section 1 m to the right of the left reaction is 9000 N. The shearing stress τ at any point in this section at a distance $y = 3$ cm from the neutral axis is

$$\tau = \frac{VQ}{Ib} = \frac{9000\,(0.06 \times 0.03) \times 0.045}{(0.06 \times 0.12^3/12) \times 0.06} = 1.41 \times 10^6 \text{ Pa}$$

The maximum shearing stress due to the vertical shear force occurs at the neutral axis where $y = 0$. Thus,

$$\tau_{max} = \frac{VQ}{Ib} = \frac{18\,000\,(0.06 \times 0.06) \times 0.03}{(0.06 \times 0.12^3/12) \times 0.06} = 3.75 \times 10^6 \text{ Pa} \qquad \text{or} \qquad 3.75 \text{ MPa}$$

7.13. Consider the cantilever beam subject to the concentrated load shown in Fig. 7-26. Determine the maximum shearing stress due to V in the beam and also determine the shearing stress 25 mm from the top surface of the beam at a section adjacent to the supporting wall.

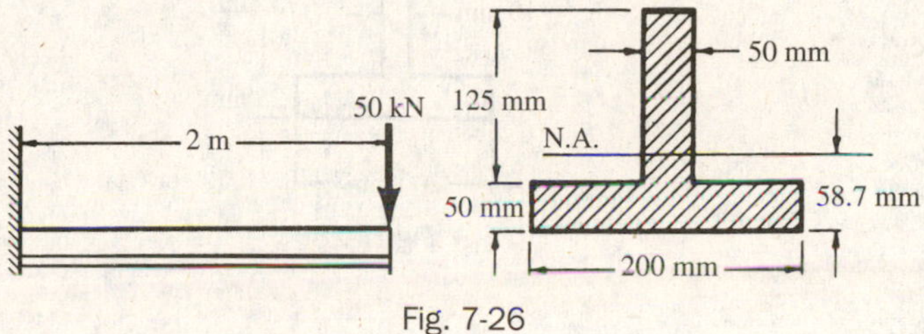

Fig. 7-26

SOLUTION: The shear force has a constant value of 50 kN at all points along the length of the beam. Because of this simple, constant value the shear diagram need not be drawn.

The location of the centroid and the moment of inertia about the centroidal axis for this particular cross section are found as usual (see Problem 7.9). The centroid is found to be 58.7 mm above the lower surface of the beam and the centroidal moment of inertia is found to be 40×10^6 mm^4.

The shearing stress is maximum at the neutral axis and is

$$\tau = \frac{VQ}{Ib} = \frac{50\,000 \times (0.1163 \times 0.05) \times 0.1163/2}{[(40 \times 10^6) \times 10^{-12}] \times 0.05} = 8.45 \times 10^6 \text{ Pa} \qquad \text{or} \qquad 8.45 \text{ MPa}$$

where, once again, Q represents the first moment of the area between the neutral axis and the outer fibers of the beam; this area is represented by the shaded region in Fig. 7-27. The value of the integral could also, of course, be found by taking the first moment of the unshaded area below the neutral axis, but that calculation would be somewhat more difficult.

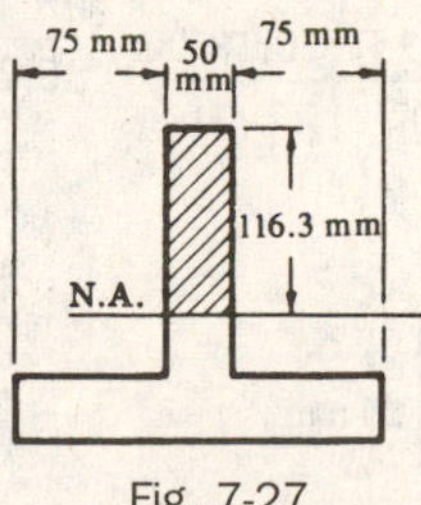

Fig. 7-27

The width b was taken to be 50 mm, since that is the width of the beam at the point where the shearing stress is being calculated. Thus the maximum shearing stress due to V is 8.45 MPa and it occurs at all points on the neutral axis along the entire length of the beam, since the shearing force has a constant value along the entire length of the beam.

The shearing stress 25 mm from the top surface of the beam is again given by the formula

$$\tau = \frac{VQ}{Ib} = \frac{50\,000(0.025 \times 0.05) \times 0.1038}{[(40 \times 10^6) \times 10^{-12}] \times 0.05} = 3.24 \times 10^6 \text{ Pa} \qquad \text{or} \qquad 3.24 \text{ MPa}$$

where 103.8 mm is the distance to the centroid of the 25 × 50 mm area.

7.14. Consider a beam having an I-type cross section as shown in Fig. 7-28. A shearing force V of 150 kN acts over the section. Determine the maximum and minimum values of the shearing stress in the vertical web of the section.

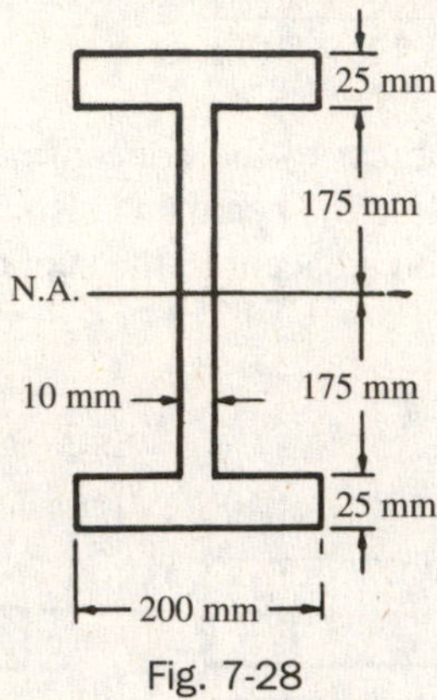

Fig. 7-28

SOLUTION: The moment of inertia I is calculated by dividing the section into rectangles, and we have

$$I = \frac{1}{12}(10)(350)^3 + 2\left[\frac{1}{12}(200)(25)^3 + 200(25)(187.5)^2\right] = 389 \times 10^6 \text{mm}^4$$

The shearing stress has a maximum value at the neutral axis. The first moment about the neutral axis of the shaded area of Fig. 7-29 is

$$Q = 175(10)(87.5) + 200(25)(187.5) = 1.1 \times 10^6 \text{mm}^3$$

Consequently the maximum shearing stress at the section a-a is found to be

$$\tau_{max} = \frac{VQ}{Ib} = \frac{150\,000(1.1 \times 10^6) \times 10^{-9}}{[(389 \times 10^6) \times 10^{-12}] \times 0.01} = 42.4 \times 10^6 \text{ Pa} \qquad \text{or} \qquad 42.4 \text{ MPa}$$

The minimum shearing stress in the web occurs at that point in the web farthest from the neutral axis, i.e., across the section b-b of Fig. 7-30. It is

$$\tau_{min} = \frac{VQ}{Ib} = \frac{150\,000(0.2 \times 0.025) \times 0.1875}{[(389 \times 10^6) \times 10^{-12}] \times 0.01} = 36.2 \times 10^6 \text{ Pa} \qquad \text{or} \qquad 36.2 \text{ MPa}$$

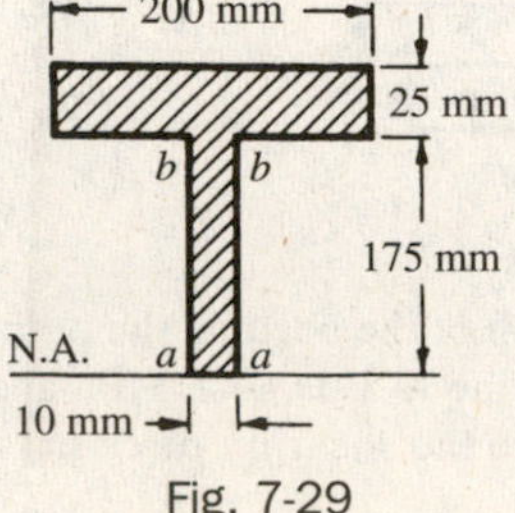

Fig. 7-29

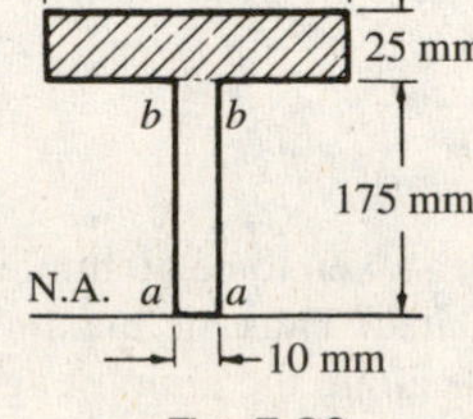

Fig. 7-30

It is to be noted that there is not too great a difference between the maximum and minimum values of shearing stress in the web of the beam. In fact, it is customary to calculate only an approximate value of the shearing stress in the web of such an I-beam. This value is obtained by dividing the total shearing force V by the cross-sectional area of the web alone. This approximate value becomes

$$\tau_{av} = \frac{150\,000}{(0.350 \times 0.01)} = 42.9 \times 10^6 \text{ Pa} \qquad \text{or} \qquad 42.9 \text{ MPa}$$

A more advanced analysis of shearing stresses in an I-beam reveals that the vertical web resists nearly all of the shearing force V and that the horizontal flanges resist only a small portion of this force. The shear stress in the web of an I-beam is specified by various codes at rather low values. Thus some codes specify 70 MPa, others 90 MPa.

7.15. A hollow cylindrical shell of outer radius 140 mm and inner radius 125 mm is subject to an axial compressive force of 168 kN and a torque of 35 kN · m. Determine the principal stresses and the maximum shearing stress acting in the shell. One end is rigidly attached to a wall.

SOLUTION: The axial compressive stress is

$$\sigma = \frac{P}{A} = \frac{168\,000}{\pi(0.14^2 - 0.125^2)} = 13.5 \times 10^6 \text{ Pa} \qquad \text{or} \qquad 13.5 \text{ MPa}$$

The shearing stress at the outerer most fibers is

$$\tau = \frac{T\rho}{J} = \frac{35\,000 \times 0.14}{\pi(0.28^4 - 0.25^4)/32} = 22.3 \times 10^6 \text{ Pa} \qquad \text{or} \qquad 22.3 \text{ MPa}$$

The stresses are displayed on an element in Fig. 7-31, and on Mohr's circle in Fig. 7-32. The principal stresses and maximum shearing stress are

$$\sigma_{max/min} = -\frac{13.5}{2} \pm \sqrt{6.75^2 + 22.3^2} = 16.6 \text{ MPa} \qquad \text{and} \qquad -30.0 \text{ MPa}$$

$$\tau_{max} = \sqrt{6.75^2 + 22.3^2} = 23.3 \text{ MPa}$$

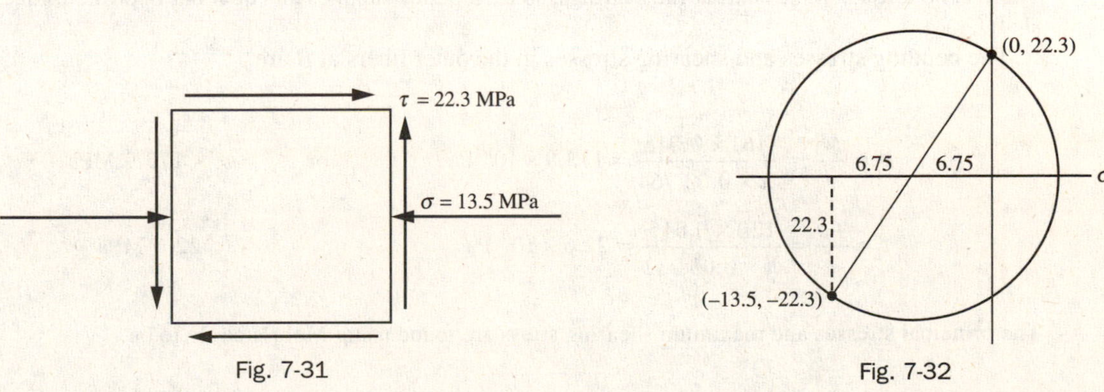

Fig. 7-31

Fig. 7-32

7.16. The 3-cm-diameter shaft in Fig. 7-33 rotates with constant angular velocity. The belt-pulls on the pulleys create a state of combined bending and torsion. The weights of the shaft and pulleys are negligible and the bearings exert only concentrated force reactions. Determine the principal stresses and the maximum shearing stress in the shaft.

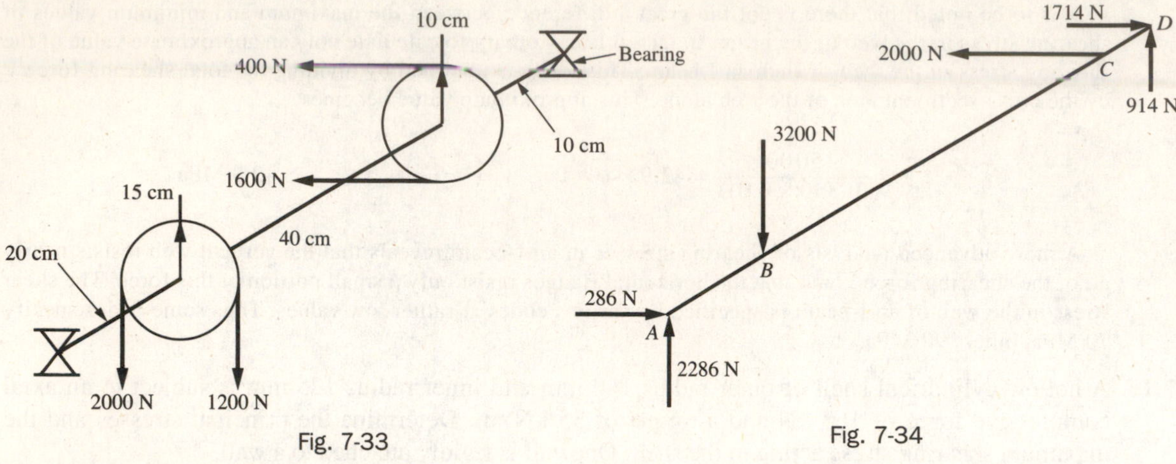

Fig. 7-33

Fig. 7-34

SOLUTION: The free-body diagram of the shaft is shown in Fig. 7-34. The transverse forces acting on the shaft due to one pulley are perpendicular to those due to the other pulley. The bending moments must be added vectorially to obtain the resultant bending moment at a particular location. This is done at B and C where the moments are the greatest. The bending moments at B and C are

$$\left.\begin{aligned}(M_B)_1 &= 2286 \times 0.2 = 457.2 \text{ N}\cdot\text{m}\\ (M_B)_2 &= 286 \times 0.2 = 57.2 \text{ N}\cdot\text{m}\end{aligned}\right\} \qquad \therefore M_B = \sqrt{457.2^2 + 57.2^2} = 461 \text{ N}\cdot\text{m}$$

$$\left.\begin{aligned}(M_C)_1 &= 914 \times 0.1 = 91.4 \text{ N}\cdot\text{m}\\ (M_C)_2 &= 1714 \times 0.1 = 171.4 \text{ N}\cdot\text{m}\end{aligned}\right\} \qquad \therefore M_C = \sqrt{91.4^2 + 171.4^2} = 194.2 \text{ N}\cdot\text{m}$$

The torque between the two pulleys is constant and equal to

$$T = (2000 - 1200) \times 0.015 = 120 \text{ N}\cdot\text{m}$$

There is no torque on the shaft at the bearings, so each pulley supplies an equal but opposite torque on the shaft.

The bending stresses and shearing stresses in the outer fibers at B are

$$\sigma = \frac{My}{I} = \frac{461 \times 0.015}{\pi \times 0.03^4/64} = 173.9 \times 10^6 \text{ Pa} \qquad \text{or} \qquad 173.9 \text{ MPa}$$

$$\tau = \frac{T\rho}{J} = \frac{120 \times 0.015}{\pi \times 0.03^4/32} = 22.6 \times 10^6 \text{ Pa} \qquad \text{or} \qquad 22.6 \text{ MPa}$$

The principal stresses and maximum shearing stress are found using Mohr's circle to be

$$\sigma_{\text{max/min}} = \frac{173.9}{2} \pm \sqrt{(173.9/2)^2 + 22.6^2} = 176.8 \text{ MPa} \qquad \text{and} \qquad -2.85 \text{ MPa}$$

$$\tau_{\text{max}} = \sqrt{(173.9/2)^2 + 22.6^2} = 89.8 \text{ MPa}$$

7.17. A 1-m-long rectangular beam is attached to a 2-m-long circular shaft and loaded as shown in Fig. 7-35. Calculate the maximum normal and shearing stresses at points A, B, and C.

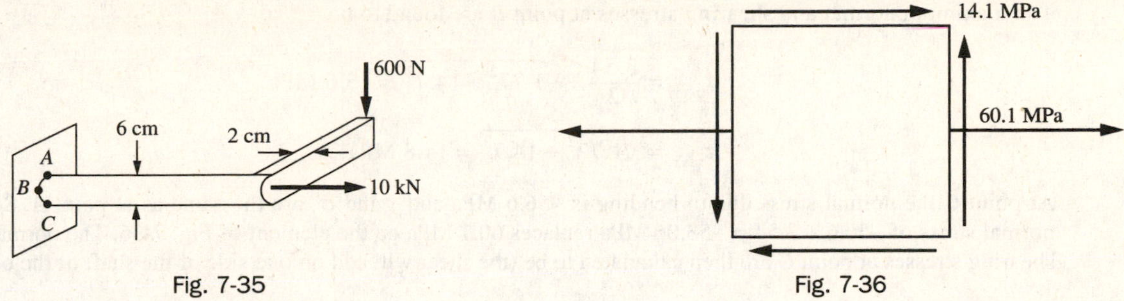

Fig. 7-35 Fig. 7-36

SOLUTION: The normal stress due to bending in the top fiber at point A is due to the moment $M = 600 \times 2 = 1200$ N · m. There is also a shearing stress due to the torque $T = 600 \times 1 = 600$ N · m in the outer fiber at point A. In addition, the tensile force of 10 kN produces a normal stress at point A. The stresses are

$$\tau = \frac{T\rho}{J} = \frac{600 \times 0.03}{\pi \times 0.06^4/32} = 14.1 \times 10^6 \text{ Pa} \quad \text{or} \quad 14.1 \text{ MPa}$$

$$\left.\begin{aligned} \sigma_1 &= \frac{Mc}{I} = \frac{1200 \times 0.03}{\pi \times 0.06^4/64} = 56.6 \times 10^6 \text{ Pa} \quad \text{or} \quad 56.6 \text{ MPa} \\ \sigma_2 &= \frac{P}{A} = \frac{10\,000}{\pi \times 0.06^2/4} = 3.54 \times 10^6 \text{ Pa} \quad \text{or} \quad 3.54 \text{ MPa} \end{aligned}\right\} \quad \therefore \sigma = 60.1 \text{ MPa}$$

The vertical shearing stress VQ/Ib is zero in the top fiber. An element at point A would be stressed as shown in Fig. 7-36. From Mohr's circle, sketched in Fig. 7-37, the maximum normal and shearing stresses at point A are calculated to be

$$\sigma_{max} = \frac{60.1}{2} + \sqrt{30.05^2 + 14.1^2} = 63.2 \text{ MPa}$$

$$\tau_{max} = \sqrt{30.05^2 + 14.1^2} = 33.2 \text{ MPa}$$

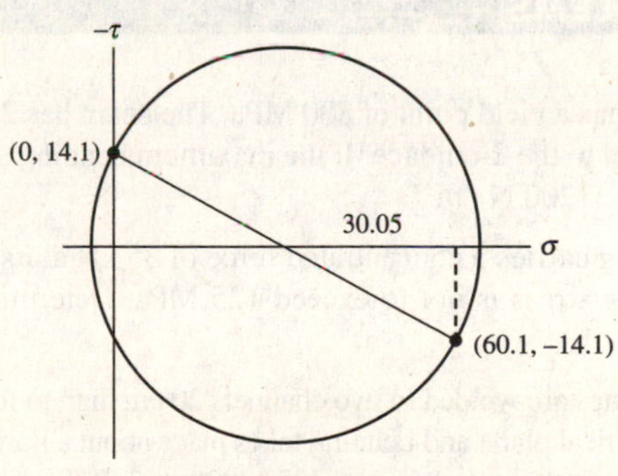

Fig. 7-37

At point B the normal stress due to bending is zero and σ_2 is the same as at point A. So, the normal stress of 3.54 MPa replaces 60.1 MPa on the element of Fig. 7-36. The normal and shearing stresses at point B are then calculated to be (the VQ/Ib shear will add on one side of the shaft or the other)

$$\left.\begin{aligned} \tau_1 &= \frac{T\rho}{J} = \frac{600 \times 0.03}{\pi \times 0.06^4/32} = 14.1 \times 10^6 \text{ Pa} \quad \text{or} \quad 14.1 \text{ MPa} \\ \tau_2 &= \frac{VQ}{Ib} = \frac{600 \times \pi \times 0.03^2 \times (4 \times 0.03/3\pi)}{(\pi \times 0.06^4/64) \times 0.06} = 0.566 \times 10^6 \text{ Pa} \end{aligned}\right\} \quad \therefore \tau = 14.7 \text{ MPa}$$

$$\sigma_2 = \frac{P}{A} = \frac{10\,000}{\pi \times 0.06^2/4} = 3.54 \times 10^6 \text{ Pa} \quad \text{or} \quad 3.54 \text{ MPa}$$

The maximum normal and shearing stresses at point B are found to be

$$\sigma_{max} = \frac{3.54}{2} + \sqrt{1.77^2 + 14.7^2} = 16.6 \text{ MPa}$$

$$\tau_{max} = \sqrt{1.77^2 + 14.7^2} = 14.8 \text{ MPa}$$

At point C the normal stress due to bending is –56.6 MPa and τ and σ_2 are the same as at point A. So, the normal stress of –56.6 + 3.54 = –53.06 MPa replaces 60.1 MPa on the element of Fig. 7-36. The normal and shearing stresses at point C are then calculated to be (the shear will add on one side of the shaft or the other)

$$\tau = \frac{T\rho}{J} = \frac{600 \times 0.03}{\pi \times 0.06^4/32} = 14.1 \times 10^6 \text{ Pa} \quad \text{or} \quad 14.1 \text{ MPa}$$

$$\sigma_1 = -\frac{Mc}{I} = -\frac{1200 \times 0.03}{\pi \times 0.06^4/64} = -56.6 \times 10^6 \text{ Pa} \quad \text{or} \quad -56.6 \text{ MPa}$$

$$\sigma_2 = \frac{P}{A} = \frac{10\,000}{\pi \times 0.06^2/4} = 3.54 \times 10^6 \text{ Pa} \quad \text{or} \quad 3.54 \text{ MPa}$$

$$\therefore \sigma = -53.04 \text{ MPa}$$

The normal stress with maximum magnitude and the maximum shearing stresses at point C are

$$\sigma_{max} = \frac{-53.04}{2} \pm \sqrt{26.52^2 + 14.1^2} = -56.5 \text{ MPa}$$

$$\tau_{max} = \sqrt{26.52^2 + 14.1^2} = 30.0 \text{ MPa}$$

Obviously, the normal and shearing stresses are maximum in the shaft at point A.

SUPPLEMENTARY PROBLEMS

7.18. A beam made of titanium has a yield point of 800 MPa. The beam has 2 cm × 6 cm rectangular cross section and bends about an axis parallel to the 2-cm face. If the maximum bending stress is 600 MPa, find the corresponding bending moment. *Ans.* 7200 N · m

7.19. A cantilever beam 3 m long carries a concentrated force of 35 kN at its free end. The material is structural steel and the maximum bending stress is not to exceed 125 MPa. Determine the required diameter if the beam is circular. *Ans.* 204 mm

7.20. Two 1 cm × 16 cm cover plates are welded to two channels 20 cm high to form the cross section of the beam shown in Fig. 7-38. Loads are in a vertical plane and bending takes place about a horizontal axis. The moment of inertia of each channel about a horizontal axis through the centroid is 3000 cm^4. If the maximum allowable elastic bending stress is 120 MPa, determine the maximum bending moment that may be developed in the beam. *Ans.* 17.1 kN · m

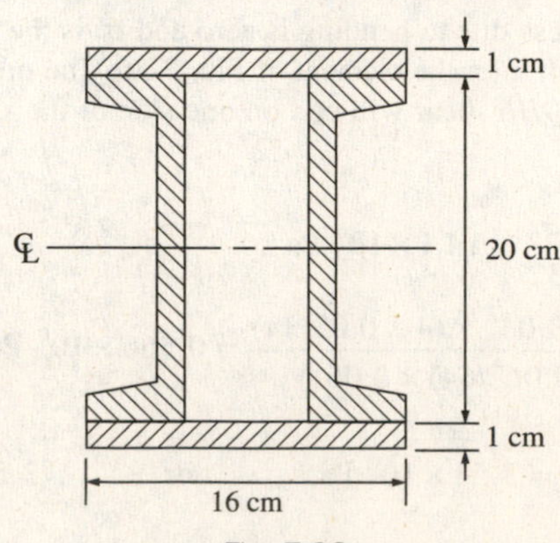

Fig. 7-38

7.21. A 250-mm-deep wide-flange section with $I = 61 \times 10^6$ mm^4 is used as a cantilever beam. The beam is 2 m long and the allowable bending stress is 125 MPa. Determine the maximum allowable intensity w of uniform load that may be carried along the entire length of the beam. *Ans.* 30.5 kN/m

7.22. The beam shown in Fig. 7-39 is simply supported at the ends and carries the two symmetrically placed loads of 60 kN each. If the working stress in either tension or compression is 125 MPa, what is the required moment of inertia of area required for a 250-mm-deep beam? *Ans.* 60×10^6 mm^4

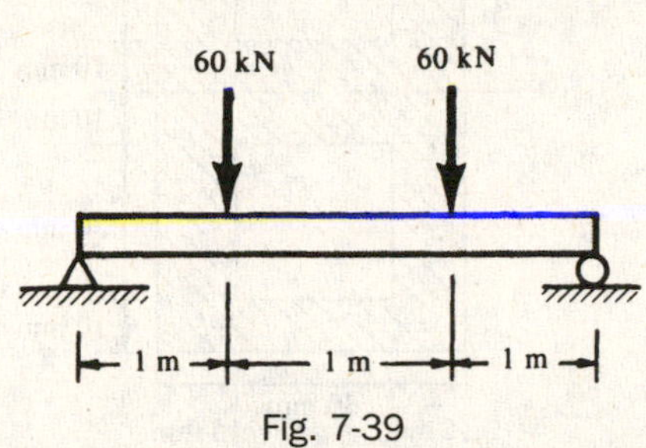

Fig. 7-39

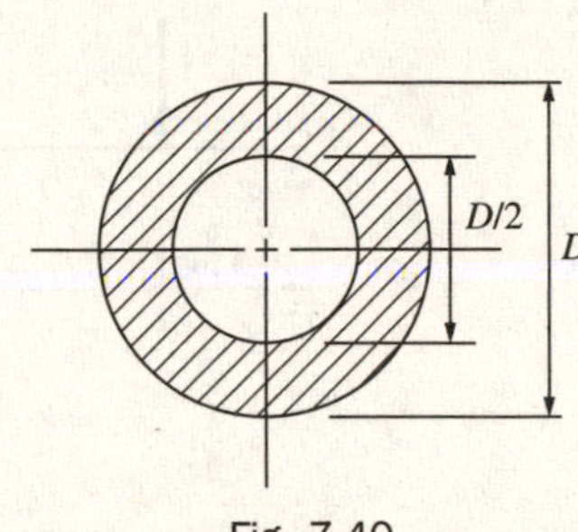

Fig. 7-40

7.23. Consider the simply supported beam subject to the two concentrated forces (60 kN each) shown in Fig. 7-39. Now, the beam is of hollow circular cross section as shown in Fig. 7-40, with an allowable working stress in either tension or compression of 125 MPa. Determine the necessary outer diameter of the beam. *Ans.* 17.4 mm

7.24. Select a suitable rectangular section twice as high as it is wide to act as a cantilever beam 2 m long that carries a uniformly distributed load of 40 kN/m. The working stress in either tension or compression is 150 MPa. *Ans.* 92.8 mm × 185.6 mm

7.25. A beam 3 m long is simply supported at each end and carries a uniformly distributed load of 10 kN/m. The beam is of rectangular cross section, 75 mm × 150 mm. Determine the magnitude and location of the peak bending stress. Also, find the magnitude of the bending stress at a point 25 mm below the upper surface at the section midway between supports. *Ans.* 40 MPa, 26.8 MPa

7.26. Reconsider the steel beam of Problem 7.25. Determine the maximum bending stress if now the weight of the beam is considered in addition to the load of 10 kN/m. The weight of steel is 77.0 kN/m^3. *Ans.* 43.6 MPa

7.27. Consider a simply supported beam carrying the concentrated and uniform loads shown in Fig. 7-41. Select the height of a 2-cm-wide rectangular section to resist these loads based upon a working stress in either tension or compression of 120 MPa. *Ans.* 130 mm

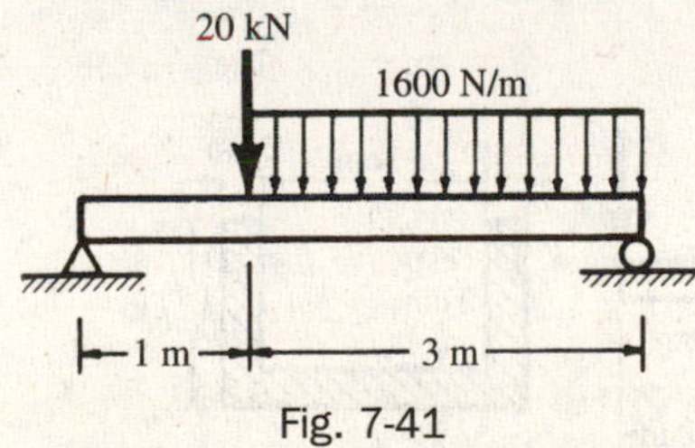

Fig. 7-41

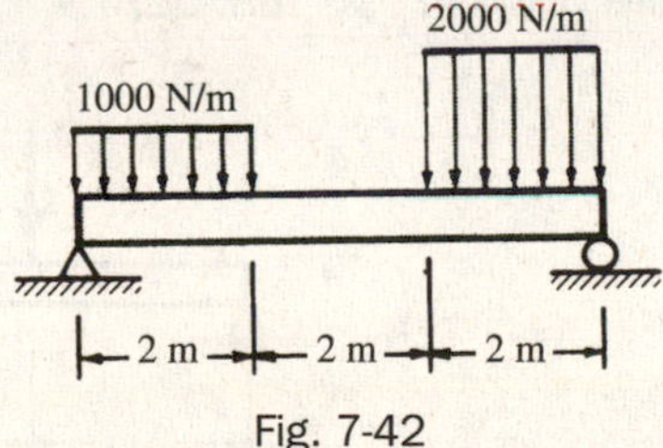

Fig. 7-42

7.28. The two distributed loads are carried by the simply supported beam as shown in Fig. 7-42. The beam has the hollow circular cross section shown in Fig. 7-40 with $D = 20$ cm. Determine the magnitude and location of the maximum bending stress in the beam. *Ans.* 4.57 MPa, 1.834 m from the right support

7.29. A T-beam having the cross section shown in Fig. 7-43 projects 2 m from a wall as a cantilever beam and carries a uniformly distributed load of 8 kN/m. Determine the maximum tensile and compressive bending stresses. *Ans.* +38.5 MPa, –81 MPa

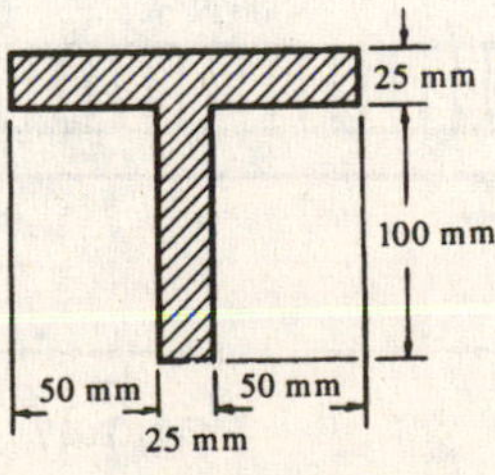

Fig. 7-43

7.30. The simply supported beam AC shown in Fig. 7-44 supports a concentrated load P. The beam section is rectangular, with two square cutouts as shown. If the allowable working stress is 120 MPa, determine the maximum value of P. *Ans.* 1.80 kN

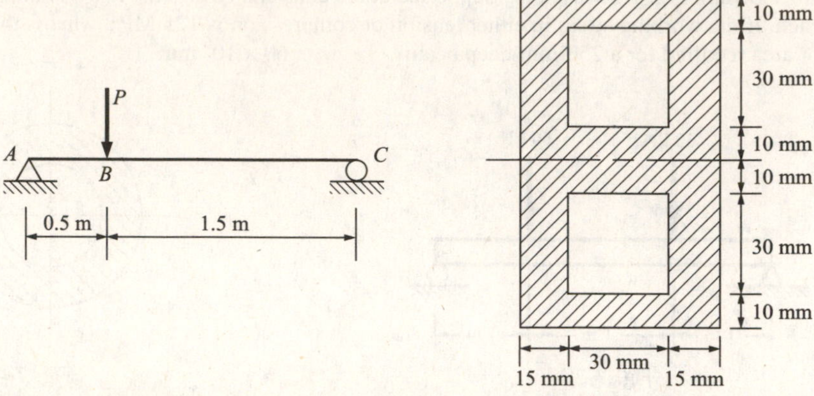

Fig. 7-44

7.31. A simply supported steel beam of channel-type cross section is loaded by both the uniformly distributed load and the couple shown in Fig. 7-45. Determine the maximum tensile and compressive stresses.

Ans. 31.2 MPa, –56.8 MPa

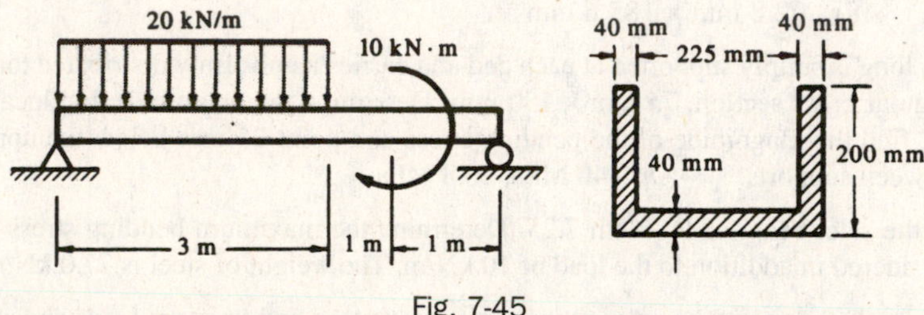

Fig. 7-45

7.32. A channel-shaped beam with an overhanging end is loaded as shown in Fig. 7-46. The material is gray cast iron having an allowable working stress of 30 MPa in tension and 120 MPa in compression. Determine the maximum allowable value of P. *Ans.* 2220 N

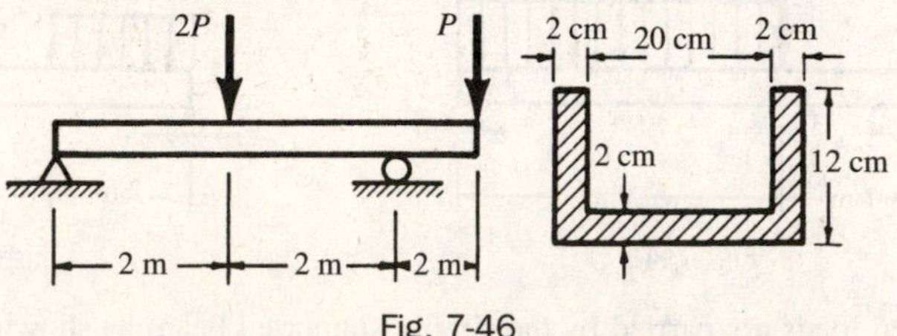

Fig. 7-46

7.33. A simply supported beam is shown in Fig. 7-47. Find (*a*) the maximum normal stress in the beam, (*b*) the maximum shearing stress in the beam due to V, and (*c*) the shearing stress at a point 1 m to the right of R_1 and 2 cm below the top surface of the beam. *Ans.* (*a*) 3.94 MPa (*b*) 337 kPa (*c*) 121 kPa

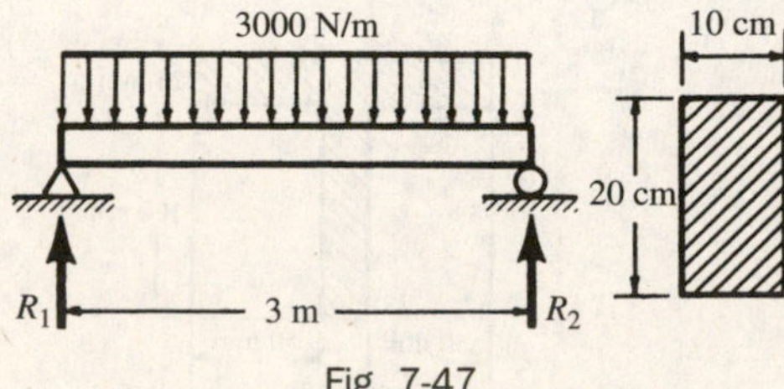

Fig. 7-47

7.34. Determine (*a*) the maximum bending stress and (*b*) the maximum shearing stress due to V in the simply supported beam shown in Fig. 7-48. *Ans.* (*a*) 240 MPa (*b*) 12 MPa

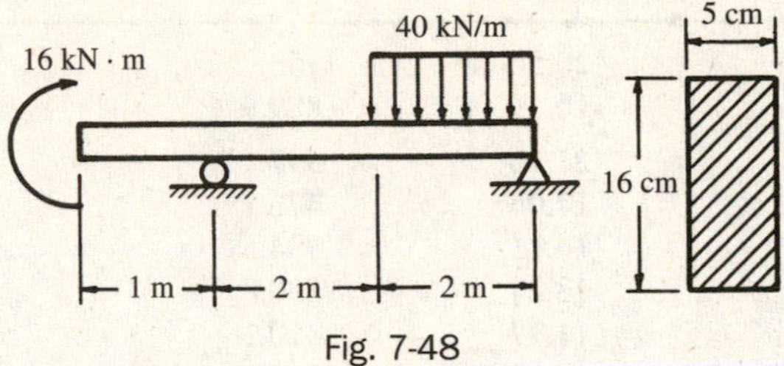

Fig. 7-48

7.35. Consider a hollow circular shaft whose outside diameter is 8 cm and whose inside diameter is equal to one-half the outside diameter. The shaft is subject to a twisting moment of 2000 N · m as well as a bending moment of 3000 N · m. Determine the principal stresses in the body. Also, determine the maximum shearing stress.

Ans. 70.1 MPa, –6.4 MPa, 38.3 MPa

7.36. The shaft shown in Fig. 7-49 is supported by bearings that exert only concentrated reactions. Two pulleys carry belts that provide the forces indicated. Determine the maximum normal stresses and the maximum shearing stress in the 10-cm-diameter shaft. *Ans.* 82.5 MPa, 41.8 MPa.

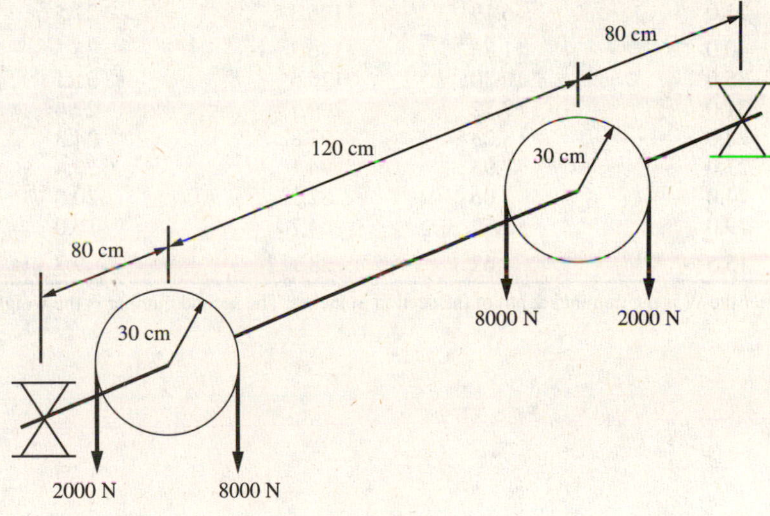

Fig. 7-49

7.37. A 50-cm-long rectangular beam is attached to an 80-cm-long circular shaft, as shown in Fig. 7-50. Calculate the maximum normal and shearing stresses at points A, B, and C. A is on the top, B is on the bottom, and C is on the neutral axis. *Ans.* At A: 88.1 MPa, 46.3 MPa; at B: 7.7 MPa, 29.6 MPa; at C 32.2 MPa, 22.2 MPa.

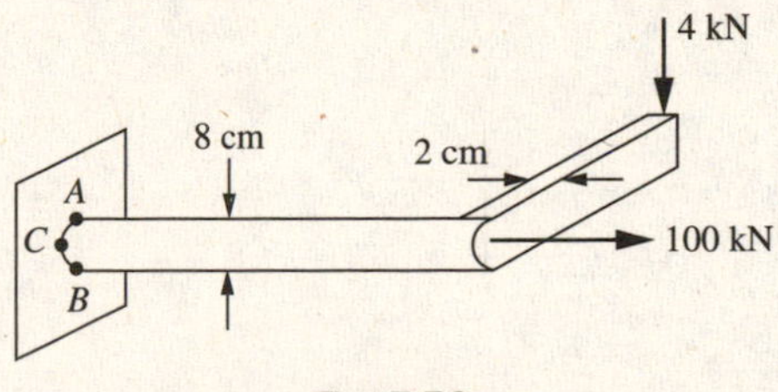

Fig. 7-50

Table 7-1. Properties of Selected Wide-Flange Sections, USCS Units

Designation*	Weight per foot, lb/ft	Area, in^2	I (about x-x axis), in^4	S, in^3	I (about y-y axis), in^4
W 18 × 70	70.0	20.56	1153.9	128.2	78.5
W 18 × 55	55.0	16.19	889.9	98.2	42.0
W 12 × 72	72.0	21.16	597.4	97.5	195.3
W 12 × 58	58.0	17.06	476.1	78.1	107.4
W 12 × 50	50.0	14.71	394.5	64.7	56.4
W 12 × 45	45.0	13.24	350.8	58.2	50.0
W 12 × 40	40.0	11.77	310.1	51.9	44.1
W 12 × 36	36.0	10.59	280.8	45.9	23.7
W 12 × 32	32.0	9.41	246.8	40.7	20.6
W 12 × 25	25.0	7.39	183.4	30.9	14.5
W 10 × 89	89.0	26.19	542.4	99.7	180.6
W 10 × 54	54.0	15.88	305.7	60.4	103.9
W 10 × 49	49.0	14.40	272.9	54.6	93.0
W 10 × 45	45.0	13.24	248.6	49.1	53.2
W 10 × 37	37.0	10.88	196.9	39.9	42.2
W 10 × 29	29.0	8.53	157.3	30.8	15.2
W 10 × 23	23.0	6.77	120.6	24.1	11.3
W 10 × 21	21.0	6.19	106.3	21.5	9.7
W 8 × 40	40.0	11.76	146.3	35.5	49.0
W 8 × 35	35.0	10.30	126.5	31.1	42.5
W 8 × 31	31.0	9.12	109.7	27.4	37.0
W 8 × 28	28.0	8.23	97.8	24.3	21.6
W 8 × 27	27.0	7.93	94.1	23.4	20.8
W 8 × 24	24.0	7.06	82.5	20.8	18.2
W 8 × 19	19.0	5.59	64.7	16.0	7.9
W 6 × $15\frac{1}{2}$	15.5	4.62	28.1	9.7	9.7

*The first number after the W is the nominal depth of the section in inches. The second number is the weight in pounds per foot of length.

Table 7-2. Properties of Selected Wide-Flange Sections, SI Units

Designation*	Mass per meter, kg/m	Area, mm^2	I (about x-x axis), $10^6\ mm^4$	S, $10^3 mm^3$	I (about y-y axis), $10^6\ mm^4$
W 460 × 103	102.9	13 200	479	2100	32.6
W 460 × 81	80.9	10 400	369	1610	17.4
W 305 × 106	105.8	13 600	248	1590	81.0
W 305 × 85	85.3	11 000	198	1280	44.6
W 305 × 74	73.5	9480	164	1060	23.4
W 305 × 66	66.2	8530	146	952	20.7
W 305 × 59	58.8	7580	129	849	18.3
W 305 × 53	52.9	6820	117	750	9.83
W 305 × 47	47.0	6060	102	665	8.55
W 305 × 37	36.8	4760	76.1	505	6.02
W 254 × 131	130.8	16 900	225	1630	74.9
W 254 × 79	79.4	10 200	127	988	43.1
W 254 × 72	72.0	9280	113	893	38.6
W 254 × 66	66.2	8530	103	803	22.1
W 254 × 54	54.4	7010	81.7	652	17.5
W 254 × 43	42.6	5490	65.3	504	6.31
W 254 × 34	33.8	4360	50.0	394	4.69
W 254 × 31	30.9	3990	44.1	352	4.02
W 203 × 59	58.8	7580	60.7	580	20.3
W 203 × 51	51.4	6630	52.5	508	17.6
W 203 × 46	45.6	5870	45.5	448	15.4
W 203 × 41	41.2	5300	40.6	397	8.96
W 203 × 40	39.7	5110	39.0	383	8.63
W 203 × 35	35.3	4550	34.2	340	7.55
W 203 × 28	27.9	3600	26.8	262	3.28
W 152 × 23	22.8	2980	11.7	159	4.02

*The first number after the W is the nominal depth of the section in millimeters. The second number is the mass in kilograms per meter of length.

Deflection of Beams

8.1 Basics

In Chapter 7 it was stated that lateral loads applied to a beam not only give rise to internal bending and shearing stresses in the bar, but also cause the bar to deflect in a direction perpendicular to its longitudinal axis. The stresses were examined in Chapter 7 and it is the purpose of this chapter to examine methods for calculating the deflections.

The deformation of a beam is expressed in terms of the deflection of the beam from its original unloaded position. The deflection is measured from the original neutral surface to the neutral surface of the deformed beam. The configuration assumed by the deformed neutral surface is known as the *elastic curve* of the beam. Figure 8-1 represents the beam in its original undeformed state and Fig. 8-2 represents the beam in the deformed configuration it has assumed under the action of the load.

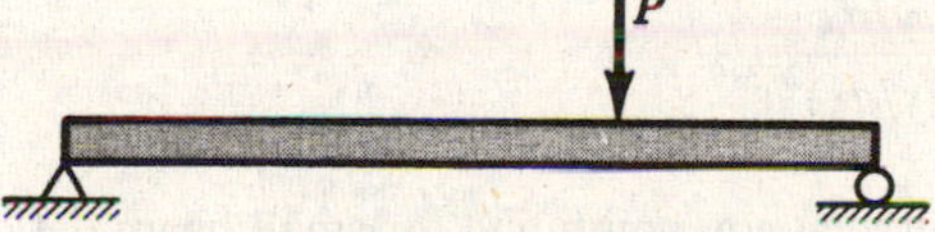

Fig. 8-1 Undeformed position of a beam.

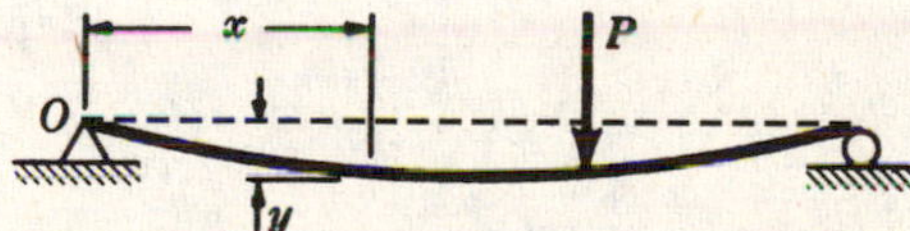

Fig. 8-2 Deformed position of the beam.

The displacement y is defined as the *deflection* of the beam. Often it will be necessary to determine the deflection y for every value of x along the beam. This relation is the *elastic curve* or *deflection curve* of the beam.

Specifications for the design of beams frequently impose limitations upon the deflections as well as the stresses. Consequently, in addition to the calculation of stresses as outlined in Chapter 7, it is essential that the designer be able to determine deflections. For example, in many building codes the maximum allowable deflection of a beam is not to exceed $\frac{1}{300}$ of the length of the beam. Components of aircraft usually are designed so that deflections do not exceed some preassigned value, so that the aerodynamic characteristics are not altered. Thus, a well-designed beam must not only be able to carry the loads to which it will be subjected but it must also not undergo undesirably large deflections. Also, the evaluation of reactions of statically indeterminate beams involves the use of various deformation relationships, to be examined in detail in Chapter 9.

Methods of Determining Beam Deflections

Numerous methods are available for the determination of beam deflections. The most commonly used are the double-integration method and the method of singularity functions. It is to be carefully noted that all methods apply *only* if all portions of the beam are acting in the *elastic range of action*.

8.2 Differential Equation of the Elastic Curve

Let us derive the differential equation of the deflection curve of a beam loaded by lateral forces.

In Chapter 7 the relationship

$$M = \frac{EI}{\rho} \tag{8.1}$$

was derived as Eq. (7.7). In this expression M denotes the bending moment acting at a particular cross section of the beam, ρ the radius of curvature of the neutral surface of the beam at this same section, E the modulus of elasticity, and I the moment of the cross-sectional area about the neutral axis passing through the centroid of the cross section. In this book we will be concerned with those beams for which E and I are constant along the entire length of the beam, but in general both M and ρ could be functions of x.

Let the heavy line in Fig. 8-3 represent the deformed neutral surface of the bent beam. Originally, the beam coincided with the x-axis prior to loading and the coordinate system that is usually found to be most convenient is shown in the sketch. The deflection y is taken to be positive in the upward direction; hence for the particular beam shown, all deflections are negative.

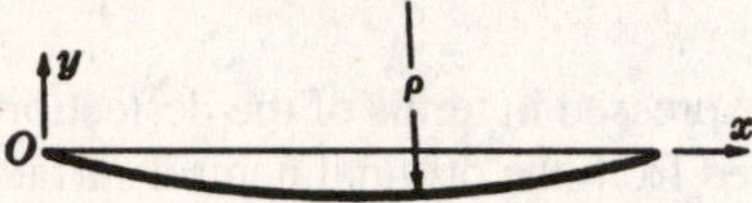

Fig. 8-3 The neutral surface of the beam.

An expression for the *curvature* at any point along the curve representing the deformed beam is readily available from differential calculus. It is

$$\frac{1}{\rho} = \frac{d^2y/dx^2}{[1 + (dy/dx)^2]^{3/2}} \tag{8.2}$$

In this expression, dy/dx represents the slope of the curve at any point; and for small beam deflections this quantity and in particular its square are small in comparison to unity and may reasonably be neglected. This assumption of small deflections simplifies the expression for curvature into

$$\frac{1}{\rho} \approx \frac{d^2y}{dx^2} \tag{8.3}$$

Hence for small deflections, Eq. (8.1) becomes

$$EI\frac{d^2y}{dx^2} = M \tag{8.4}$$

This is the differential equation of the deflection curve of a beam loaded by lateral forces. It is called the Euler-Bernoulli equation of bending of a beam. In any problem it is necessary to integrate this equation to obtain an algebraic relationship between the deflection y and the coordinate x along the length of the beam. This will be carried out in the problems.

8.3 Deflection by Integration

The double-integration method for calculating deflections of beams merely consists of integrating Eq. (8.4). The first integration yields the slope dy/dx at any point in the beam and the second integration gives the deflection y for any value of x. The bending moment M must, of course, be expressed as a function of the coordinate x before the equation can be integrated. For the cases to be studied here the integrations are straightforward.

Since the differential equation (8.4) is of the second order, its solution must contain two constants of integration. These two constants must be evaluated from known conditions concerning the slope or deflection at certain points in the beam. For example, in the case of a cantilever beam the constants would be determined from the conditions of zero change of slope as well as zero deflection at the built-in end of the beam, and at a pin or roller the deflection would be zero.

Frequently two or more equations are necessary to describe the bending moment in the various regions along the length of a beam. This was emphasized in Chapter 6. In such a case, Eq. (8.4) must be written for each region of the beam and integration of these equations yields two constants of integration for each region. These constants must then be determined so as to impose conditions of continuous deformations and slopes at the points common to adjacent regions. Problems will illustrate.

The sign conventions for bending moment adopted in Chapter 6 will be retained here. The quantities E and I appearing in Eq. (8.4) are, of course, positive. Thus, from this equation, if M is positive for a certain value of x, then d^2y/dx^2 is also positive. With the above sign convention for bending moments, it is necessary to consider the coordinate x along the length of the beam to be positive to the right and the deflection y to be positive upward. With these algebraic signs the integration of Eq. (8.4) may be carried out to yield the deflection $y(x)$ with the understanding that upward beam deflections are positive and downward deflections negative.

In the derivation of Eq. (8.4) it was assumed that the slope of the elastic curve was much less than unity and that all portions of the beam were acting in the elastic range. Equation (8.4) is derived on the basis of the beam being straight prior to the application of loads.

8.4 Deflections Using Singularity Functions

In Section 8.3 we described how the elastic deflections of transversely loaded beams are found through direct integration of the second-order Euler-Bernoulli equation. As we saw, the approach is direct but may become very lengthy even for relatively simple engineering situations in which several loads are imposed.

A more expedient approach is based upon the use of the singularity functions introduced in Chapter 6. The method is direct and may be applied to a beam subject to any combination of concentrated forces, moments, and distributed loads. One must only remember the definition of the singularity function given in Chapter 6: the quantity $\langle x - a \rangle$ vanishes if $x < a$ but is equal to $(x - a)$ if $x > a$.

There are several possible approaches for using singularity functions for the determination of beam deflections. Perhaps the simplest is to employ the approach of Chapter 6 in which the bending moment is written in terms of singularity functions in the form of one equation valid along the entire length of the beam. Two integrations of Eq. (8.1) lead to the equation for the deflected beam in terms of two constants of integration which must be determined from boundary conditions. As noted in Chapter 6, integration of the singularity functions proceeds directly and in the same manner as simple power functions. Thus, the approach is direct and avoids the problem of the determination of a pair of constants corresponding to each region of the beam (between loads), as in the case of double integration. Problems 8.11 to 8.14 and 8.27 to 8.34 will illustrate.

It should be noted that the singularity function approach leads directly into a computerized approach for the determination of beam deflections.

8.5 Deflections Using Superposition

The equations that describe the deflection of a beam and the stresses due to the applied loads are all linear (i.e., if a load is doubled, the stresses and deflections are also doubled). Thus, we are able to superpose the contributions from each separate load to obtain the resultant effect of several loads. The contributions of each separate load are typically available from previous work for deflections and stresses. If only the maximum deflection is of interest, as is often the case, the results of several simple beams are presented in Table 8-1 for quick reference. Problems 8.15, 8.16, and 8.17 illustrate the procedure.

Table 8-1. Beam Deflection Formulas

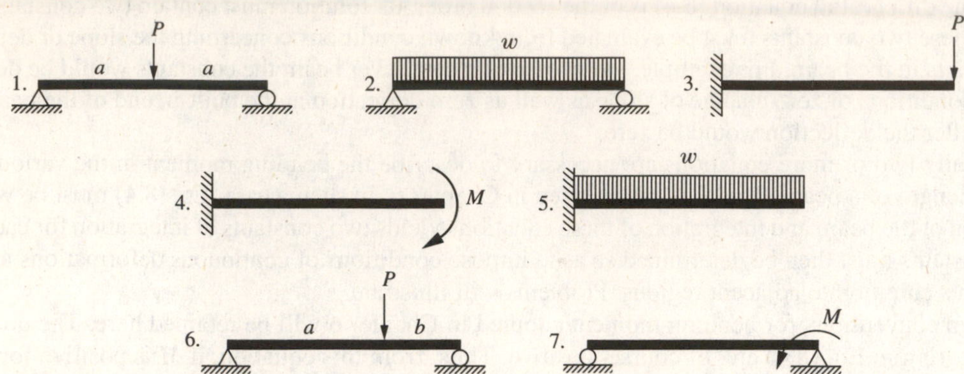

	Max Shear	Max Moment	Max Deflection	Max Beam Slope
1	$P/2$	$PL/4$	$PL^3/48EI$	$PL^2/16EI$
2	$wL/2$	$wL^2/8$	$5wL^4/384EI$	$wL^3/24EI$
3	P	PL	$PL^3/3EI$	$PL^2/2EI$
4	0	M	$ML^2/2EI$	ML/EI
5	wL	$wL^2/2$	$wL^4/8EI$	$wL^3/6EI$
6	$\dfrac{Pa}{L}$	$\dfrac{Pab}{L}$	$y_{\max} = \dfrac{Pb(L^2 - b^2)^{3/2}}{9\sqrt{3}LEI}$ at $x = \sqrt{\dfrac{L^2 - b^2}{3}}$ $y_{\text{center}} = \dfrac{Pb(3L^2 - 4b^2)}{48EI}$	$\theta_{\text{left}} = -\dfrac{Pb(L^2 - b^2)}{6LEI}$ $\theta_{\text{right}} = \dfrac{Pa(L^2 - a^2)}{6LEI}$
7	$\dfrac{M}{L}$	M	$y_{\max} = \dfrac{\sqrt{3}ML^2}{27EI}$ at $x = \dfrac{L}{\sqrt{3}}$ $y_{\text{center}} = \dfrac{ML^2}{16EI}$	$\theta_{\text{left}} = -\dfrac{ML}{6EI}$ $\theta_{\text{right}} = \dfrac{ML}{3EI}$

SOLVED PROBLEMS

8.1. Determine the deflection at every point of the cantilever beam subject to the single concentrated force P, as shown in Fig. 8-4.

SOLUTION: The x-y coordinate system shown is introduced, where the x-axis coincides with the original unbent position of the beam. The deformed beam has the appearance indicated by the heavy line in Fig. 8-5. It is first necessary to find the reactions exerted by the supporting wall upon the bar, and these are easily found from statics to be a vertical force reaction P and a moment PL, as shown.

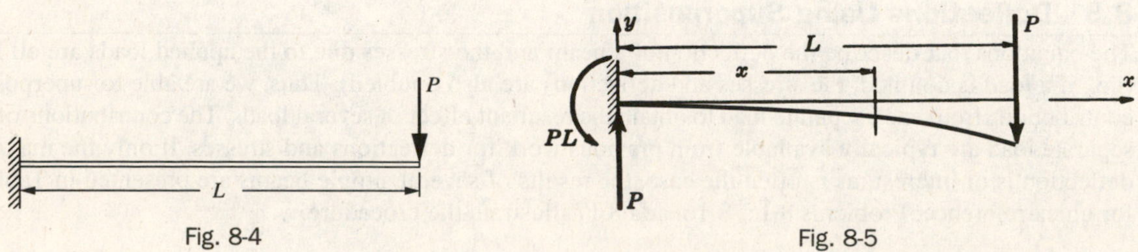

Fig. 8-4

Fig. 8-5

According to the sign convention of Chap. 6, the bending moment M at the section x is

$$M = -PL + Px$$

The differential equation (8.4) of the bent beam is then

$$EI\frac{d^2y}{dx^2} = -PL + Px \tag{1}$$

This equation is readily integrated once to yield

$$EI\frac{dy}{dx} = -PLx + \frac{Px^2}{2} + C_1 \tag{2}$$

which represents the equation of the slope, where C_1 denotes a constant of integration. This constant may be evaluated by use of the condition that the slope dy/dx of the beam at the wall is zero since the beam is rigidly clamped there. Equation (2) is true for all values of x and y, and if the condition $x = 0$ is substituted we obtain $0 = 0 + 0 + C_1$ or $C_1 = 0$.

Next, integration of Eq. (2) yields

$$EIy = -PL\frac{x^2}{2} + \frac{Px^3}{6} + C_2 \tag{3}$$

where C_2 is a second constant of integration. Again, the condition at the supporting wall will determine this constant. At $x = 0$, the deflection y is zero since the bar is rigidly clamped. We find $0 = 0 + 0 + C_2$ or $C_2 = 0$.

Thus Eqs. (2) and (3) with $C_1 = C_2 = 0$ give the slope dy/dx and deflection y at any point x in the beam. The deflection is maximum at the right end of the beam ($x = L$), under the load P, and from Eq. (3),

$$EIy_{max} = \frac{-PL^3}{3} \tag{4}$$

where the negative value denotes that this point on the deflection curve lies below the x-axis. If only the magnitude of the maximum deflection at $x = L$ is desired, it is usually denoted by Δ_{max} and we have

$$\Delta_{max} = \frac{PL^3}{3EI} \tag{5}$$

This is the maximum deflection formula for Beam 3 of Table 8-1.

8.2. The cantilever beam shown in Fig. 8-4 is 3 m long and loaded by an end force of 20 kN. The cross section is a W 203 × 59 steel section, which according to Table 7-2 has $I = 60.7 \times 10^{-6}\ \text{m}^4$ and $S = 580 \times 10^{-6}\ \text{m}^3$. Find the maximum deflection of the beam. Take $E = 200$ GPa. Neglect the weight of the beam.

SOLUTION: The maximum deflection occurs at the free end of the beam under the concentrated force and was found in Problem 8.1 to be, by Eq. (5),

$$\Delta_{max} = \frac{PL^3}{3EI} = \frac{(20\,000)(3)^3}{3(200 \times 10^9)(60.7 \times 10^{-6})} = 0.0148\,\text{m} \qquad \text{or} \qquad 14.8\ \text{mm}$$

In the derivation of this deflection formula it was assumed that the material of the beam follows Hooke's law. Actually, from the above calculation alone there is no assurance that the material is not stressed beyond the proportional limit. If it were, then the basic beam-bending Eq. (8.4) would no longer be valid and the above numerical value would be meaningless. Consequently, in every problem involving beam deflections it is to be emphasized that the maximum bending stress in the beam is below the proportional limit of the material. The maximum bending stress is

$$\sigma = \frac{Mc}{I} = \frac{M}{I/c} = \frac{M}{S} = \frac{60\,000}{580 \times 10^{-6}} = 103 \times 10^6\,\text{Pa} \qquad \text{or} \qquad 103\ \text{MPa}$$

Since this value is below the proportional limit of steel, which is approximately 200 MPa, the use of the beam deflection equation was justifiable.

8.3. Determine the slope of the right end of the cantilever beam loaded as shown in Fig. 8-4 using the beam and loads described in Problem 8.2.

SOLUTION: In Problem 8.1 the equation of the slope was found to be

$$EI\frac{dy}{dx} = -PLx + \frac{Px^2}{2}$$

At the free end, $x = L$, and

$$EI\left(\frac{dy}{dx}\right)_{x=L} = -PL^2 + \frac{PL^2}{2}$$

The slope at the end is thus

$$\left(\frac{dy}{dx}\right)_{x=L} = \frac{-PL^2}{2EI} = \frac{-20\,000 \times 3^2}{2(200 \times 10^9)(60.7 \times 10^{-6})} = -0.00741 \qquad \text{or} \qquad -0.425°$$

Note: $dy/dx = \tan\theta \cong \theta$ since the slope of beams is very small.

8.4. Determine the deflection curve of a cantilever beam subject to the uniformly distributed load w, shown in Fig. 8-6.

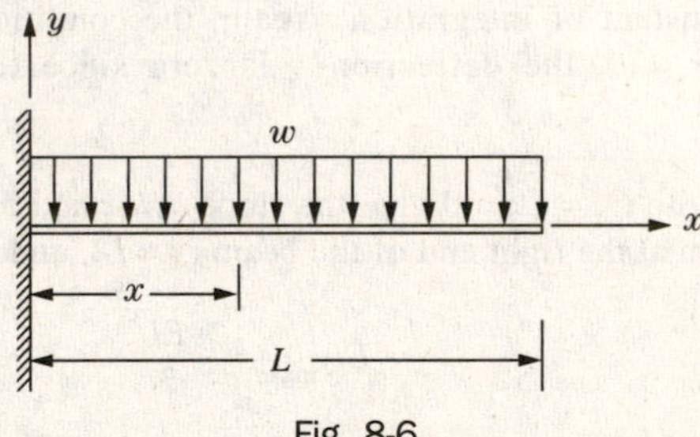

Fig. 8-6

SOLUTION: The equation for the bending moment could be determined in a manner analogous to that used in Problem 8.1, but instead let us determine the bending moment at the section at a distance x from the wall by considering the forces to the right of this section, rather than those to the left.

The force due to the distributed load w is $w(L - x)$. This force acts at the midpoint of this length of beam to the right of x and thus its moment arm from x is $\frac{1}{2}(L - x)$. The bending moment at the section x is thus given by

$$M = -\frac{w}{2}(L - x)^2$$

the negative sign being necessary since downward loads produce negative bending.

The differential equation describing the bent beam is thus

$$EI\frac{d^2y}{dx^2} = -\frac{w}{2}(L - x)^2 \tag{1}$$

The first integration yields

$$EI\frac{dy}{dx} = \frac{w}{2}\frac{(L - x)^3}{3} + C_1 \tag{2}$$

The constant may be evaluated by using $(dy/dx)_{x=0} = 0$. We find $C_1 = -wL^3/6$. We thus have

$$EI\frac{dy}{dx} = \frac{w}{6}(L - x)^3 - \frac{wL^3}{6} \tag{3}$$

The next integration yields

$$EIy = -\frac{w}{6}\frac{(L - x)^4}{4} - \frac{wL^3}{6}x + C_2 \tag{4}$$

At the clamped end, the deflection is zero so that

$$0 = \frac{-wL^4}{24} + C_2 \qquad \text{or} \qquad C_2 = \frac{wL^4}{24}$$

The final form of the deflection curve of the beam is thus

$$EIy = -\frac{w}{24}(L-x)^4 - \frac{wL^3}{6}x + \frac{wL^4}{24} \tag{5}$$

The deflection is maximum at the right end of the bar ($x = L$) and there we have from Eq. (5)

$$EIy_{\max} = -\frac{wL^4}{6} + \frac{wL^4}{24} = -\frac{wL^4}{8}$$

where the negative value denotes that this point on the deflection curve lies below the x-axis. The magnitude of the maximum deflection is

$$\Delta_{\max} = \frac{wL^4}{8EI} \tag{6}$$

This is the maximum deflection formula for Beam 5 of Table 8-1.

8.5. Obtain an expression for the deflection curve of the simply supported beam of Fig. 8-7 subject to the uniformly distributed load w.

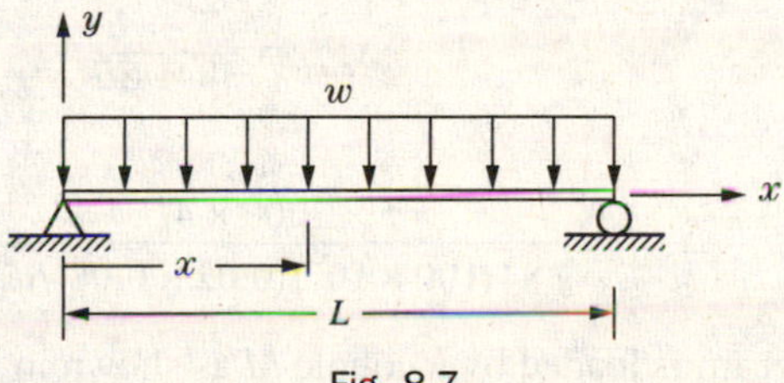

Fig. 8-7

SOLUTION: The total load acting on the beam is wL and, because of symmetry, each of the end reactions is $wL/2$. Because of the symmetry of loading, it is evident that the deflected beam is symmetric about the midpoint $x = L/2$.

The portion of the uniform load to the left of the section at a distance x from the left support is replaced by its resultant acting at the midpoint of the section of length x. The resultant is wx acting downward and hence giving rise to a negative bending moment. The reaction $wL/2$ gives rise to a positive bending moment. Consequently, for any value of x, the bending moment is

$$M = \frac{wL}{2}x - wx\frac{x}{2}$$

The differential equation of the loaded beam is thus

$$EI\frac{d^2y}{dx^2} = \frac{wL}{2}x - \frac{wx^2}{2} \tag{1}$$

Integrating,

$$EI\frac{dy}{dx} = \frac{wL}{2}\frac{x^2}{2} - \frac{w}{2}\frac{x^2}{3} + C_1 \tag{2}$$

Since the deflected beam is symmetric about the center of the span, i.e., about $x = L/2$, it is evident that the slope must be zero there. Substituting $(dy/dx)_{x=L/2} = 0$ in Eq. (2) we have

$$0 = \frac{wL}{2}\frac{L^2}{8} - \frac{w}{2}\frac{L^3}{24} + C_1 \qquad \text{or} \qquad C_1 = -\frac{wL^3}{24}$$

The slope dy/dx at any point is thus given by

$$EI\frac{dy}{dx} = \frac{wL}{4}x^2 - \frac{w}{6}x^3 - \frac{wL^3}{24} \tag{3}$$

Integrating again, we find

$$EIy = \frac{wL}{4}\frac{x^3}{3} - \frac{w}{6}\frac{x^4}{4} - \frac{wL^3}{24}x + C_2 \tag{4}$$

The deflection y is zero at the left support. Substituting $y_{x=0} = 0$ in Eq. (3), we find $C_2 = 0$. The final form of the deflection curve of the beam is thus

$$EIy = \frac{wL}{12}x^3 - \frac{w}{24}x^4 - \frac{wL^3}{24}x \tag{5}$$

The maximum deflection of the beam occurs at the center because of symmetry. Substituting $x = L/2$ in Eq. (5), we obtain

$$EIy_{max} = -\frac{5wL^4}{384} \qquad \text{or} \qquad \Delta_{max} = \frac{5wL^4}{384EI} \tag{6}$$

which is the maximum deflection formula for Beam 2 in Table 8-1.

8.6. A simply supported beam of length 4 m and rectangular cross section 2 cm × 8 cm carries a uniform load of 2000 N/m. The beam is titanium, having $E = 100$ GPa. Determine the maximum deflection of the beam if the 8-cm dimension is vertical.

SOLUTION: From Problem 8.5, the maximum deflection is

$$\Delta_{max} = \frac{5}{384}\frac{wL^4}{EI}$$

Substituting,

$$\Delta_{max} = \frac{5 \times 2000 \times 4^4}{384(100 \times 10^9)(0.02 \times 0.08^3/12)} = 0.0781 \text{ m}$$

8.7. A simply supported beam is loaded by a couple M as shown in Fig. 8-8. The beam is 2 m long and of square cross section 50 mm on a side. If the maximum permissible deflection in the beam is 5 mm, and the allowable bending stress is 150 MPa, find the maximum allowable load M. Use $E = 200$ GPa.

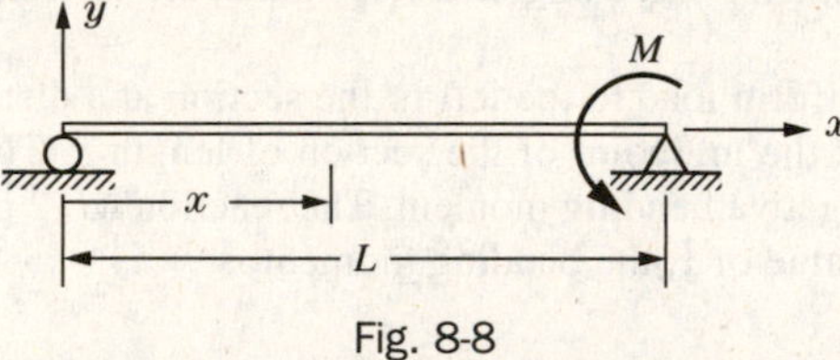

Fig. 8-8

SOLUTION: It is perhaps simplest to determine two values of M: one based upon the assumption that the deflection of 5 mm is realized, the other based on the assumption that the maximum bending stress in the bar is 150 MPa. The true value of M is then the minimum of these two values.

First, the reactions at the two ends are determined. Moments about the ends provide

$$R_L = R_R = \frac{M}{L} \qquad \text{and thus} \qquad M_x = \frac{M}{L}x$$

The differential equation describing the bent beam is thus

$$EI\frac{d^2y}{dx^2} = \frac{M}{L}x \tag{1}$$

Integrating twice

$$EI\frac{dy}{dx} = \frac{M}{2L}x^2 + C_1 \tag{2}$$

$$EIy = \frac{M}{6L}x^3 + C_1x + C_2 \tag{3}$$

We may now determine the two constants of integration through use of the fact that the beam deflection is zero at each end. When $x = 0$, $y = 0$, so from Eq. (3) we have

$$0 = 0 + 0 + C_2 \qquad C_2 = 0$$

Next, when $x = L$, $y = 0$, so we have from Eq. (3)

$$0 = \frac{ML^2}{6} + C_1 L \qquad \therefore C_1 = -\frac{ML}{6}$$

The desired equation of the deflection curve and the equation for the slope is

$$EIy = \frac{Mx^3}{6L} - \frac{MLx}{6} \tag{4}$$

and

$$EI\frac{dy}{dx} = \frac{Mx^2}{2L} - \frac{ML}{6} \tag{5}$$

The point of peak deflection occurs when the slope given by Eq. (5) is zero. This provides

$$x = \frac{L}{\sqrt{3}} \tag{6}$$

Returning to Eq. (4), the maximum deflection is

$$y_{max} = \frac{M}{6LEI}\left(\frac{L}{\sqrt{3}}\right)^3 - \frac{ML}{6EI}\left(\frac{L}{\sqrt{3}}\right) = -\frac{ML^2\sqrt{3}}{27EI} \tag{7}$$

Let us now consider that the maximum deflection in the beam is 5 mm. According to Eq. (7), we have

$$0.005 = \frac{M(2)^2\sqrt{3}}{27(200\times 10^9)[(0.05)(0.05)^3/12]} \qquad \text{or} \qquad M = 2030 \text{ N}\cdot\text{m} \tag{8}$$

We shall now assume that the allowable bending stress of 150 MPa is set up in the outer fibers of the beam at the section of maximum bending moment. The maximum bending moment occurs at the right end with magnitude M. There

$$\sigma_{max} = \frac{Mc}{I} \qquad 150\times 10^6 = \frac{M(0.025)}{(0.05)(0.05)^3/12} \qquad \text{or} \qquad M = 3125 \text{ N}\cdot\text{m} \tag{9}$$

Thus the maximum allowable moment is given by Eq. (8) and is 2030 N · m.

8.8. A simply supported beam, loaded at the midpoint, is 4 m long and of circular cross section of 10 cm in diameter. If the maximum permissible deflection is 5 mm, determine the maximum value of the load P. The material is steel for which $E = 200$ GPa. Use Table 8-1.

SOLUTION: The maximum deflection, is $\Delta_{max} = PL^3/48EI$. Thus,

$$0.005 = \frac{P\times 4^3}{48(200\times 10^9)(\pi\times 0.1^4/64)} \qquad \therefore P = 3680 \text{ N}$$

8.9. Consider the simply supported Beam 6 of Table 8.1. If the cross section is rectangular, 50 mm × 100 mm and $P = 20$ kN with $a = 1$ m and $b = 0.5$ m, determine the maximum deflection of the beam. The beam is steel, for which $E = 200$ GPa.

SOLUTION: Since $a > b$, the maximum deflection occurs to the left of the load P. It occurs at that point where the slope of the beam is zero and is given by

$$\Delta_{max} = \frac{Pb(L^2-b^2)^{3/2}}{9\sqrt{3}\,LEI} = \frac{20\,000\times 0.5(1.5^2-0.5^2)}{9\sqrt{3}\times 1.5(200\times 10^9)\times 0.05\times 0.1^3/12} = 0.00145 \text{ m} \qquad \text{or} \qquad 1.45 \text{ mm}$$

8.10. Determine the equation of the deflection curve for a cantilever beam loaded by a uniformly distributed load w per unit length, and a concentrated force P at the free end (see Fig. 8-9).

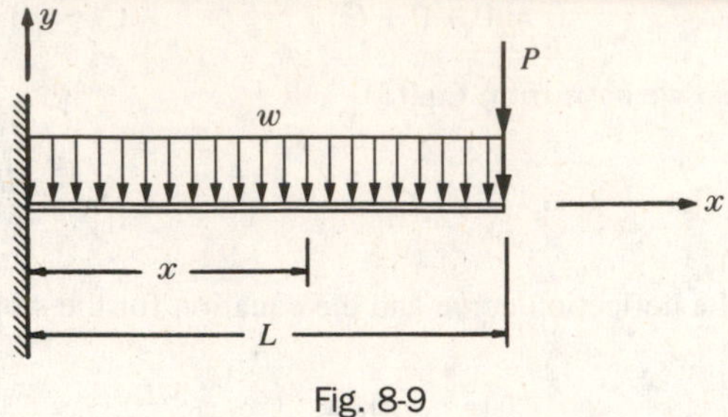

Fig. 8-9

SOLUTION: One logical approach to this problem is to determine the reactions at the wall, then write the differential equation of the bent beam, integrate this equation twice, and determine the constant of integration from the conditions of zero slope and zero deflection at the wall.

Actually this procedure has already been carried out in Problem 8.1 for the concentrated load, and in Problem 8.4 for the uniformly distributed load. For the concentrated force alone the deflection y was found in Eq. (3) of Problem 8.1 to be

$$EIy = -PL\frac{x^2}{2} + \frac{Px^3}{6} \tag{1}$$

For the uniformly distributed load alone the deflection y was found in Eq. (5) of Problem 8.4 to be

$$EIy = -\frac{w}{24}(L-x)^4 - \frac{wL^3}{6}x + \frac{wL^4}{24} \tag{2}$$

It is possible to obtain the resultant effect of these two loads when they act simultaneously merely by adding together the effects of each as they act separately. This is called the *method of superposition*. It is useful in determining deflections of beams subject to a combination of loads, such as we have here. Essentially it consists of utilizing the results of simpler beam-deflection problems to build up the solutions of more complicated problems. Thus it is not an independent method of determining beam deflections.

For the present beam the final deflection equation is given by adding Eqs. (1) and (2):

$$EIy = -PL\frac{x^2}{2} + \frac{Px^3}{6} - \frac{w}{24}(L-x)^4 - \frac{wL^3}{6}x + \frac{wL^4}{24} \tag{3}$$

The slope dy/dx at any point in the beam is merely found by differentiating both sides of Eq. (3) with respect to x.

The method of superposition is valid in all cases where there is a linear relationship between each separate load and the separate deflection which it produces.

8.11. Using singularity functions, determine the deflection curve of the cantilever beam subject to the loads shown in Fig. 8-10.

SOLUTION: In this case it is not necessary to determine the reactions of the wall supporting the beam at C. From the techniques of Chapter 6 we find the bending moment along the entire length of the beam to be given by

$$M = -P\langle x\rangle^1 - 2P\left\langle x - \frac{L}{4}\right\rangle^1 \tag{1}$$

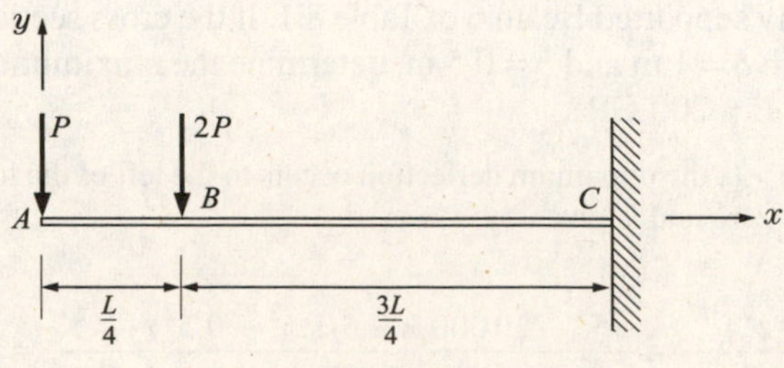

Fig. 8-10

where the angular brackets have the meanings given in the section "Singularity Functions" of Chap. 6. Thus, the differential equation for the bent beam is

$$EI\frac{d^2y}{dx^2} = -P\langle x\rangle^1 - 2P\left\langle x - \frac{L}{4}\right\rangle^1 \tag{2}$$

The first integration yields

$$EI\frac{dy}{dx} = -P\frac{\langle x\rangle^2}{2} - 2P\frac{\left\langle x - \frac{L}{4}\right\rangle^2}{2} + C_1 \tag{3}$$

where C_1 is a constant of integration. The next integration leads to

$$EIy = -\frac{P}{2}\frac{\langle x\rangle^3}{3} - 2P\frac{\left\langle x - \frac{L}{4}\right\rangle^3}{2(3)} + C_1\langle x\rangle + C_2 \tag{4}$$

where C_2 is a second constant of integration. These two constants may be determined from the two boundary conditions.

When $x = L$, $dy/dx = 0$, so from Eq. (3):

$$0 = -\frac{PL^2}{2} - P\left(\frac{3L}{4}\right)^2 + C_1 \tag{5}$$

When $x = L$, $y = 0$, so from Eq. (4):

$$0 = -\frac{PL^3}{6} - \frac{P}{3}\left(\frac{3L}{4}\right)^3 + C_1L + C_2 \tag{6}$$

Solving Eqs. (5) and (6),

$$C_1 = \frac{17}{16}PL^2 \qquad C_2 = -\frac{145}{192}PL^3 \tag{7}$$

The desired deflection curve is thus

$$EIy = -\frac{P}{6}\langle x\rangle^3 - \frac{P}{3}\left\langle x - \frac{L}{4}\right\rangle^3 + \frac{17}{16}PL^2\langle x\rangle - \frac{145}{192}PL^3 \tag{8}$$

For example, the deflection at point B where $x = L/4$ is found from Eq. (8) to be

$$EIy]_{x=L/4} = -\frac{P}{6}\left(\frac{L}{4}\right)^3 - 0 + \frac{17}{16}PL^2\left(\frac{L}{4}\right) - \frac{145}{192}PL^3 \qquad \therefore\ y]_{x=L/4} = -\frac{94.5PL^3}{192EI}$$

8.12. The cantilever beam ABC shown in Fig. 8-11 is subject to a uniform load w per unit length distributed over its right half, together with a concentrated couple $wL^2/2$ applied at B. Using singularity functions, determine the deflection curve of the beam, and the maximum deflection.

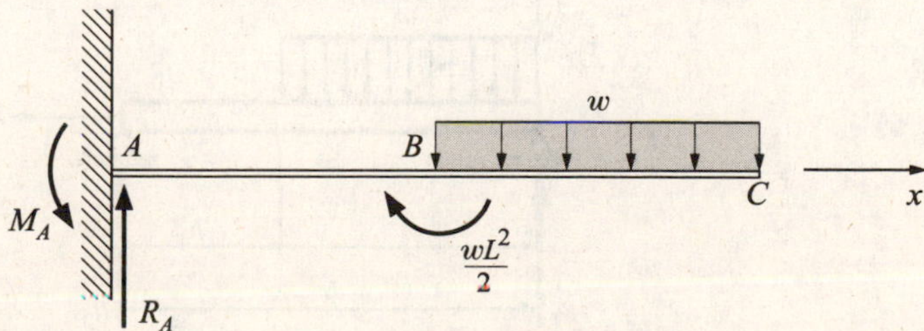

Fig. 8-11

SOLUTION: From statics we have

$$\Sigma M_A = M_A - \frac{wL^2}{2} - w\left(\frac{L}{2}\right)\left(\frac{3L}{4}\right) = 0 \qquad \therefore M_A = \frac{7wL^2}{8}$$

$$\Sigma F_y = R_A - w\left(\frac{L}{2}\right) = 0 \qquad \therefore R_A = \frac{wL}{2}$$

Using singularity functions, we may write the bending moment as

$$M = \frac{wL}{2}\langle x\rangle^1 - \frac{7wL^2}{8}\langle x\rangle^0 + \frac{wL^2}{2}\left\langle x - \frac{L}{2}\right\rangle^0 - w\left\langle x - \frac{L}{2}\right\rangle^1 \frac{\left\langle x - \frac{L}{2}\right\rangle^1}{2} \tag{1}$$

Thus, the differential equation of the loaded beam is

$$EI\frac{d^2y}{dx^2} = \frac{wL}{2}\langle x\rangle^1 - \frac{7wL^2}{8}\langle x\rangle^0 + \frac{wL^2}{2}\left\langle x - \frac{L}{2}\right\rangle^0 - w\left\langle x - \frac{L}{2}\right\rangle^1 \frac{\left\langle x - \frac{L}{2}\right\rangle^1}{2} \tag{2}$$

Integrating,

$$EI\frac{dy}{dx} = \frac{wL}{2}\frac{\langle x\rangle^2}{2} - \frac{7wL^2}{8}\langle x\rangle + \frac{wL^2}{2}\frac{\left\langle x - \frac{L}{2}\right\rangle^1}{1} - \frac{w}{2}\frac{\left\langle x - \frac{L}{2}\right\rangle^3}{3} + C_1 \tag{3}$$

The first boundary condition is: When $x = 0$, $dy/dx = 0$. Substituting in Eq. (3), we find $C_1 = 0$.
Integrating again,

$$EIy = \frac{wL}{4}\frac{\langle x\rangle^3}{3} - \frac{7wL^2}{8}\frac{\langle x\rangle^2}{2} + \frac{wL^2}{2}\frac{\left\langle x - \frac{L}{2}\right\rangle^2}{2} - \frac{w}{6}\frac{\left\langle x - \frac{L}{2}\right\rangle^4}{4} + C_2 \tag{4}$$

The second boundary condition is: When $x = 0$, $y = 0$. Substituting in Eq. (4), we find $C_2 = 0$.
Thus, the desired deflection equation is

$$EIy = \frac{wL}{12}\langle x\rangle^3 - \frac{7wL^2}{16}\langle x\rangle^2 + \frac{wL^2}{4}\left\langle x - \frac{L}{2}\right\rangle^2 - \frac{w}{24}\left\langle x - \frac{L}{2}\right\rangle^4 \tag{5}$$

This yields the deflection at the tip to be

$$EIy]_{x=L} = \frac{wL^4}{12} - \frac{7wL^4}{16} + \frac{wL^2}{4}\left(\frac{L}{2}\right)^2 - \frac{w}{24}\left(\frac{L}{2}\right)^4 \qquad \therefore \Delta_{max} = \frac{113wL^4}{384EI}$$

8.13. Consider a simply supported beam subject to a uniform load distributed over a portion of its length, as indicated in Fig. 8-12. Use singularity functions to determine the deflection curve of the beam.

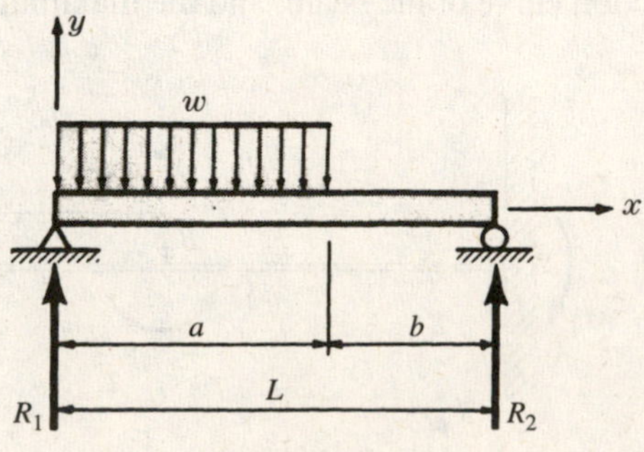

Fig. 8-12

SOLUTION: From statics the reactions are found to be

$$R_1 = \frac{w_0}{2L}(L^2 - b^2) \qquad R_2 = w_0 a - \frac{w_0}{2L}(L^2 - b^2)$$

The bending moment at any point x along the length of the beam is

$$M = R_1 x - \frac{w}{2}\langle x\rangle^2 + \frac{w}{2}\langle x - a\rangle^2 \tag{1}$$

Note that the last term on the right is required to cancel the distributed load represented by the term on the right section of the beam for values of x greater than $x = a$. Thus,

$$EI\frac{d^2y}{dx^2} = M = R_1\langle x\rangle^1 - \frac{w}{2}\langle x\rangle^2 + \frac{w}{2}\langle x - a\rangle^2 \tag{2}$$

Integrating,

$$EI\frac{dy}{dx} = \frac{R_1}{2}\langle x\rangle^2 - \frac{w}{6}\langle x\rangle^3 + \frac{w}{6}\langle x - a\rangle^3 + C_1 \tag{3}$$

Finally,

$$EIy = \frac{R_1}{6}\langle x\rangle^3 - \frac{w}{24}\langle x\rangle^4 + \frac{w}{24}\langle x - a\rangle^4 + C_1 x + C_2 \tag{4}$$

To determine C_1 and C_2, we impose the boundary conditions that $y = 0$ at $x = 0$ and $x = L$. From Eq. (4) we thus find

$$C_1 = \frac{wL^3}{24} - \frac{wb^4}{24L} - \frac{wL}{12}(L^2 - b^2) \qquad C_2 = 0$$

The deflection curve is then

$$EIy = \frac{w}{12L}(L^2 - b^2)\langle x\rangle^3 - \frac{w}{24}\langle x\rangle^4 + \frac{w}{24}\langle x - a\rangle^4 + \left[-\frac{wL^3}{24} - \frac{wb^4}{24L} + \frac{wLb^2}{12}\right]x \tag{5}$$

8.14. Consider the overhanging beam shown in Fig. 8-13. Determine the equation of the deflection curve using singularity functions.

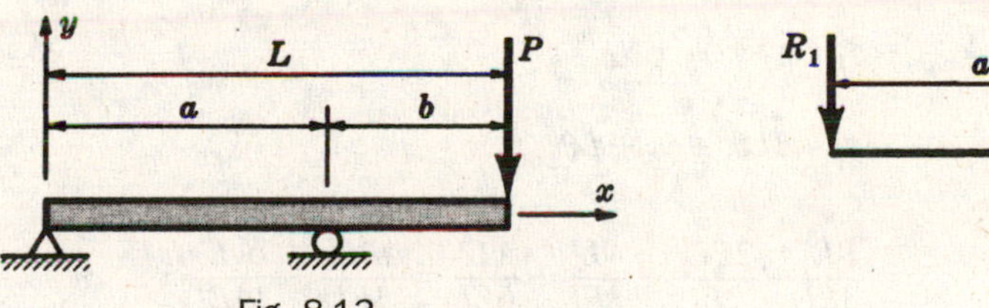

Fig. 8-13

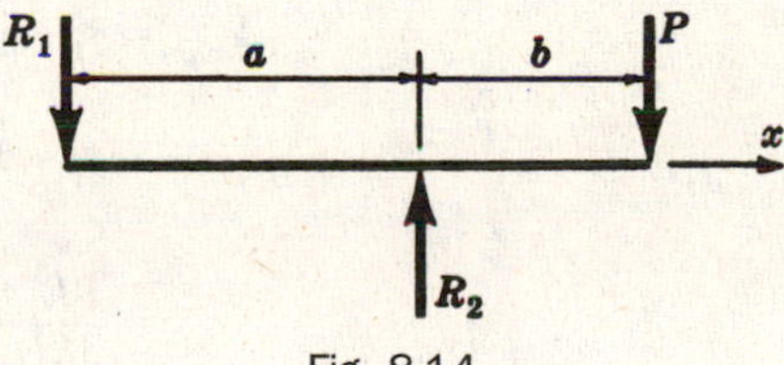

Fig. 8-14

From statics the reactions are first found to be $R_1 = Pb/a$ and $R_2 = P[1 + (b/a)]$, acting as indicated in Fig. 8-14. The bending moment at any point x along the entire length of the beam is

$$M(x) = -R_1\langle x\rangle^1 + R_2\langle x - a\rangle^1 \tag{1}$$

Thus

$$EI\frac{d^2y}{dx^2} = M = -R_1\langle x\rangle^1 + R_2\langle x - a\rangle^1 \tag{2}$$

from which

$$EI\frac{dy}{dx} = -\frac{R_1}{2}\langle x\rangle^2 + \frac{R_2}{2}\langle x - a\rangle^2 + C_1 \tag{3}$$

$$EIy = -\frac{R_1}{6}\langle x\rangle^3 + \frac{R_2}{6}\langle x - a\rangle^3 + C_1 x + C_2 \tag{4}$$

The boundary conditions are $y = 0$ at $x = 0$ and $x = a$. From these conditions, C_1 and C_2 are found from (4) to be

$$C_2 = 0, \qquad C_1 = \frac{Pab}{6}$$

The deflection curve is thus

$$EIy = -\frac{Pb}{6a}\langle x\rangle^3 + \frac{P}{6}\left(1 + \frac{b}{a}\right)\langle x - a\rangle^3 + \frac{Pabx}{6} \tag{5}$$

8.15. In many situations, it is the maximum deflection of a beam that is of primary interest. Using the results of the simple beams of Table 8-1, determine an expression for the deflection of the right end of the beam shown in Fig. 8-15(*a*).

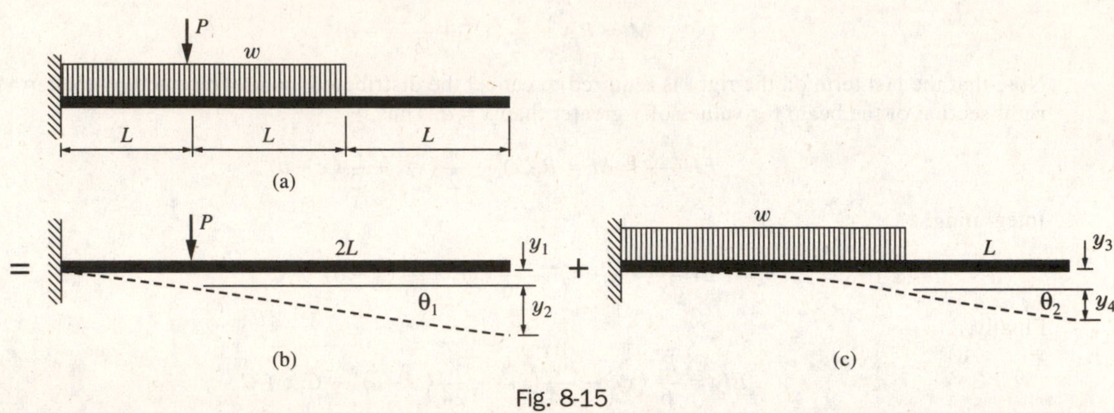

Fig. 8-15

SOLUTION: Using superposition, the beam can be replaced by the sum of two beams, as shown in Figs. 8-15(*b*) and (*c*). Using the formulas from Table 8-1, we have

$$y_1 = \frac{PL^3}{3EI} \qquad \theta_1 = \frac{PL^2}{2EI} \qquad y_3 = \frac{wL^4}{8EI} \qquad \theta_2 = \frac{wL^3}{6EI}$$

Consequently, superposition allows us to obtain

$$\begin{aligned}\Delta_{max} &= y_1 + y_2 + y_3 + y_4 \\ &= y_1 + 2L\theta_1 + y_3 + L\theta_2 \\ &= \frac{PL^3}{3EI} + \frac{2PL^3}{2EI} + \frac{wL^4}{8EI} + \frac{wL^4}{6EI} = \frac{4PL^3}{3EI} + \frac{7wL^4}{24EI}\end{aligned}$$

8.16. Using superposition and the formulas of Table 8-1, calculate the maximum deflection of the beam shown in Fig. 8-16(*a*). Use $E = 200$ GPa and $I = 3500\ \text{cm}^4$.

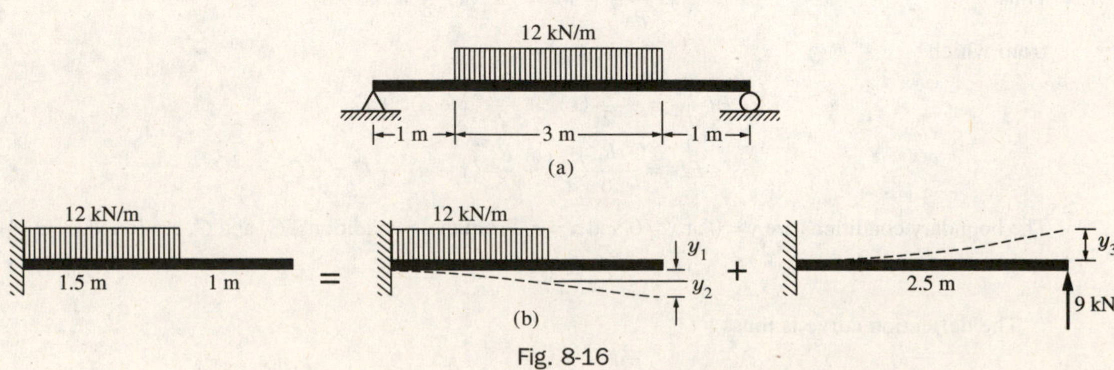

Fig. 8-16

SOLUTION: Observe that the beam is loaded symmetrically about its center where the slope is zero. So, the center of the beam can be considered to be the end of a cantilever beam. The right half of the beam is then composed of the two beams shown in Fig. 8-16(*b*). Using superposition and Table 8-1, we have

$$\Delta_{max} = y_3 - y_1 - y_2$$

$$= \frac{PL_1^3}{3EI} - \frac{wL_2^4}{8EI} - \frac{wL_2^3}{6EI} \times L_3$$

$$= \frac{1}{(200 \times 10^9)(3500 \times 10^{-8})}\left(\frac{9000 \times 2.5^3}{3} - \frac{12\,000 \times 1.5^4}{8} - \frac{12\,000 \times 1.5^3}{6} \times 1\right)$$

$$= 0.00465 \text{ m} \quad \text{or} \quad 4.65 \text{ mm}$$

This displacement is upward but we know the right end does not deflect. Hence, this is the downward displacement of the center of the beam.

8.17. Using superposition and the formulas in Table 8-1, find an expression for the deflection of the beam of Fig. 8-17 at (a) point *B* and (b) point *D*.

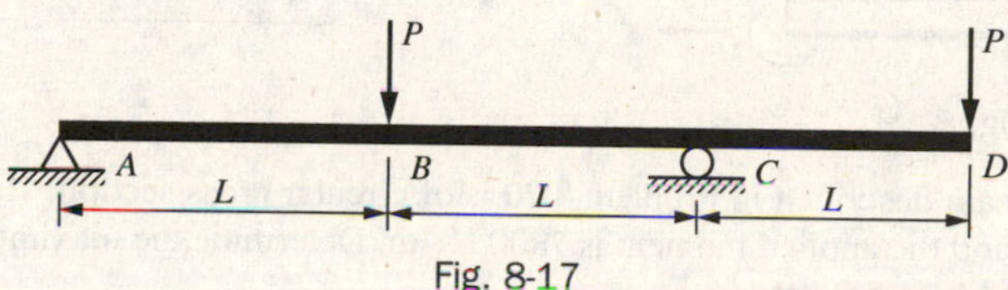

Fig. 8-17

SOLUTION: (a) To find the deflection of point *B*, move the force *P* on the right end to point *C* and consider the beam of Fig. 8-18(*a*). It is composed of the two beams of Figs. 8-18(*b*) and (*c*). The force over *C* produces no deflection. Consequently, using the formulas of Table 8-1, the deflection of the midpoint *B* is

$$y_B = y_1 + y_3$$

$$= -\frac{P(2L)^3}{48EI} + \frac{ML^2}{16EI} = -\frac{PL^3}{6EI} + \frac{PL^3}{16EI} = -\frac{5PL^3}{32EI}$$

(b) The deflection of the right end at *D* results from the rotation of the cross section at *C* and the deflection caused by the force on the overhang [see Fig. 8-18(*d*)]. It is

$$y_D = y_2 - y_4 - y_5$$

$$= \theta_1 L - \theta_2 L - y_5$$

$$= \frac{PL^2}{16EI} \times L - \frac{ML}{6EI} \times L - \frac{PL^3}{3EI} = -\frac{7PL^3}{16EI}$$

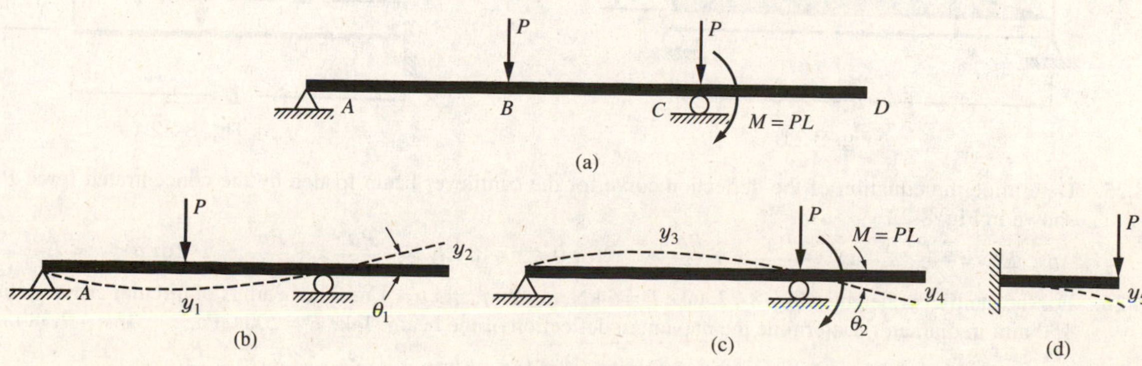

Fig. 8-18

SUPPLEMENTARY PROBLEMS

8.18. The cantilever beam loaded as shown in Problem 8.1 is made of a titanium alloy, having $E = 105$ GPa. The load P is 20 kN, $L = 4$ m, and the moment of inertia of the beam cross section is 104×10^6 mm^4. Find the maximum deflection of the beam. *Ans.* –39 mm

8.19. Consider the simply supported beam loaded as shown by Beam 6 in Table 8-1. The length of the beam is 6 m, $a = 4$ m, the load $P = 4000$ N, and $I = 6000$ cm^4. Determine the deflection at the center of the beam. Use $E = 200$ GPa. *Ans.* –1.28 mm

8.20. Refer to Fig. 8-19. Determine the deflection at every point of the cantilever beam subject to the single moment M shown. *Ans.* $y = -Mx^2/2EI$

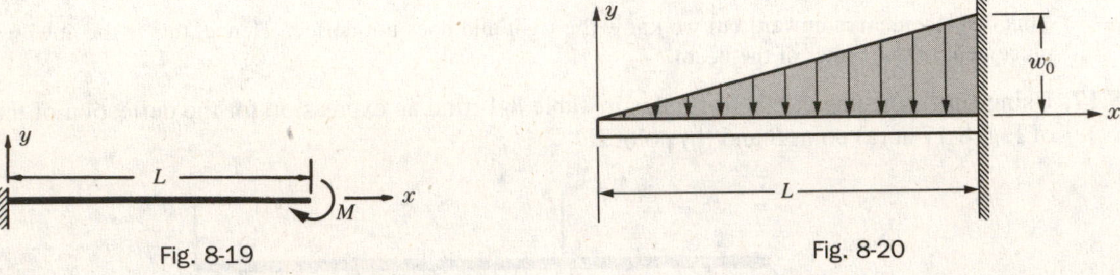

Fig. 8-19 Fig. 8-20

8.21. The cantilever beam described in Problem 8.20 is of circular cross section, 10 cm in diameter. The length of the beam is 4 m and the applied moment is 7000 N · m. Determine the maximum deflection of the beam. Use $E = 200$ GPa. *Ans.* 5.7 cm

8.22. Refer to Fig. 8-20. Find the equation of the deflection curve for the cantilever beam subject to the varying load shown.

Ans. $y(x) = -\dfrac{w_0}{6LEI}\left(\dfrac{x^5}{20} - \dfrac{xL^4}{4} + \dfrac{L^5}{5}\right)$

8.23. The cross section of the cantilever beam loaded as shown in Fig. 8-20 is rectangular, 50 × 75 mm. The bar, 1 m long, is aluminum for which $E = 65$ GPa. Determine the permissible maximum intensity of loading if the maximum deflection is not to exceed 5 mm and the maximum stress is not to exceed 50 MPa.

Ans. $w_0 = 14.1$ kN/m and 17.1 kN/m. Select 14.1 kN/m.

8.24. Refer to Fig. 8-21. Determine the equation of the deflection curve for the simply supported beam supporting the load of uniformly varying intensity.

Ans. $y = \dfrac{w_0 L}{2EI}\left(-\dfrac{x^5}{60L^2} + \dfrac{x^3}{18} - \dfrac{7L^2 x}{180}\right)$

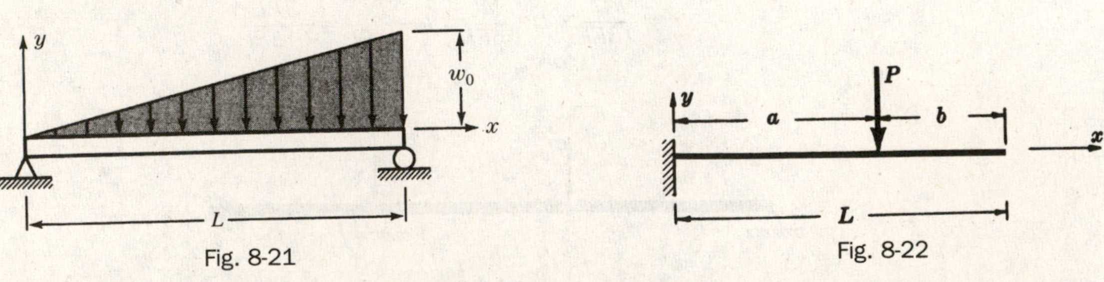

Fig. 8-21 Fig. 8-22

8.25. Determine the equation of the deflection curve for the cantilever beam loaded by the concentrated force P as shown in Fig. 8-22.

Ans. $EIy = -\dfrac{P}{6}(a-x)^3 - \dfrac{Pa^2}{2}x + \dfrac{Pa^3}{6}$ for $0 < x < a$; $EIy = -\dfrac{Pa^2}{2}x + \dfrac{Pa^3}{6}$ for $a < x < L$

8.26. For the cantilever beam of Fig. 8-22, take $P = 5$ kN, $a = 2$ m, and $b = 1$ m. The beam is of circular cross section, 150 mm in diameter. Determine the maximum deflection of the beam. Take $E = 200$ GPa. *Ans.* –7.38 mm

8.27. Repeat Problem 8.25 using singularity functions. *Ans.* $EIy = \dfrac{P}{6}\langle x\rangle^3 - \dfrac{Pa}{2}\langle x\rangle^2 - \dfrac{P}{6}\langle x-a\rangle^3$

8.28. The cantilever beam shown in Fig. 8-23 is subject to a uniform load w per unit length over its right half BC. Determine the deflection curve as well as the maximum deflection. Use singularity functions. Note: $\langle x\rangle^2 = x^2$, since x is never negative.

Ans. $EIy = \dfrac{wL\langle x\rangle^3}{12} - \dfrac{3wL^2\langle x\rangle^2}{16} - \dfrac{w}{24}\left\langle x - \dfrac{L}{2}\right\rangle^4 \qquad \Delta_{max} = \dfrac{41}{384}\left(\dfrac{wL^4}{EI}\right)$

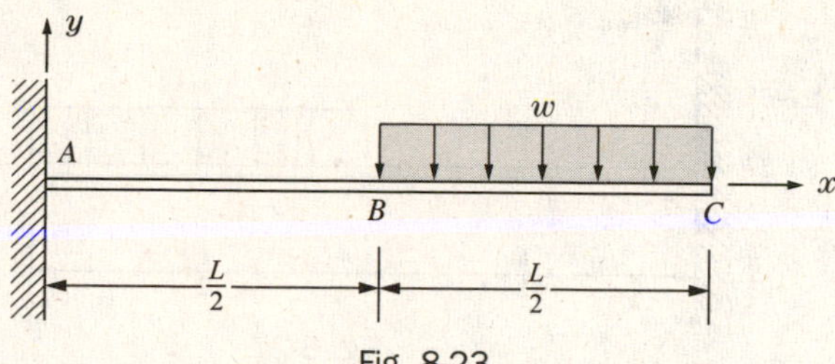

Fig. 8-23

8.29. The simply supported overhanging beam supports the load w per unit length as shown in Fig. 8-24. Find the deflection curve of the beam. Use singularity functions.

Ans. $EIy = -\dfrac{wx^4}{24} + wa\langle x - a\rangle^3 + \dfrac{wa}{2}\langle x - 3a\rangle^2 + \dfrac{5wa^3}{3}x - \dfrac{4wa}{3}x + \dfrac{4a^2w}{3} - \dfrac{13wa^4}{8}$

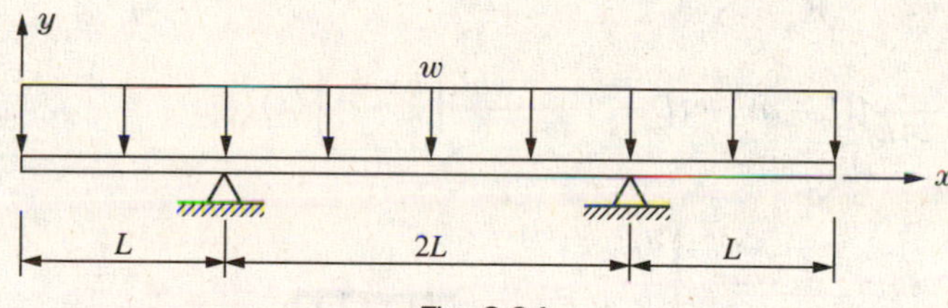

Fig. 8-24

8.30. A simply supported beam with overhanging ends is loaded by the uniformly distributed loads shown in Fig. 8-25. Determine the deflection of the midpoint of the beam with respect to origin at the level of the supports.

Ans. $\dfrac{wa^4}{4EI}$

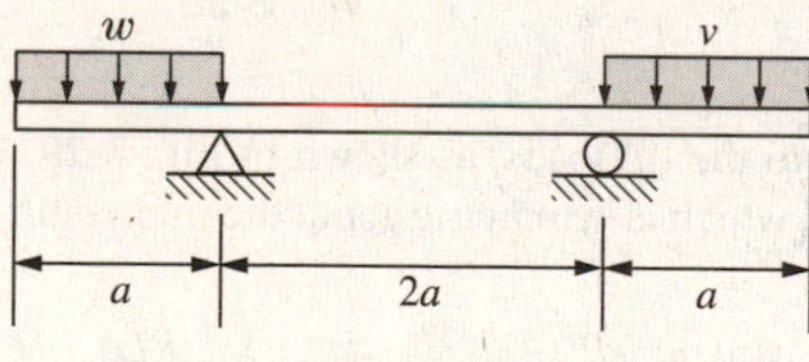

Fig. 8-25

8.31. For the beam described in Problem 8.30, determine the deflection of one end of the beam with respect to origin at the level of the supports. Use singularity functions.

Ans. $-\dfrac{5wa^4}{8EI}$

8.32. The overhanging beam is loaded by the uniformly distributed load as well as the concentrated force shown in Fig. 8-26. Determine the deflection of point A of the beam. Use singularity functions.

Ans. $-\dfrac{5wa^4}{8EI}$

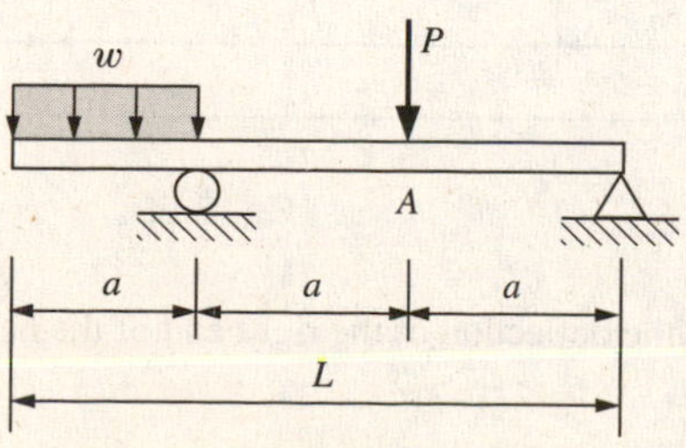

Fig. 8-26

8.33. The cantilever beam ABC is loaded by a uniformly distributed load over the right half BC as shown in Fig. 8-27. Use singularity functions to determine the deflection curve of the bent beam. Also, determine the deflection at the tip C.

Ans. $EIy = \frac{wL}{12}\langle x\rangle^3 - \frac{3}{8}wL^2\frac{\langle x\rangle^2}{2} - \frac{w}{24}\left\langle x - \frac{L}{2}\right\rangle^4 \qquad \Delta_{max} = -\frac{41wL^4}{384EI}$

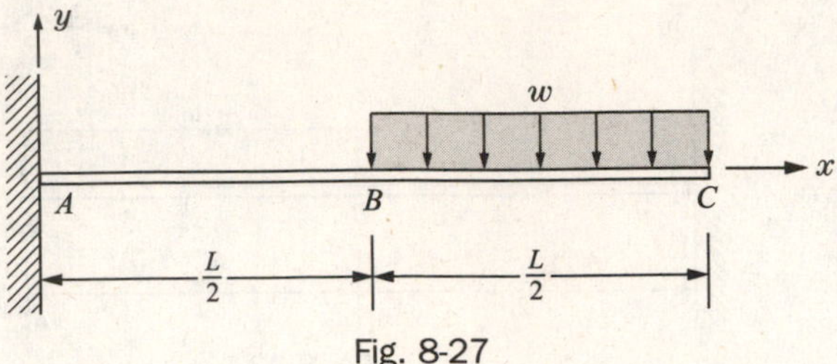

Fig. 8-27

8.34. Consider a simply supported beam subject to a uniform load acting over a portion of the beam as indicated in Fig. 8-28. Use singularity functions to determine the equation of the deflection curve.

Ans. $$EIy = \frac{wb}{6L}\left(\frac{b}{2}+c\right)\langle x\rangle^3 - \frac{w}{24}\langle x-a\rangle^4 + \frac{w}{24}\langle x-a-b\rangle^4$$
$$+\left\{\frac{w}{24L}[(L-a)^4-(L-c)^4] - \frac{wbL}{6}\left(\frac{b}{2}+c\right)\right\}\langle x\rangle$$

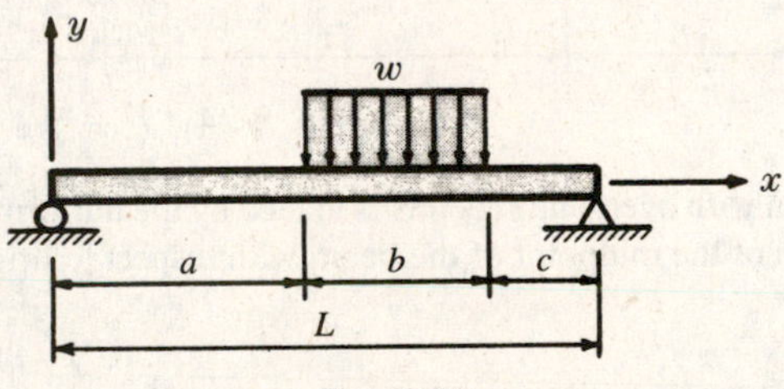

Fig. 8-28

8.35. The beam $ABCD$ is subject to the tip loads, as shown in Fig. 8-29. Use singularity functions to determine the deflection curve of the beam, which is symmetric about the midlength of the beam. Also, determine the deflection at point A.

Ans. $EIy = -\frac{P}{6}\langle x\rangle^3 + \frac{P}{6}\langle x-a\rangle^3 + \frac{P}{6}\langle x-(a+L_1)\rangle^3 + \left(\frac{PLa}{2} - \frac{Pa^2}{2}\right)\langle x\rangle,\ \Delta_A = \frac{Pa^2}{EI}\left(\frac{L}{2} - \frac{2a}{3}\right)$

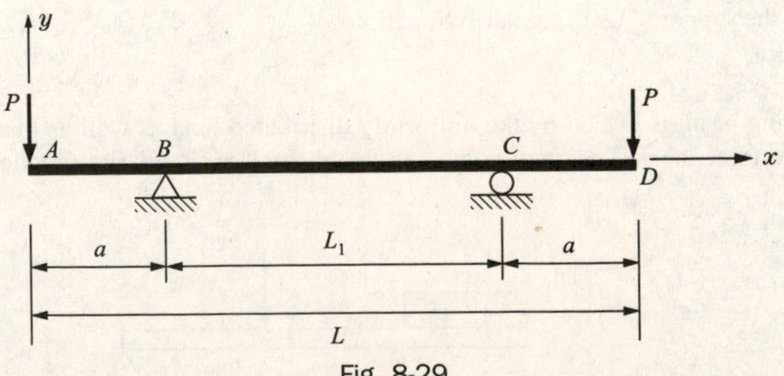

Fig. 8-29

8.36. Determine an expression for the deflection of the right end of the beam shown in Fig. 8-30. Use the formulas of Table 8-1.

Ans. $\frac{7wa^4}{24EI} + \frac{8Pa^3}{3EI}$

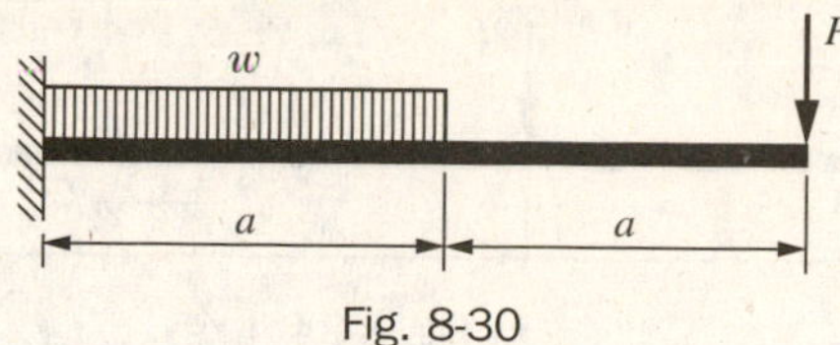

Fig. 8-30

8.37. Determine an expression for the deflection of the right end of the beam shown in Fig. 8-31. Use the formulas of Table 8-1.

Ans. $\dfrac{2Ma^2}{EI} + \dfrac{41wa^4}{24EI}$

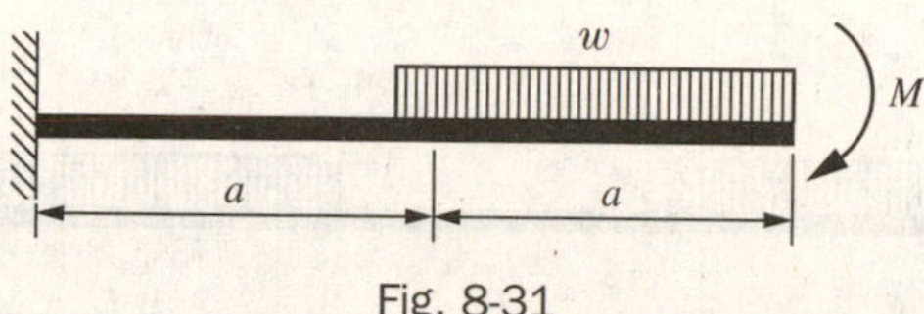

Fig. 8-31

8.38. Determine an expression for the deflection of the left end of the beam shown in Fig. 8-32. Use the formulas of Table 8-1.

Ans. $\dfrac{5Pa^3}{6EI} + \dfrac{7wa^4}{24EI} - \dfrac{2Ma^2}{EI}$

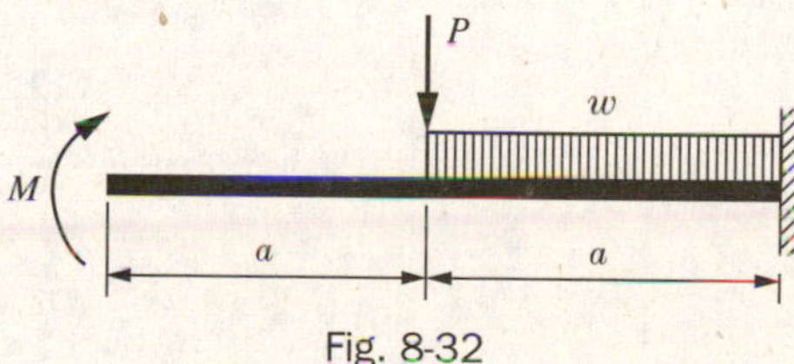

Fig. 8-32

8.39. Using $a = 120$ cm, $P = 2000$ N, $w = 2400$ N/m, $M = 1600$ N · m, $E = 200$ GPa, and $I = 15 \times 10^6$ mm^4, calculate the deflection of the free end of the beam shown in (a) Fig. 8-30, (b) Fig. 8-31, and (c) Fig. 8-32.
Ans. (a) 3.56 mm ↓, (b) 4.37 mm ↓, (c) 0.294 mm ↑

8.40. Determine an expression for the deflection of the center of the beam shown in Fig. 8-33. Use the formulas of Table 8-1.

Ans. $\dfrac{11Pa^3}{12EI} + \dfrac{10wa^4}{3EI}$

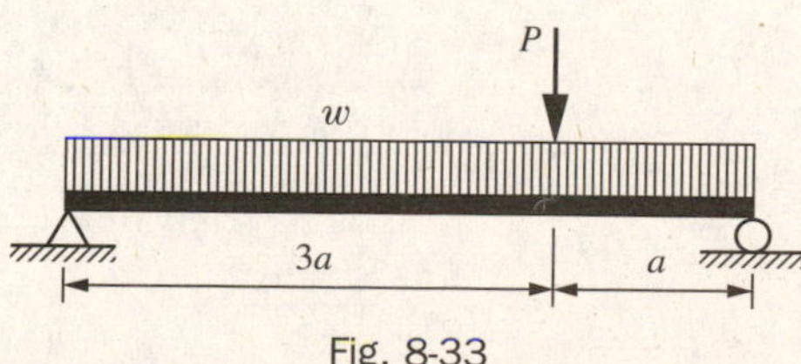

Fig. 8-33

8.41. Determine an expression for the deflection of the right end of the beam shown in Fig. 8-34. Use the formulas of Table 8-1.

Ans. $\dfrac{5Pa^3}{12EI}$ ↓

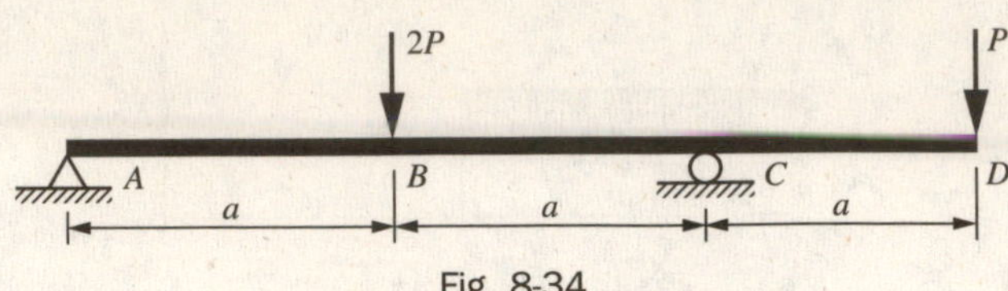

Fig. 8-34

8.42. Determine an expression for the deflection of the right end of the beam shown in Fig. 8-35. Use the formulas of Table 8-1.

Ans. $\dfrac{29wa^4}{4EI}$

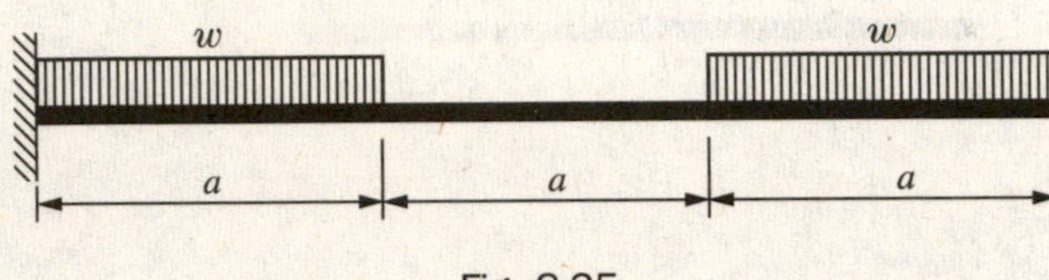

Fig. 8-35

CHAPTER 9

Statically Indeterminate Elastic Beams

9.1 Basics

In Chapters 7 and 8 the stresses and deflections were determined for beams having various conditions of loading and support. In the cases treated it was always possible to completely determine the reactions exerted upon the beam merely by applying the equations of static equilibrium. In these cases the beams are said to be *statically determinate*.

In this chapter we shall consider those beams where the number of unknown reactions exceeds the number of equilibrium equations available for the system. In such a case it is necessary to supplement the equilibrium equations with additional equations stemming from the deformations of the beam. In these cases the beams are said to be *statically indeterminate*.

Types of Statically Indeterminate Beams

Several common types of statically indeterminate beams are illustrated below. Although a wide variety of such structures exists in practice, the following four diagrams (Figs. 9-1, 9-2, and 9-3), will illustrate the nature of an indeterminate system. For the beams shown below the reactions of each constitute a parallel force system and hence there are two equations of static equilibrium available. Thus the determination of the reactions in each of these cases necessitates the use of additional equations arising from the deformation of the beam.

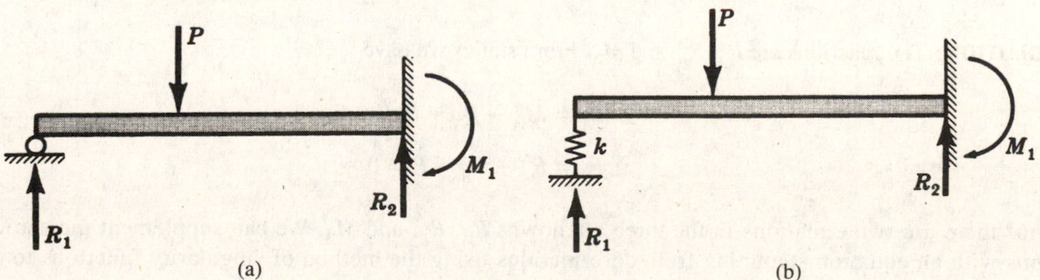

Fig. 9-1 Examples of statically indeterminate beams.

In the case [Fig. 9-1(*a*)] of a beam fixed at one end and supported at the other, sometimes termed a *supported cantilever*, we have as unknown reactions R_1, R_2, and M_1. The two statics equations must be supplemented by one equation based upon deformations.

In Fig. 9-1(*b*) the beam is fixed at one end and has a flexible springlike support at the other. In the case of a simple linear spring the flexible support exerts a force proportional to the beam deflection at that point. The unknown reactions are again R_1, R_2, and M_1. Again, the two statics equations must be supplemented by one equation stemming from deformations.

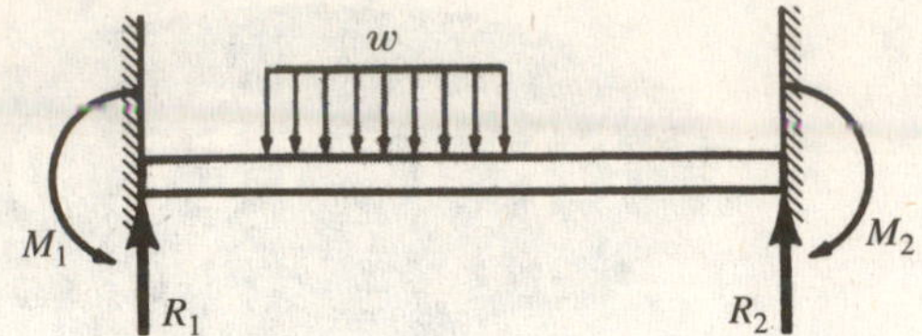

Fig. 9-2 A beam cantilevered at both ends.

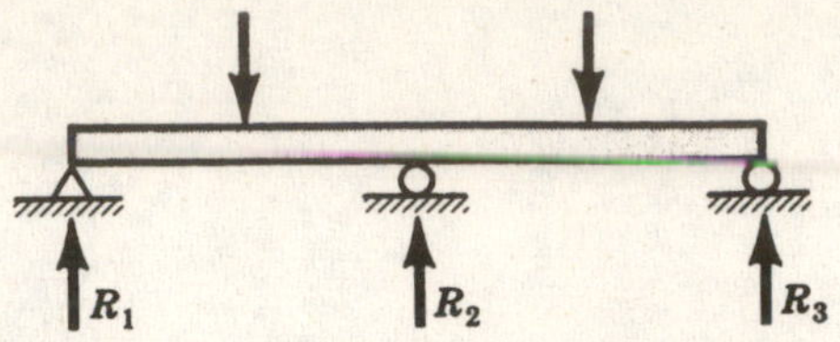

Fig. 9-3 A continuous beam with three supports.

As shown in Fig. 9-2, a beam fixed or clamped at both ends has the unknown reactions R_1, R_2, M_1, and M_2. The two statics equations must be supplemented by two equations arising from the deformations.

In Fig. 9-3 the beam is supported on three supports at the same level. The unknown reactions are R_1, R_2, and R_3. The two statics equations must be supplemented by one equation based upon deformations. A beam of this type that rests on more than two supports is called a *continuous beam*.

The following problems will illustrate the solution technique for solving problems involving statically indeterminate beams.

SOLVED PROBLEMS

9.1. A beam is clamped at A, simply supported at B, and subject to the concentrated force shown in Fig. 9-4. Determine all reactions assuming all dimensions are known.

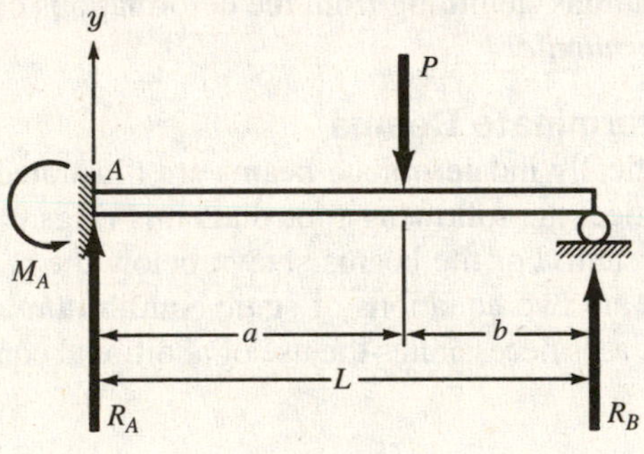

Fig. 9-4

SOLUTION: The reactions are R_A, R_B, and M_A. From statics we have

$$\Sigma M_A = M_A - Pa + R_b L = 0 \tag{1}$$

$$\Sigma F_y = R_A + R_B - P = 0 \tag{2}$$

Thus there are two equations in the three unknowns R_A, R_B, and M_A. We can supplement the statics equations with an equation stemming from deformations using the method of singularity functions to describe the bent beam. That is,

$$EI\frac{d^2y}{dx^2} = R_A\langle x\rangle - M_A\langle x\rangle^0 - P\langle x-a\rangle \tag{3}$$

Integrating the first time, we have

$$EI\frac{dy}{dx} = R_A\frac{\langle x\rangle^2}{2} - M_A\langle x\rangle - \frac{P}{2}\langle x-a\rangle^2 + C_1 \tag{4}$$

The first boundary condition is that at $x = 0$, $dy/dx = 0$, and thus $C_1 = 0$. Integrating again,

$$Ely = \frac{R_A}{2}\frac{\langle x\rangle^3}{3} - M_A\frac{\langle x\rangle^2}{2} - \frac{p}{2}\frac{\langle x-a\rangle^3}{3} + C_2 \tag{5}$$

The second boundary condition is that at $x = 0$, $y = 0$, and we find $C_2 = 0$.

The third boundary condition is that at $x = L$, $y = 0$. Substituting in Eq. (5), we have

$$0 = \frac{R_A L^3}{6} - \frac{M_A L^2}{2} - \frac{Pb^3}{6} \tag{6}$$

Simultaneous solution of the three equations (1), (2), and (6) leads to

$$R_A = \frac{Pb}{2L^3}(3L^2 - b^2)$$

$$R_B = \frac{Pa^2}{2L^3}(2L + b)$$

$$M_A = \frac{Pb}{2L^2}(L^2 - b^2)$$

9.2. The beam AB in Fig. 9-5 is clamped at A, spring supported at B, and loaded by the uniformly distributed load w per unit length. Prior to application of the load, the spring is just stress free. To determine the flexural rigidity EI of the beam, an experiment is conducted without the uniform load w and also without the spring being present. In this experiment it is found that a vertical force of 10 000 N applied at end B deflects that point 50 mm. The vertical force is removed, the spring is then attached to the beam at B and a uniform load of magnitude 5 kN/m is applied between A and B. Determine the deflection of point B under these conditions.

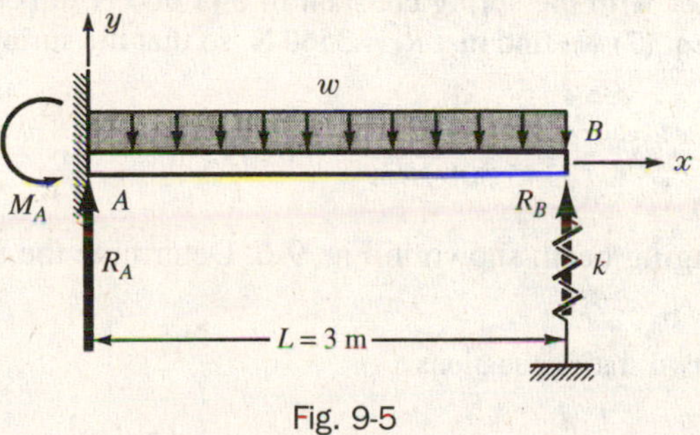

Fig. 9-5

SOLUTION: The forces acting on the beam when it is uniformly loaded as well as spring supported at its tip are shown in Fig. 9-5. The force R_B represents the force exerted by the spring on the beam. The differential equation of the bent beam is

$$EI\frac{d^2y}{dx^2} = -M_A + R_A x - \frac{w}{2}x^2 \tag{1}$$

Integrating the first time, we find

$$EI\frac{dy}{dx} = -M_A x + \frac{R_A}{2}x^2 - \frac{w}{6}x^3 + C_1 \tag{2}$$

Now, invoking the boundary condition that when $x = 0$, $dy/dx = 0$, we find from Eq. (2) that $C_1 = 0$. The second integration yields

$$EIy = -\frac{M_A}{2}x^2 + \frac{R_A}{6}x^3 - \frac{w}{24}x^4 + C_2 \tag{3}$$

and the second boundary condition is that $x = 0$ when $y = 0$, so from Eq. (3) we have $C_2 = 0$. From Eq. (3) we have the deflection at B due to the uniform load plus the presence of the spring to be given by

$$EIy_L = -\frac{M_A L^2}{2} + \frac{R_A L^3}{6} - \frac{wL^4}{24} \tag{4}$$

But for linear action of the spring we have the usual relation

$$R_B = -ky_L = k\Delta_B \tag{5}$$

Also, from statics for this parallel force system we have the two equilibrium equations

$$\sum M_A = M_A + R_B L - \frac{wL^2}{2} = 0 \tag{6}$$

$$\sum F_y = R_A + R_B - (5000\ \text{N/m})(3\ \text{m}) = 0 \tag{7}$$

Simultaneous solution of Eqs. (4), (5), (6), and (7) provides

$$R_A\left(\frac{EI}{k} + \frac{L^3}{3}\right) = \frac{EIwL}{k} + \frac{5wL^4}{24} \tag{8}$$

The flexural rigidity EI is easily found by consideration of the experimental evidence. The tip deflection of a tip-loaded cantilever beam is $PL^3/3EI$ which becomes, for this experiment,

$$0.050\ \text{m} = \frac{(10\,000\ \text{N})(3\ \text{m})^3}{3EI}$$

from which

$$EI = 1.8 \times 10^6\ \text{N} \cdot \text{m}^2 \tag{9}$$

If this value together with the spring constant of 345 000 N/m is substituted in Eq. (8), we find that $R_A = 11\,440$ N. From Eq. (7) we find that $R_B = 3560$ N, so that the spring Eq. (5) indicates the displacement of point B to be

$$\Delta_B = \frac{3560\ \text{N}}{345\,000\ \text{N/m}} = 0.01032\ \text{m} \quad \text{or} \quad 10.3\ \text{mm} \tag{10}$$

9.3. Consider the overhanging beam shown in Fig. 9-6. Determine the magnitude of the supporting force at B.

SOLUTION: There are two statics equations

$$\sum M_A = M_1 + R_2 a - \frac{w(a+b)^2}{2} = 0 \tag{1}$$

$$\sum F_y = R_1 + R_2 - w(a+b) = 0 \tag{2}$$

Let us employ the method of singularity functions to write the differential equation of the bent beam

$$EI\frac{d^2y}{dx^2} = -M_1\langle x\rangle^0 + R_1\langle x\rangle^1 - \frac{w}{2}\langle x\rangle^2 + R_2\langle x-a\rangle^1 \tag{3}$$

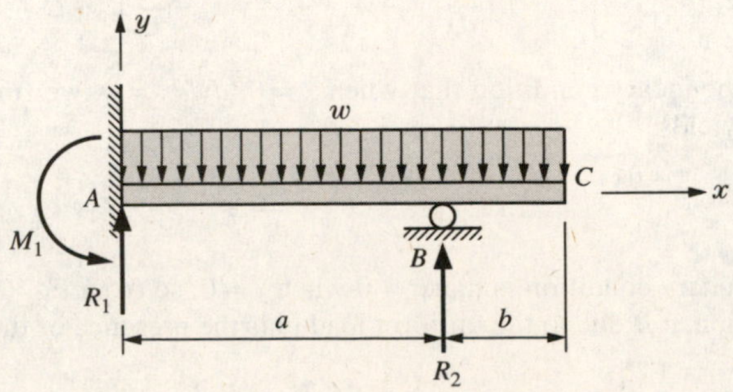

Fig. 9-6

Note that in Eq. (3) a negative sign is assigned to M_1 since, as we work from left to right starting at the origin A, the reactive moment M_1 tends to bend the portion of the beam to the right of A into a configuration having curvature concave downward, which is negative according to the bending moment sign convention given in Chapter 6.

Integrating,

$$EI\frac{dy}{dx} = -M_1\langle x\rangle^1 + \frac{R_1}{2}\langle x\rangle^2 - \frac{w}{6}\langle x\rangle^3 + \frac{R_2}{2}\langle x-a\rangle^2 + C_1 \tag{4}$$

But when $x = 0$, $dy/dx = 0$; hence $C_1 = 0$. Integrating again,

$$EIy = -\frac{M_1}{2}\langle x\rangle^2 + \frac{R_1}{6}\langle x\rangle^3 - \frac{w}{24}\langle x\rangle^4 + \frac{R_2}{6}\langle x-a\rangle^3 + C_2 \tag{5}$$

Also, when $x = 0$, $y = 0$, so that $C_2 = 0$.

Since the support at point B is unyielding, y must vanish in Eq. (5) when $x = a$. Substituting, we find

$$0 = -\frac{M_1a^2}{2} + \frac{R_1a^3}{6} - \frac{wa^4}{24} \qquad \text{from which } M_1 = R_1\frac{a}{3} - \frac{wa^2}{12}$$

Solving this in conjunction with the statics equations, we find

$$R_1 = \frac{5}{8}wa - \frac{3wb^2}{4a} \qquad R_2 = \frac{3}{8}wa + wb + \frac{3wb^2}{4a}$$

9.4. The clamped end beam is loaded as shown in Fig. 9-7 by a couple M. Determine all reactions.

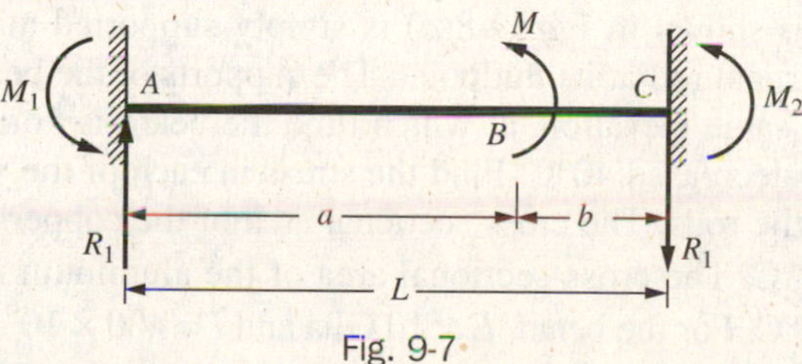

Fig. 9-7

SOLUTION: Under the action of the couple, the initially straight beam bends into a configuration in which tangents to the deformed configuration remain horizontal at ends A and B, and of course there is zero vertical displacement at each of these ends. This gives rise to the reactions shown in which the vertical (shear) reactions are of equal magnitude for vertical equilibrium. This leaves only one equation from statics, namely,

$$\Sigma M_A = -M_1 - M_2 - M + R_1(a+b) = 0 \tag{1}$$

This equation contains R_1, M_1, and M_2 as unknowns. Since there are no more statics equations available, we must supplement Eq. (1) with two additional equations stemming from deformations of the system. We employ the method of singularity functions and write the bending moment at any point along the length of the beam as

$$M = -M_1\langle x\rangle^0 + R_1\langle x\rangle - M\langle x-a\rangle^0 \tag{2}$$

The differential equation of the bent beam is thus

$$EI\frac{d^2y}{dx^2} = -M_1\langle x\rangle^0 + R_1\langle x\rangle - M\langle x-a\rangle^0 \tag{3}$$

Integrating the first time, we obtain

$$EI\frac{dy}{dx} = -M_1\langle x\rangle + R_1\frac{\langle x\rangle^2}{2} - M\frac{\langle x-a\rangle^1}{1} + C_1 \tag{4}$$

When $x = 0$, $dy/dx = 0$; hence from Eq. (4) we have $C_1 = 0$. Integrating again

$$EIy = -M_A \frac{\langle x\rangle^2}{2} + \frac{R_1}{2}\cdot\frac{\langle x\rangle^3}{3} - M\frac{\langle x-a\rangle^2}{2} + C_2 \tag{5}$$

When $x = 0$, $y = 0$. Substituting these values in Eq. (5), we find $C_2 = 0$.
Also, when $x = L$, $dy/dx = 0$. Thus from Eq. (4) we have

$$0 = -M_1 L + \frac{R_1 L^2}{2} - Mb \tag{6}$$

The last boundary condition is that when $x = L$, $y = 0$. From Eq. (5) we obtain

$$0 = -\frac{M_1}{2}L^2 + \frac{R_1}{2}\cdot\frac{L^3}{6} - M\frac{b^2}{2} \tag{7}$$

It is now possible to solve Eqs. (1), (6), and (7) simultaneously to obtain the desired reactions

$$R_1 = \frac{6Mab}{L^3} \qquad M_1 = \frac{M(2ab - b^2)}{L^2} \qquad M_2 = \frac{M(2ab - a^2)}{L^2} \tag{8}$$

There may have been a temptation to say that the deflection under the point of application of the couple, at B, is zero. There is no reason for making such an assumption and, in fact, we may now return to the deflection Eq. (5) and calculate the deflection at $x = a$ and find that it is

$$EI[y]_{x=a} = \frac{Ma^2(2ab - b^2)}{2L^2} + \frac{Ma^4 b}{L^3} \tag{9}$$

which is clearly nonzero.

9.5. The horizontal beam shown in Fig. 9-8(a) is simply supported at the ends and is connected to a composite elastic vertical rod at its midpoint. The supports of the beam and the top of the copper rod are originally at the same elevation, at which time the beam is horizontal. The temperature of both vertical rods is then decreased 40°C. Find the stress in each of the vertical rods. Neglect the weight of the beam and of the rods. The cross-sectional area of the copper rod is 500 mm^2, $E_{cu} = 100$ GPa, and $\alpha_{cu} = 20 \times 10^{-6}$/°C. The cross-sectional area of the aluminum rod is 1000 mm^2, $E_{al} = 70$ GPa, and $\alpha_{al} = 25 \times 10^{-6}$/°C. For the beam, $E = 10$ GPa and $I = 400 \times 10^6$ mm^4.

SOLUTION: A free-body diagram of the horizontal beam appears as in Fig. 9-8(b). Here, P denotes the force exerted upon the beam by the copper rod. Since this force is initially unknown, there are three forces acting upon the beam, but only two equations of equilibrium for a parallel force system; hence the problem is statically indeterminate. It will thus be necessary to consider the deformations of the system.

A free-body diagram of the two vertical rods appears as in Fig. 9-8(c). The simplest procedure is temporarily to cut the connection between the beam and the copper rod, and then allow the vertical rods to contract freely because of the decrease in temperature. If the horizontal beam offers no restraint, the copper rod will contract an amount

$$\Delta_{cu} = (20 \times 10^{-6})(10^3)(40) = 0.8 \text{ mm}$$

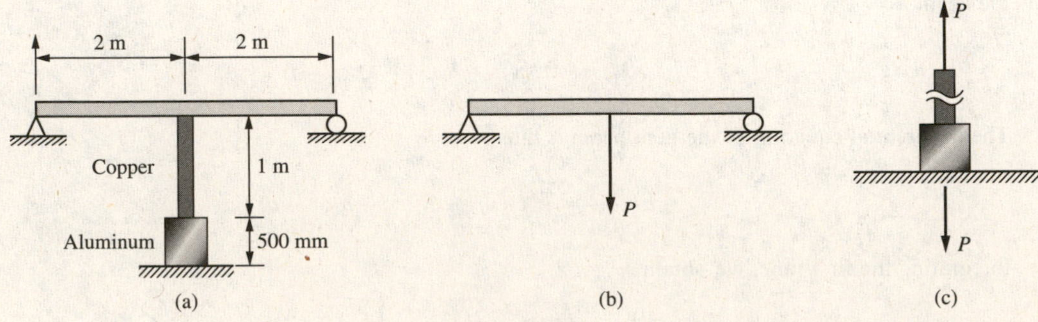

Fig. 9-8

and the aluminum rod will contract by an amount

$$\Delta_{al} = (25 \times 10^{-6})(500)(40) = 0.5 \text{ mm}$$

However, the beam exerts a force P upon the copper rod and the same force acts in the aluminum rod as shown in Fig. 9-8(c). These forces elongate the vertical rods by an amount of

$$\left(\frac{PL}{AE}\right)_{cu} + \left(\frac{PL}{AE}\right)_{al} = \frac{P \times 1}{(500 \times 10^{-6})(100 \times 10^9)} + \frac{P \times 0.5}{(1000 \times 10^{-6})(70 \times 10^9)} = 2.714 \times 10^{-8} P \text{ m}$$

The downward force P exerted by the copper rod upon the horizontal beam causes a vertical deflection of the beam. In Table 8-1 this central deflection is $\Delta = PL^3/48EI$.

Actually, there is a connection between the copper rod and the horizontal beam and the resultant shortening of the vertical rods is exactly equal to the downward vertical deflection of the midpoint of the beam. For the shortening of the rods to be equal to the deflection of the beam we must have

$$0.0008 + 0.0005 - 2.714 \times 10^{-8} P = \frac{P \times 4^3}{48(10 \times 10^9)(400 \times 10^6 \times 10^{-12})}$$

Solving, $P = 3610$ N; then,

$$\sigma_{cu} = \frac{3610}{500 \times 10^{-6}} = 7.22 \times 10^6 \text{ Pa} \qquad \sigma_{al} = \frac{3610}{1000 \times 10^{-6}} = 3.61 \times 10^6 \text{ Pa}$$

9.6. The beam of flexural rigidity EI shown in Fig. 9-9 is clamped at both ends and subject to a uniformly distributed load extending along the region BC of length $0.6L$. Determine all reactions.

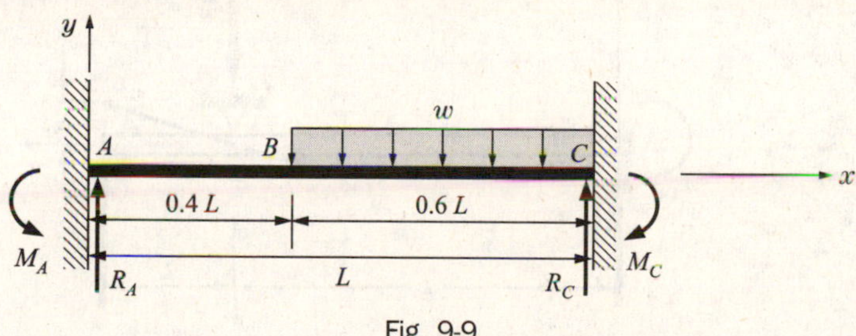

Fig. 9-9

SOLUTION: At end A as well as C the supporting walls exert bending moments M_A and M_C plus shearing forces R_A and R_C as shown. For such a plane, parallel force system there are two equations of static equilibrium and we must supplement these equations with additional relations stemming from beam deformations. The bending moment along the length ABC is conveniently written in terms of singularity functions:

$$EI\frac{d^2y}{dx^2} = -M_A\langle x\rangle^0 + R_A\langle x\rangle - \frac{w\langle x-0.4L\rangle^2}{2} \tag{1}$$

Integrating,

$$EI\frac{dy}{dx} = -M_A\langle x\rangle^1 + R_A\frac{\langle x\rangle^2}{2} - \frac{w}{2}\frac{\langle x-0.4L\rangle^3}{3} + C_1 \tag{2}$$

When $x = 0$, the slope $dy/dx = 0$. Substituting in Eq. (2), we have $C_1 = 0$. When $x = L$, $dy/dx = 0$. Substituting in Eq. (2), we find

$$0 = -M_AL + \frac{R_AL^3}{2} - \frac{w}{6}(0.6L)^3 \tag{3}$$

Next, integrating Eq. (2), we find

$$EIy = -M_A\frac{\langle x\rangle^2}{2} + \frac{R_A}{2}\frac{\langle x\rangle^3}{3} - \frac{w}{6}\frac{\langle x-0.4L\rangle^4}{4} + C_2 \tag{4}$$

When $x = 0$, $y = 0$, so from Eq. (4) we have $C_2 = 0$. The last boundary condition is: when $x = L$, $y = 0$, so from Eq. (4) we have

$$0 = -\frac{M_A L^2}{2} + \frac{R_A L^3}{6} - \frac{w}{24}\langle 0.6L\rangle^4 \tag{5}$$

The expressions for M_A given in Eqs. (3) and (5) may now be equated to obtain a single equation containing R_A as an unknown. Solving this equation, we find

$$R_A = wL\left\{(0.6)^3 - \frac{(0.6)^4}{2}\right\} = 0.1512wL$$

Substituting this value in Eq. (3), we find $M_A = 0.0396wL^2$.

From statics we have the two equations,

$$\Sigma F_y = -(0.6L)w + 0.1512wL + R_C = 0 \qquad \therefore R_C = 0.4488wL$$

$$\Sigma M_A = -0.0396wL^2 - M_C + (0.4488wL)(L) - [w(0.6L)](0.7L) = 0 \qquad \therefore M_C = 0.0684wL^2$$

9.7. The beam in Fig. 9-10 of flexural rigidity EI is clamped at A, supported between knife edges at B, and loaded by a vertical force P at the unsupported tip C. Determine the deflection at C.

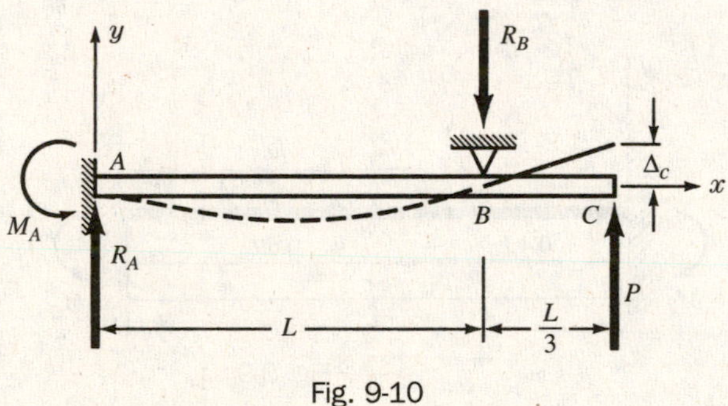

Fig. 9-10

SOLUTION: The reactions at A are the moment M_A and shear force R_A as shown in Fig. 9-10. From statics we have

$$\Sigma M_A = M_A + P\left(\frac{4L}{3}\right) - R_B(L) = 0 \tag{1}$$

$$\Sigma F_y = R_A + P - R_B = 0 \tag{2}$$

These two equations contain the three unknowns M_A, R_A, and R_B. Thus, we must supplement these two statics equations with another equation arising from deformation of the beam. Using the x-y coordinate system shown, the differential equation of the deformed beam in terms of singularity functions is

$$EI\frac{d^2y}{dx^2} = -M_A\langle x\rangle^0 + R_A\langle x\rangle^1 - R_B\langle x - L\rangle^1 \tag{3}$$

The first integration yields

$$EI\frac{dy}{dx} = -M_A\langle x\rangle^1 + R_A\frac{\langle x\rangle^2}{2} - R_B\frac{\langle x - L\rangle^2}{2} + C_1 \tag{4}$$

When $x = 0$, $dy/dx = 0$; hence from Eq. (4), $C_1 = 0$. The next integration yields

$$EIy = -M_A\frac{\langle x\rangle^2}{2} + \frac{R_A}{2}\frac{\langle x\rangle^3}{2} - \frac{R_B}{2}\frac{\langle x - L\rangle^3}{3} + C_2 \tag{5}$$

When $x = 0$, $y = 0$, so that $C_2 = 0$. Also, when $x = L$, $y = 0$. Substituting in Eq. (5), we find

$$0 = -\frac{M_A L^2}{2} + \frac{R_A L^3}{6} - 0 \qquad (6)$$

Solving Eqs. (1), (2), and (6) simultaneously, we have

$$R_A = \frac{3M_A}{L} = \frac{P}{2} \qquad M_A = \frac{PL}{6} \qquad R_B = \frac{3P}{2} \qquad (7)$$

If we now introduce these values into Eq. (5) and also set $x = 4L/3$ (point C), we have

$$\Delta_C = \frac{0.0401PL^3}{EI} \qquad (8)$$

9.8. In Problem 9.7 if the beam is a steel section with $I = 1400\ \text{cm}^4$ with length $L = 4.2$ m, determine the force P required to deflect the tip C 5 mm.

SOLUTION: From Eq. (8) of Problem 9.7, we have the tip deflection Δ_C as

$$\Delta_C = 0.0401\frac{PL^3}{EI} \qquad 0.005 = 0.0401\frac{P \times 4.2^3}{(200 \times 10^9)(1400 \times 10^{-8})} \qquad \therefore P = 4710\ \text{N}$$

9.9. The beam of flexural rigidity EI in Fig. 9-11 is clamped at end A, supported at C, and loaded by the couple at B together with the load uniformly distributed over the region BC. Determine all reactions.

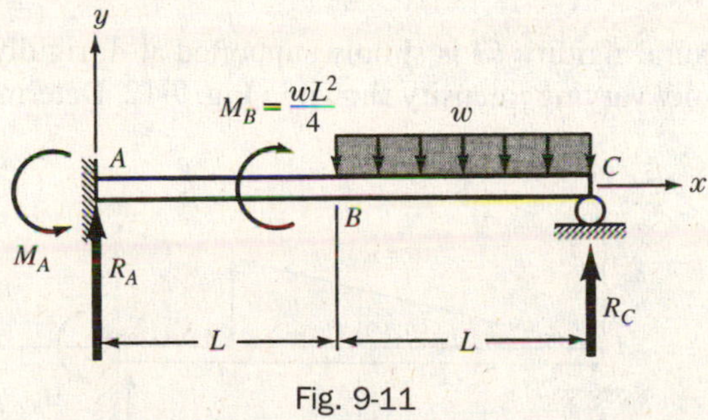

Fig. 9-11

SOLUTION: The reactions at the left support A consist of moment M_A and force R_A. From statics, we have

$$\Sigma M_A = M_A - \frac{wL^2}{4} - (wL)\left(\frac{3L}{2}\right) + R_C(2L) = 0 \qquad (1)$$

$$\Sigma F_y = R_A + R_C - wL = 0 \qquad (2)$$

These two equations contain the three unknowns M_A, R_A, and R_C. Accordingly we must supplement the two statics equations with another equation stemming from deformations of the system.

For the x-y coordinate system shown, the differential equation of the bent beam is

$$EI\frac{d^2y}{dx^2} = -M_A\langle x\rangle^0 + R_A\langle x\rangle^1 + \frac{wL^2}{4}\langle x - L\rangle^0 - \frac{w}{2}\langle x - L\rangle^2 \qquad (3)$$

Integrating the first time, this becomes

$$EI\frac{dy}{dx} = -M_A\langle x\rangle^1 + R_A\frac{\langle x\rangle^2}{2} + \frac{wL^2}{4}\langle x - L\rangle - \frac{w}{2}\frac{\langle x - L\rangle^3}{3} + C_1 \qquad (4)$$

When $x = 0$, $dy/dx = 0$, so $C_1 = 0$. Integrating the second time, we find

$$EIy = -M_A\frac{\langle x\rangle^2}{2} + R_A\frac{\langle x\rangle^3}{6} + \frac{wL^2}{4}\frac{\langle x - L\rangle^2}{2} - \frac{w}{2}\frac{\langle x - L\rangle^4}{12} + C_2 \qquad (5)$$

When $x = 0$, $y = 0$, $C_2 = 0$. At point C when $x = 2L$, $y = 0$. Substituting these values in Eq. (5), we have

$$-2M_A L^2 + \frac{4R_A L^3}{3} + \frac{wL^2}{4} \cdot \frac{L^2}{2} - \frac{wL^4}{24} = 0 \tag{6}$$

Solving Eqs. (1), (2), and (6) simultaneously, we find

$$M_A = \frac{3}{16} wL^2 \qquad R_A = \frac{7}{32} wL \qquad R_C = \frac{25}{32} wL \tag{7}$$

9.10. In Problem 9.9 if the beam is titanium having a Young's modulus of 110 GPa, with a rectangular cross section 20 mm × 30 mm, is 2 m long, and carries the uniform load in BC of 960 N/m, determine the deflection at the midpoint B. (Note: $L = 1$ m in Fig. 9-11.)

SOLUTION: From Eq. (5) of Problem 9.9 we have the deflection at the midpoint B as

$$EIy]_{x=1} = -\frac{M_A}{2} L^2 + \frac{R_A}{6} L^3$$

$$= -\frac{3}{16} \frac{960 \times 1^2}{2} + \frac{7}{32} \frac{960 \times 1}{6} = -55$$

The deflection at the midpoint B is then

$$\Delta_B = \frac{55}{EI} = \frac{55}{(110 \times 10^9)(20 \times 30^3/12) \times 10^{-12}} = 0.0111 \text{ m} \qquad \text{or} \qquad 11.1 \text{ mm}$$

9.11. The beam AB of flexural rigidity EI is simply supported at A, rigidly clamped at end B, and subject to the load of uniformly varying intensity shown in Fig. 9-12. Determine the reactions developed at A and B.

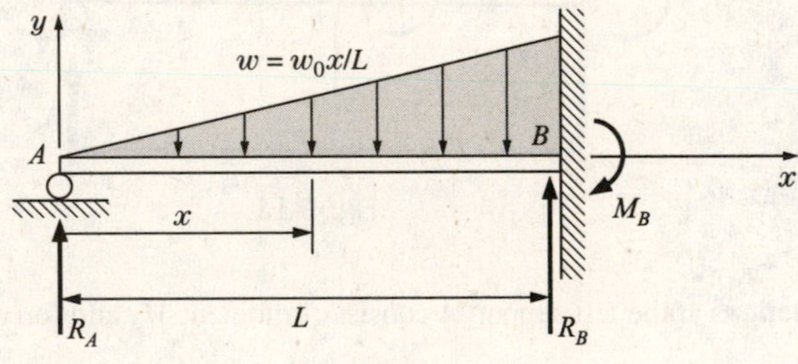

Fig. 9-12

SOLUTION: Let us denote the vertical force reaction at A by R_A, that at B by R_B, and the moment exerted by the wall on the beam at B by M_B, as shown. The contribution to bending moment of the distributed loading at any point a distance x to the right of A is

$$M = R_A x - \frac{w_0 x}{L}\left(\frac{x}{2}\right)\left(\frac{x}{3}\right)$$

Thus,

$$EI\frac{d^2y}{dx^2} = R_A x - \frac{w_0 x^3}{6} \tag{1}$$

Integrating the first time,

$$EI\frac{dy}{dx} = R_A \frac{x^2}{2} - \frac{w_0 x^4}{24L} + C_1 \tag{2}$$

When $x = L$, $dy/dx = 0$, so from Eq. (2)

$$0 = R_A \frac{L^2}{2} - \frac{w_0 L^3}{24} + C_1 \tag{3}$$

Integrating a second time,

$$EIy = R_A \frac{x^3}{6} - \frac{w_0 x^5}{120L} + C_1 x + C_2 \tag{4}$$

When $x = L$, $y = 0$, so we have from Eq. (4)

$$0 = \frac{R_A L^3}{6} - \frac{w_0 L^4}{120} + C_1 L + C_2 \tag{5}$$

Also, when $x = 0$, $y = 0$, so from Eq. (4), $C_2 = 0$. From Eqs. (3) and (5) we have

$$C_1 = \frac{w_0 L^3}{24} - \frac{R_A L^2}{2} = -\frac{R_A L^2}{6} + \frac{w_0 L^3}{120} \qquad \therefore R_A = \frac{1}{10} w_0 L \tag{6}$$

The two statics equations for such a force system are

$$\Sigma F_y = R_A + R_B - \frac{w_0 L}{2} = 0$$

$$\Sigma M_B = -R_A L - M_0 + \left(\frac{w_0}{2}\right)(L)\left(\frac{L}{3}\right) = 0$$

Solving for the reactions at B, there results

$$R_B = \frac{2}{5} w_0 L$$

$$M_B = \frac{1}{15} w_0 L^2$$

9.12. The initially horizontal beam ABC in Fig. 9-13 is clamped at C and supported on a smooth roller at B. A uniform load w acts over the entire length of the beam. After application of the load, the reaction at B is mechanically displaced upward an amount Δ. Determine the reaction R_B after this displacement has been imposed.

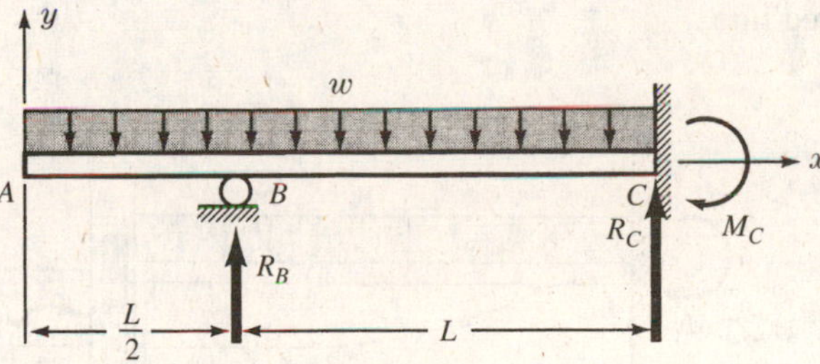

Fig. 9-13

SOLUTION: The beam reactions are R_B, R_C, and a moment M_C. Using the method of singularity functions, we have the equation of the bent beam,

$$EI \frac{d^2 y}{dx^2} = -w \frac{\langle x \rangle^2}{2} + R_B \left\langle x - \frac{L}{2} \right\rangle \tag{1}$$

Integrating the first time, we obtain

$$EI \frac{dy}{dx} = -\frac{w}{2} \frac{\langle x \rangle^3}{3} + \frac{R_B}{2} \left\langle x - \frac{L}{2} \right\rangle^2 + C_1 \tag{2}$$

We know that, when $x = 3L/2$, $dy/dx = 0$. Substituting in Eq. (2)

$$0 = -\frac{w}{6}\left(\frac{27L^3}{8}\right) + R_B \frac{L^2}{2} + C_1$$

from which

$$C_1 = \frac{9}{16}wL^3 - \frac{R_B L^2}{2} \tag{3}$$

Integrating a second time,

$$EIy = -\frac{w}{6}\frac{\langle x\rangle^4}{4} + \frac{R_B}{2}\frac{\langle x - L/2\rangle^3}{3} + \left(\frac{9}{16}wL^3 - R_B\frac{L^2}{2}\right)\langle x\rangle + C_2 \tag{4}$$

When $x = 3L/2$, $y = 0$. Substituting in Eq. (4), we have

$$wL^4\left[-\frac{27}{128} + \frac{27}{32}\right] + R_B\left[\frac{L^3}{6} - \frac{3L^3}{4}\right] + C_2 = 0$$

from which

$$C_2 = -\frac{81}{128}wL^4 + \frac{7}{12}R_B L^3$$

The last boundary condition stems from the imposed displacement at point B; that is, when $x = L/2$, $y = \Delta$. Substituting these values in Eq. (4), we have

$$EI\Delta = -\frac{w}{24}\left(\frac{L^4}{16}\right) + 0 + \left(\frac{9}{16}wL^3 - R_B\frac{L^2}{2}\right)\left(\frac{L}{2}\right) - \frac{81}{128}wL^4 + \frac{7}{12}R_B L^3$$

Solving for R_B, we obtain

$$R_B = \frac{3EI\Delta}{L^3} + \frac{17}{16}w_0 L$$

9.13. The horizontal beam AB shown in Fig. 9-14 is clamped at A, subject to a uniformly distributed load, and supported at B in such a manner that it is free to deflect vertically but is completely restrained against rotation at that point. Determine the vertical deflection at B after the beam has deflected as shown by the dotted line.

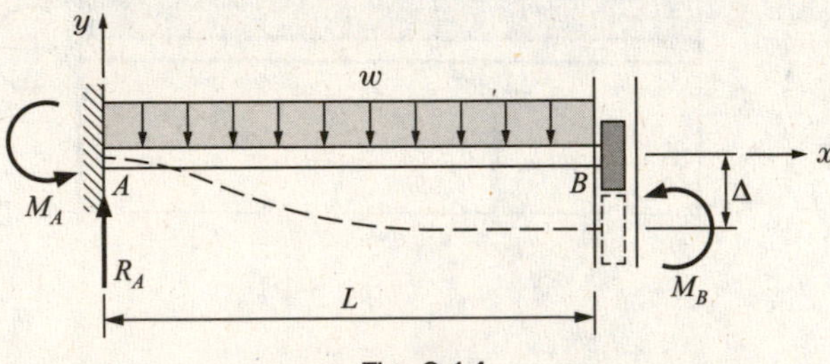

Fig. 9-14

SOLUTION: The equation of the deflected beam is

$$EI\frac{d^2y}{dx^2} = -M_A + R_A x - \frac{wx^2}{2} \tag{1}$$

Integrating the first time, we find

$$EI\frac{dy}{dx} = -M_A x + R_A\frac{x^2}{2} - \frac{w}{2}\frac{x^3}{3} + C_1 \tag{2}$$

When $x = 0$, $dy/dx = 0$, so that $C_1 = 0$. Integrating again,

$$EIy = -M_A\frac{x^2}{2} + \frac{R_A x^3}{6} - \frac{wx^4}{24} + C_2 \tag{3}$$

Imposing the boundary condition that $x = 0$ at $y = 0$, we have $C_2 = 0$.

The third boundary condition is that, when $x = L$, $dy/dx = 0$. Substituting these values in Eq. (2), we obtain the equation

$$0 = -M_A L + \frac{R_A L^2}{2} - \frac{wL^3}{6} \tag{4}$$

From statics, we have the two equilibrium equations

$$\Sigma M_A = M_A + M_B - \frac{wL^2}{2} = 0 \tag{5}$$

$$\Sigma F_y = R_A - wL = 0 \tag{6}$$

Solving Eqs. (4), (5), and (6) simultaneously, we have

$$R_A = wL \qquad M_A = \frac{2}{3}wL^2 \qquad M_B = \frac{1}{6}wL^2$$

Substitution of these values in Eq. (3) leads to

$$EI[y]_{x=L} = -\frac{wL^4}{6} + \frac{wL^4}{6} - \frac{wL^4}{24} \qquad \text{or} \qquad \therefore \Delta_B = \frac{wL^4}{24EI}$$

SUPPLEMENTARY PROBLEMS

9.14. A clamped-end beam is supported at the right end, clamped at the left, and carries the two concentrated forces shown in Fig. 9-15. Determine the reaction at the wall and the reaction at the right end of the beam. *Ans.* $4P/3$ at A, $PL/3$ acting counterclockwise at A, $2P/3$ at B

9.15. Determine the deflection under the point of application of the force P located a distance $L/3$ from the right end of the beam shown in Fig. 9-15. *Ans.* $7PL^3/486EI$

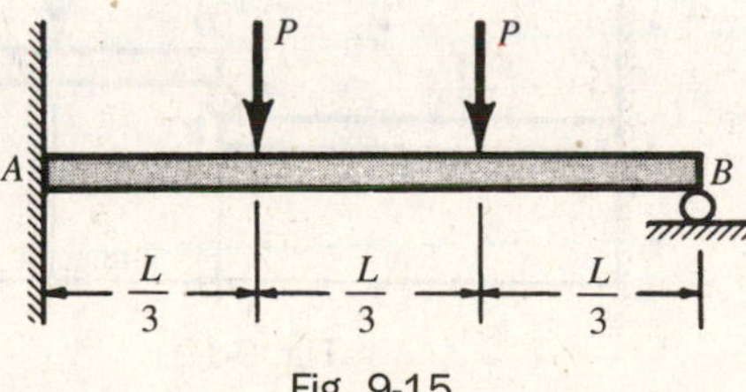

Fig. 9-15

9.16. The titanium 2-m-long beam of Problem 9.14 has an ultimate tensile strength of 1.2 GPa. If the cross section is 5 cm × 12 cm and a safety factor of 1.4 is employed, determine the maximum allowable value of each load P. *Ans.* 19 300 N

9.17. A clamped-end beam is supported at an intermediate point and loaded as shown in Fig. 9-16. Determine the various reactions. *Ans.* $R_A = wL/4$, $M_A = 0$, $R_B = 5wL/4$

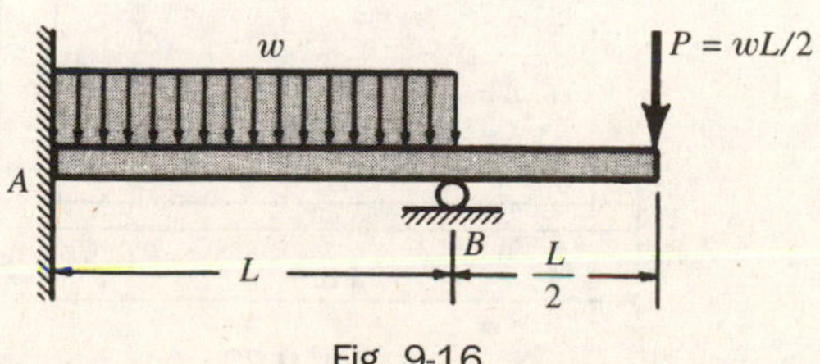

Fig. 9-16

9.18. A clamped-end beam is supported at the right end, clamped at the left, and carries the load of uniformly varying intensity, as indicated in Fig. 9-17. Determine the resisting moment at the left end. *Ans.* $7wL^2/120$

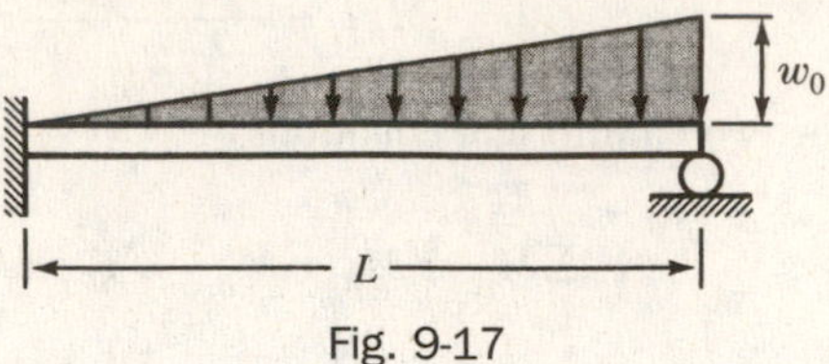

Fig. 9-17

9.19. The beam shown in Fig. 9-18 is clamped at the left end, supported at the right, and loaded by a couple M. Determine the reaction at the right support. *Ans.* $4M/9a$

9.20. For the beam shown in Fig. 9-18, determine the deflection under the point of application of the applied moment M. *Ans.* $2Ma^2/27EI$

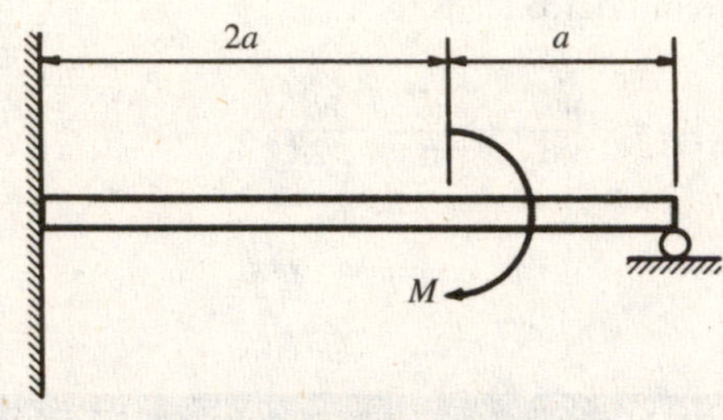

Fig. 9-18

9.21. In Fig. 9-19 AB and CD are cantilever beams with a roller E between their end points. A load of 5 kN is applied as shown. Both beams are made of steel for which $E = 200$ GPa. For beam AB, $I = 20 \times 10^6\,\text{mm}^4$; for CD, $I = 30 \times 10^6\,\text{mm}^4$. Find the reaction at E. *Ans.* 398 N

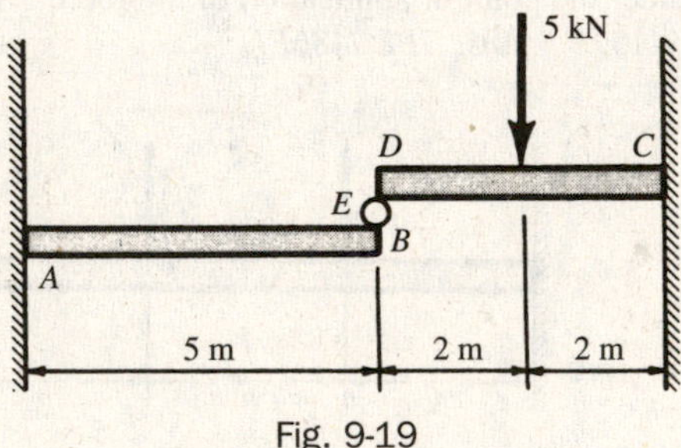

Fig. 9-19

9.22. The beam AB in Fig. 9-20 has $I = 11.7 \times 10^6\,\text{mm}^4$. Member CD is a vertical steel wire of 3-mm-diameter cross section and length 4 m. Both the beam and the wire are steel for which $E = 200$ GPa. Prior to the application of any load to the beam, due to a fabrication error, the end D of the wire is 5 mm above the tip B of the beam. The end D of the wire and the beam are then mechanically pulled together and joined. Determine the axial stress in the wire. *Ans.* 106 MPa

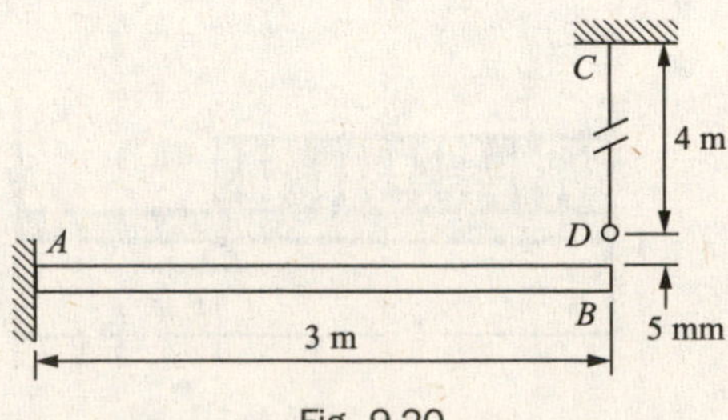

Fig. 9-20

9.23. A beam is clamped at both ends and supports a uniform load over its right half, as shown in Fig. 9-21. Determine the reactions at A and B.

Ans. $3wL/32$, $5wL^2/192$ acting counterclockwise; $13wL/32$, $11wL^2/192$ acting clockwise

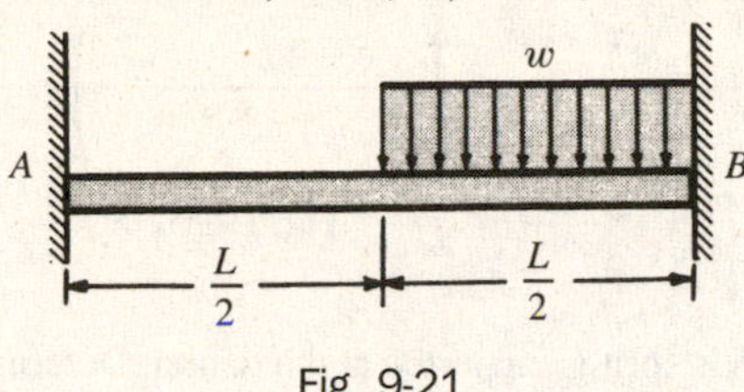

Fig. 9-21

9.24. Determine the deflection at the center of the beam shown in Fig. 9-21. *Ans.* $wL^4/768EI$

9.25. The beam of flexural rigidity EI in Fig. 9-22 is clamped at A, supported between knife edges at B, and subjected to the couple M_0 at its unsupported tip C. Determine the deflection of point C. *Ans.* $M_0L^2/4EI$

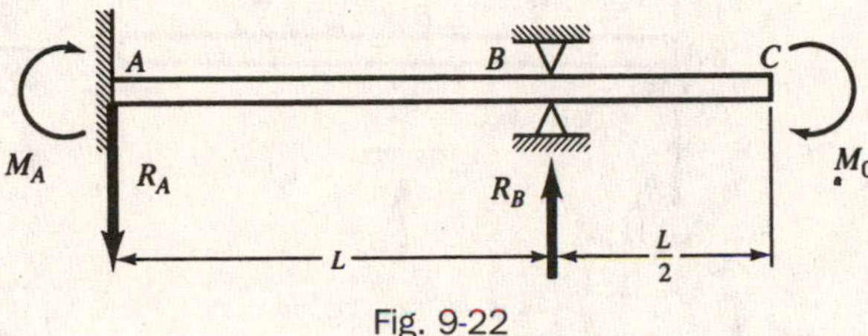

Fig. 9-22

9.26. The cantilever beam in Fig. 9-23 of length 3 m and rectangular cross section 100 mm × 200 mm has its free end (at no load) 3 mm above the top of a spring whose constant is 150 kN/m. The material of the beam has $E = 110$ GPa and a yield point of 900 MPa. A downward force P of 7000 N is applied to the tip of the beam. Find the deformation of the top of the spring under this load. *Ans.* 4.72 mm

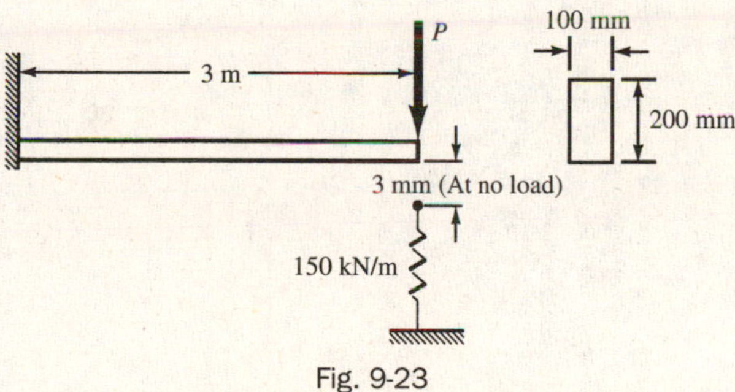

Fig. 9-23

9.27. A beam AB is clamped at each end and subject to a load of uniformly varying intensity as shown in Fig. 9-24. Determine the moment reactions developed at each end of the beam.

Ans. $w_0L^2/30$ counterclockwise at A, $w_0L^2/20$ clockwise at B

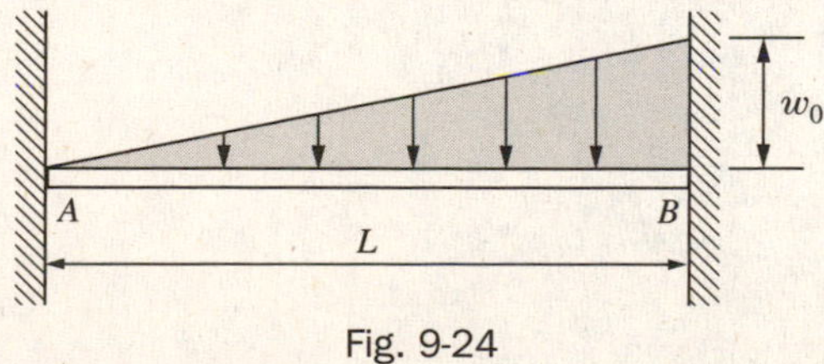

Fig. 9-24

9.28. The three-span continuous beam shown in Fig. 9-25 supports a uniformly distributed load in the left and central span, but is unloaded in the right span. Determine the reactions at A, B, C, and D.

Ans. $R_A = 0.383wL(\uparrow)$, $R_B = 1.20wL\ (\uparrow)$, $R_C = 0.450wL\ (\uparrow)$, $R_D = -0.033wL\ (\downarrow)$

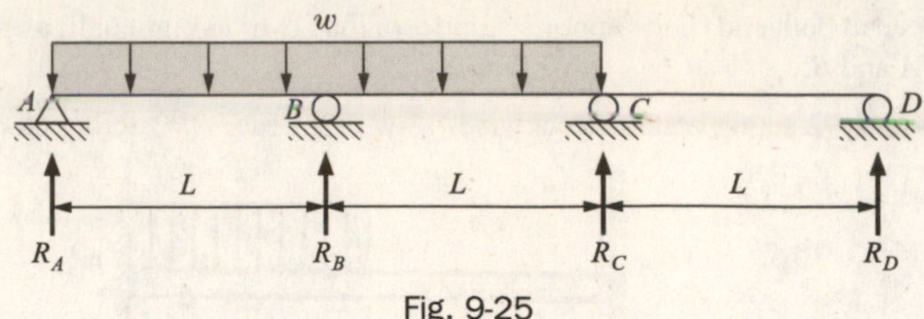

Fig. 9-25

9.29. The beam shown in Fig. 9-26 is spring supported at the center. Determine the spring constant so that the bending moment will be zero at the point where the spring supports the beam. *Ans.* $k = 16EI/L^3$

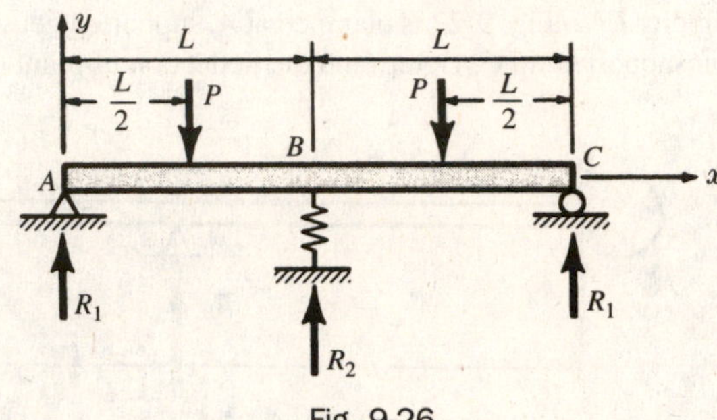

Fig. 9-26

Columns

10.1 Basics

A long slender bar subject to axial compression is called a *column*. The term "column" is frequently used to describe a vertical member, whereas the word "strut" is occasionally used in regard to inclined bars.

Many aircraft structural components, structural connections between stages of boosters for space vehicles, certain members in bridge trusses, and structural frameworks of buildings are common examples of columns.

Failure of a column occurs by buckling, i.e., by lateral deflection of the bar. In comparison it is to be noted that failure of a short compression member occurs by yielding of the material. Buckling, and hence failure, of a column may occur even though the maximum stress in the bar is much less than the yield point of the material. Linkages in oscillating or reciprocating machines may also fail by buckling.

The Critical Load of a Column

The *critical load* of a slender bar subject to axial compression is that value of the axial force that is just sufficient to keep the bar in a slightly deflected configuration. Figure 10-1 shows a pin-ended bar in a buckled configuration due to the critical load P_{cr}.

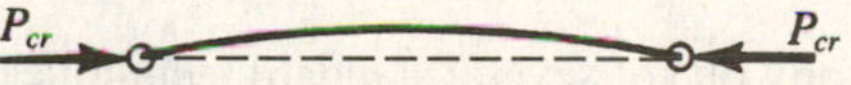

Fig. 10-1 A column critically loaded at pinned ends.

Slenderness Ratio of a Column

The ratio of the length of the column to the minimum radius of gyration of the cross-sectional area is termed the *slenderness ratio*. This ratio is dimensionless. The method of determining the radius of gyration of an area was presented in a statics course. It is related to the moment of inertia by

$$r_x = \sqrt{\frac{I_x}{A}} \qquad r_y = \sqrt{\frac{I_y}{A}} \tag{10.1}$$

If the column is free to rotate at each end, then buckling takes place about that axis for which the radius of gyration is minimum.

10.2 Critical Load of a Long Slender Column

Let us derive an expression for the critical load for a long slender pin-ended column loaded by an axial compressive force at each end. The line of action of the forces passes through the centroid of the cross section of the bar. The critical load is defined to be that axial force that is just sufficient to hold the bar in a slightly deformed configuration. Under the action of the load P_{cr} the bar has the deflected shape shown in Fig. 10-2. The column is subject to a load P with the x- and y-axes identified is shown in Fig. 10-2.

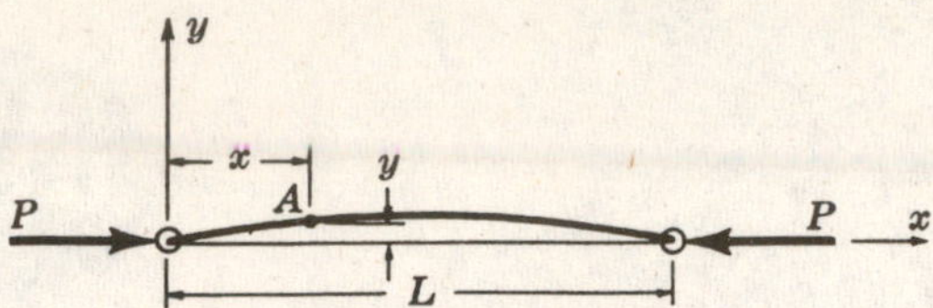

Fig. 10-2 The nomenclature used to derive P_{cr}

It is of course necessary that one end of the bar be able to move axially with respect to the other end in order that lateral deflection may take place. The differential equation of the deflection curve is the same as that presented in Chapter 8, namely,

$$EI\frac{d^2y}{dx^2} = M \tag{10.2}$$

Here the bending moment at the point A having coordinates (x, y) is merely the moment of the force P applied at the left end of the bar about an axis through the point A and perpendicular to the plane of the page. It is to be carefully noted that this force is assumed to produce curvature of the bar that is concave downward, which, according to the sign convention of Chapter 6, constitutes negative bending. Hence the bending moment is $M = -Py$. Thus we have

$$EI\frac{d^2y}{dx^2} = -Py \tag{10.3}$$

If we set

$$\frac{P}{EI} = k^2 \tag{10.4}$$

Equation (10.3) becomes

$$\frac{d^2y}{dx^2} + k^2y = 0 \tag{10.5}$$

This equation is readily solved by any one of several standard techniques discussed in works on differential equations. However, the solution is almost immediately apparent. We need merely find a function which when differentiated twice and added to itself (times a constant) is equal to zero. Evidently either sin kx or cos kx possesses this property. In fact, a combination of these terms in the form

$$y(x) = C\sin kx + D\cos kx \tag{10.6}$$

may also be taken to be a solution of Eq. (10.5). This may be readily checked by substituting $y(x)$ into Eq. (10.5).

Having obtained $y(x)$, it is next necessary to determine C and D. At the left end of the bar, $y = 0$ when $x = 0$. Substituting these values in Eq. (10.6), we obtain

$$0 = 0 + D \qquad \text{or} \qquad D = 0$$

At the right end of the bar, $y = 0$ when $x = L$. Substituting these values in Eq. (10.6) with $D = 0$, we obtain

$$0 = C\sin kL$$

Evidently either $C = 0$ or $\sin kL = 0$. But if $C = 0$ then y is everywhere zero and we have only the trivial case of a straight bar which is the configuration prior to the occurrence of buckling. Since we are not interested in that solution, we must take

$$\sin kL = 0 \tag{10.7}$$

For this to be true, we must have

$$kL = n\pi \ (n = 1, 2, 3, \ldots) \tag{10.8}$$

Substituting $k^2 = P/EI$ in Eq. (10.8), we find

$$\sqrt{\frac{P}{EI}}L = n\pi \quad \text{or} \quad P = \frac{n^2\pi^2 EI}{L^2} \tag{10.9}$$

The smallest value of this load P evidently occurs when $n = 1$. Then we have the so-called first mode of buckling where the critical load is given by

$$P_{cr} = \frac{\pi^2 EI}{L^2} \tag{10.10}$$

This is *Euler's buckling load for a pin-ended column*. The deflection shape corresponding to this load is

$$y(x) = C\sin\left(\sqrt{\frac{P}{EI}}x\right) \tag{10.11}$$

Substituting in this equation from Eq. (10.10), we obtain

$$y(x) = C\sin\frac{\pi x}{L} \tag{10.12}$$

Thus the deflected shape is a sine function. Because of the approximations introduced in the derivation of Eq. (10.2), it is not possible to obtain the amplitude of the buckled shape, denoted by C in Eq. (10.12).

As may be seen from Eq. (10.10), buckling of the bar will take place about that axis in the cross section for which I assumes a minimum value.

Equation (10.10) may be modified to the form

$$P_{cr} = \frac{\pi^2 EI}{(KL)^2} \tag{10.13}$$

where KL is an effective length of the column, defined to be a portion of the deflected bar between points corresponding to zero curvature. For example, for a column pinned at both ends, $K = 1$. If both ends are rigidly clamped, $K = 0.5$. For one end clamped and the other pinned, $K = 0.7$. In the case of a cantilever-type column loaded at its free end, $K = 2$. These four cases are displayed in Fig. 10-3.

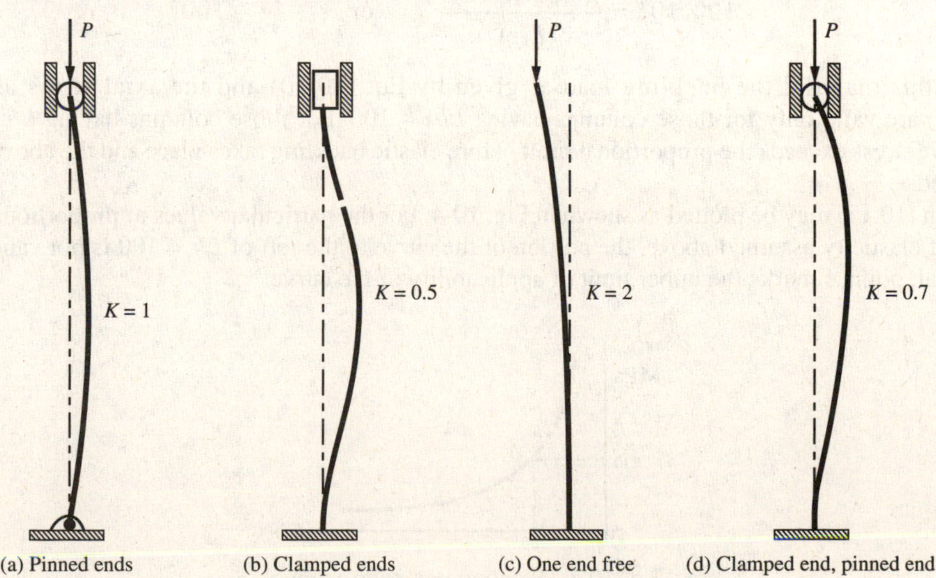

Fig. 10-3 Columns with various end conditions.

The Euler buckling load P_{cr} is not immediately applicable if the corresponding axial stress, found from the expression $\sigma_{cr} = P_{cr}/A$, where A represents the cross-sectional area of the bar, exceeds the proportional limit of the material. For example, for a steel bar having a proportional limit of 210 MPa, Eq. (10.10) is valid only for columns whose slenderness ratio exceeds 100. The value of P_{cr} represented by this formula is a failure load; consequently, a safety factor must be introduced to obtain a design load. In the derivation of the equation $EI(d^2y/dx^2) = M$ used to determine the critical load of Eq. (10.10), it was assumed that there is a linear relationship between stress and strain. Thus the critical load indicated by Eq. (10.10) is correct only if the proportional limit of the material has not been exceeded.

The axial stress in the bar immediately prior to the instant when the bar assumes its buckled configuration is given by

$$\sigma_{cr} = \frac{P_{cr}}{A} \tag{10.14}$$

where A represents the cross-sectional area of the column. Substituting for P_{cr} its value as given by Eq. (10.10), we find

$$\sigma_{cr} = \frac{\pi^2 EI}{AL^2} \tag{10.15}$$

But the moment of inertia is related to the radius of gyration r by

$$I = Ar^2 \tag{10.16}$$

Substituting this value in Eq. (10.15), we find

$$\sigma_{cr} = \frac{\pi^2 EAr^2}{AL^2} = \pi^2 E\left(\frac{r}{L}\right)^2$$

or

$$\sigma_{cr} = \frac{\pi^2 E}{(L/r)^2} \tag{10.17}$$

The ratio L/r is called the *slenderness ratio* of the column.

Let us consider a steel column having a proportional limit of 210 MPa and $E = 200$ GPa. The stress of 210 MPa marks the upper limit of stress for which Eq. (10.17) may be used. To find the value of L/r corresponding to these constants, we substitute in Eq. (10.17) and obtain

$$210 \times 10^6 = \frac{\pi^2(200 \times 10^9)}{(L/r)^2} \qquad \text{or} \qquad \frac{L}{r} \approx 100 \tag{10.18}$$

Thus, for this material the buckling load as given by Eq. (10.10) and the axial stress as given by Eq. (10.17) are valid only for those columns having $L/r \geqslant 100$. For those columns having $L/r < 100$, the compressive stress exceeds the proportional limit before elastic buckling takes place and the above equations are not valid.

Equation (10.17) may be plotted as shown in Fig. 10-4. For the particular values of proportional limit and modulus of elasticity assumed above, the portion of the curve to the left of $L/r = 100$ is not valid. Thus for this material, point A marks the upper limit of applicability of the curve.

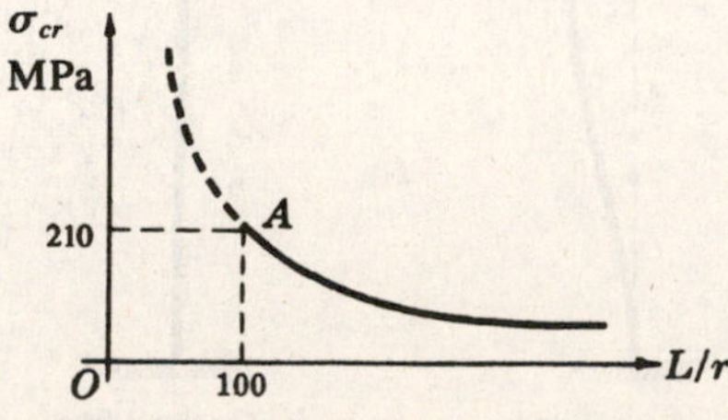

Fig. 10-4 Slenderness ratio versus critical stress.

10.3 Eccentrically Loaded Columns

The derivation of the expression leading to the Euler buckling load assumes that the column is loaded perfectly concentrically. If the axial force P is applied with an eccentricity e, as shown in Fig. 10-5, let us derive an expression for the peak compressive stress in the bar.

Assume an initially straight, pin-ended column subject to an axial compressive force applied with known eccentricity e (see Fig. 10-5).

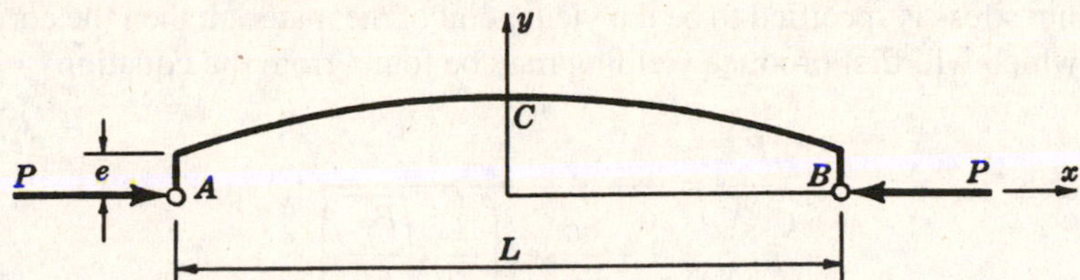

Fig. 10-5 An eccentrically loaded column.

The differential equation of the bar in its deflected configuration is

$$EI\frac{d^2y}{dx^2} = -Py \tag{10.19}$$

which has the standard solution

$$y = C_1 \sin\left(\sqrt{\frac{P}{EI}}x\right) + C_2 \cos\left(\sqrt{\frac{P}{EI}}x\right) \tag{10.20}$$

Since $y = e$ at each of the ends $x = -L/2$ and $x = L/2$, the values of the two constants of integration are readily found to be

$$C_1 = 0 \qquad C_2 = \frac{e}{\cos\left(\sqrt{\frac{P}{EI}}\frac{L}{2}\right)} \tag{10.21}$$

Thus, the deflection curve of the bent bar is

$$y = \frac{e}{\cos\left(\sqrt{\frac{P}{EI}}\frac{L}{2}\right)}\cos\left(\sqrt{\frac{P}{EI}}x\right) \tag{10.22}$$

The maximum value of deflection occurs at $x = 0$, by symmetry, and is

$$y_{max} = e\sec\left(\sqrt{\frac{P}{EI}}\frac{L}{2}\right) \tag{10.23}$$

Introducing the value of the critical load P_{cr} as given by Eq. (10.10), this becomes

$$y_{max} = e\sec\left(\frac{\pi}{2}\sqrt{\frac{P}{P_{cr}}}\right) \tag{10.24}$$

Evidently the maximum deflection, which occurs at the center of the bar, becomes very great as the load P approaches the critical value. The phenomenon is one of gradually increasing lateral deflections, not buckling. The maximum compressive stress occurs on the concave side of the bar at C and is given by

$$\sigma_{max} = \frac{P}{A} + \frac{M_{max}c}{I} = \frac{P}{A} + \frac{Pec}{I}\sec\left(\frac{\pi}{2}\sqrt{\frac{P}{P_{cr}}}\right) \tag{10.25}$$

where c denotes the distance from the neutral axis to the outer fibers of the bar. If we now introduce the radius of gyration r [see Eq. (10.1)] of the cross section, this becomes

$$\sigma_{max} = \frac{P}{A}\left[1 + \frac{ec}{r^2}\sec\left(\frac{L}{2r}\sqrt{\frac{P}{AE}}\right)\right] \tag{10.26}$$

This is the *secant formula* for an eccentrically loaded long column. In it, P/A is the average compressive stress. If the maximum stress is specified to be the yield point of the material, then the corresponding average compressive stress which will first produce yielding may be found from the equation

$$\frac{P_{yp}}{A} = \frac{\sigma_{yp}}{1 + \frac{ec}{r^2}\sec\left(\frac{L}{2r}\sqrt{\frac{P_{yp}}{AE}}\right)} \tag{10.27}$$

For any designated value of the eccentricity ratio ec/r^2, this equation may be solved by trial and error and a curve of P/A versus L/r plotted to indicate the value of P/A at which yielding first begins in the extreme fibers.

10.4 Design Formulas for Columns Having Intermediate Slenderness Ratios

The design of compression members having large values of the slenderness ratio proceeds according to the Euler formula presented above together with an appropriate safety factor. For the design of shorter compression members, it is customary to employ any one of the many semiempirical formulas giving a relationship between the yield stress and the slenderness ratio of the bar.

For steel columns, one commonly employed design expression is that due to the American Institute of Steel Construction (AISC), which states that the allowable (working) axial stress on a steel column having slenderness ratio L/r is

$$\sigma_a = \frac{[1 - (KL/r)^2]\sigma_{yp}}{\left[\frac{5}{3} + \frac{3(KL/r)}{8C_c} - \frac{(KL/r)^3}{8C_c^3}\right]} \qquad \text{for } \frac{KL}{r} < C_c \tag{10.28}$$

$$\sigma_a = \frac{12\pi^2 E}{23(KL/r)^2} \qquad \text{for } \frac{KL}{r} > C_c \tag{10.29}$$

where

$$C_c = \sqrt{\frac{2\pi^2 E}{\sigma_{yp}}}$$

and σ_{yp} is the yield point of the material and E is Young's modulus (see Problems 10.8, 10.9, and 10.10).

Another approach is in the use of the Structural Stability Research Council's (SSRC) equations which give mean axial compressive stress σ_u immediately prior to collapse:

$$\begin{aligned}
\sigma_u &= \sigma_{yp} && \text{for } 0 < \lambda < 0.15 \\
\sigma_u &= \sigma_{yp}(1.035 - 0.202\lambda - 0.222\lambda^2) && \text{for } 0.15 \leqslant \lambda \leqslant 1.0 \\
\sigma_u &= \sigma_{yp}(-0.111 + 0.636\lambda^{-1} + 0.087\lambda^{-2}) && \text{for } 1.0 \leqslant \lambda \leqslant 2.0 \\
\sigma_u &= \sigma_{yp}(0.009 + 0.877\lambda^{-2}) && \text{for } 2.0 \leqslant \lambda < 3.6 \\
\sigma_u &= \sigma_{yp}\lambda^{-2}(\text{Euler's curve}) && \text{for } \lambda \geqslant 3.6
\end{aligned} \tag{10.30}$$

where

$$\lambda = \frac{L}{\pi r}\sqrt{\frac{\sigma_{yp}}{E}} \tag{10.31}$$

No safety factor is present in these equations but of course one must be introduced by the designer (see Problem 10.11). Also, either SI or English units may be used.

SOLVED PROBLEMS

10.1. Determine the critical load of a long, slender bar clamped at each end and subject to axial thrust as shown in Fig. 10-6.

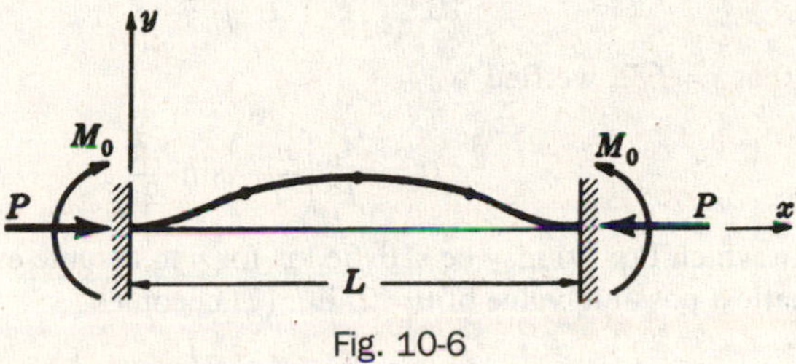

Fig. 10-6

SOLUTION: Let us introduce the x–y coordinate system shown in Fig. 10-6 and let (x, y) represent the coordinates of an arbitrary point on the bar. The bending moment at this point is found as the sum of the moments of the forces to the left of this section about an axis through this point and perpendicular to the plane of the page. Hence at this point we have $M = -Py + M_0$. The differential equation for the bending of the bar is then $EId^2y/dx^2 = -Py + M_0$, or

$$\frac{d^2y}{dx^2} + \frac{P}{EI}y = \frac{M_0}{EI} \tag{1}$$

As discussed in texts on differential equations, the solution to Eq. (1) consists of two parts. The first part is merely the solution of the so-called homogeneous equation obtained by setting the right-hand side of Eq. (1) equal to zero. We must then solve the equation

$$\frac{d^2y}{dx^2} + \frac{P}{EI}y = 0 \tag{2}$$

But the solution to this equation has already been found [see Eq. (10.6)] to be, with $k^2 = P/EI$,

$$y = C\cos kx + D\sin kx \tag{3}$$

The second part of the solution of Eq. (1) is given by a particular solution, i.e., any function satisfying Eq. (1). Evidently one such function is given by

$$y = \frac{M_0}{P} \tag{4}$$

The general solution of Eq. (1) is given by the sum of the solutions represented by Eqs. (3) and (4), or

$$y = C\cos kx + D\sin kx + \frac{M_0}{P} \tag{5}$$

Consequently,

$$\frac{dy}{dx} = -Ck\sin kx + Dk\cos kx \tag{6}$$

At the left end of the bar we have $y = 0$ when $x = 0$. Substituting these values in Eq. (5), we find $0 = C + M_0/P$. Also, at the left end of the bar we have $dy/dx = 0$ when $x = 0$; substituting in Eq. (6), we obtain

$$0 = 0 + Dk \qquad \therefore D = 0$$

At the right end of the bar we have $dy/dx = 0$ when $x = L$; substituting in Eq. (6), with $D = 0$, we find

$$0 = -Ck \sin kL$$

But $C = -M_0/P$ and since this ratio is not zero, then $\sin kL = 0$. This occurs only when $kL = n\pi$ where $n = 1, 2, 3, \ldots$. Consequently, substituting $k = \sqrt{P/EI}$, we have

$$P_{cr} = \frac{n^2\pi^2 EI}{L^2} \qquad n = 1, 2, 3 \ldots \tag{7}$$

For the so-called first mode of buckling illustrated in Fig. 10-6, the deflection curve of the bent bar has a horizontal tangent at $x = L/2$; that is, $dy/dx = 0$ there. Equation (6) now takes the form

$$\frac{dy}{dx} = \frac{M_0}{P}\left(\frac{n\pi}{L}\right)\sin\frac{n\pi x}{L} \tag{8}$$

and since $dy/dx = 0$ at $x = L/2$, we find

$$0 = \frac{M_0}{P}\left(\frac{n\pi}{L}\right)\sin\frac{n\pi}{2} \tag{9}$$

The only manner in which Eq. (9) may be satisfied is for n to assume even values; that is, $n = 2, 4, 6, \ldots$. Thus for the smallest possible value of $n = 2$, Eq. (7) becomes

$$P_{cr} = \frac{4\pi^2 EI}{L^2}$$

10.2. Determine the critical load for a long slender column clamped at one end, free at the other, and loaded by an axial compressive force applied at the free end.

SOLUTION: The critical load is that axial compressive force P that is just sufficient to keep the bar in a slightly deformed configuration, as shown in Fig. 10-7. The moment M_0 represents the effect of the support in preventing any angular rotation of the left end of the bar.

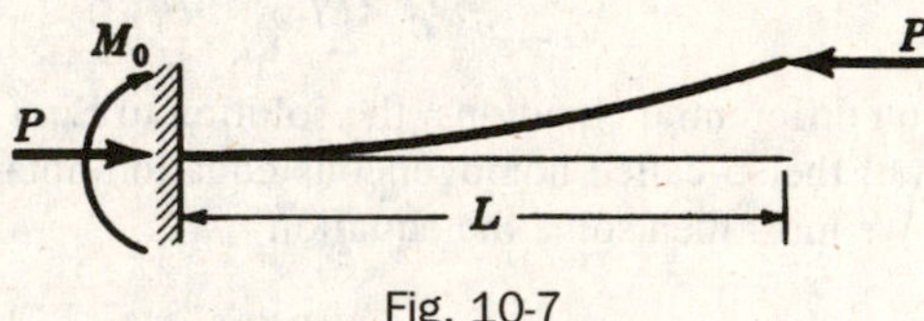

Fig. 10-7

Inspection of the above deflection curve for the buckled column indicates that the entire bar corresponds to one-half of the deflected pin-ended bar discussed in Section 10.2. Thus for the column under consideration, the length L corresponds to $L/2$ for the pin-ended column. Hence the critical load for the present column may be found from Eq. (10.10), by replacing L by $2L$. This yields

$$P_{cr} = \frac{\pi^2 EI}{(2L)^2} = \frac{\pi^2 EI}{4L^2}$$

10.3. A steel bar of rectangular cross section 40 mm × 50 mm and pinned at each end is subject to axial compression. If the proportional limit of the material is 230 MPa and $E = 200$ GPa, determine the minimum length for which Euler's equation may be used to determine the buckling load.

SOLUTION: The minimum second moment is $I = \frac{1}{12}bh^3 = \frac{1}{12}(50)(40)^3 = 2.67 \times 10^5\,\text{mm}^4$. Hence the least radius of gyration is

$$r = \sqrt{\frac{I}{A}} = \sqrt{\frac{2.67 \times 10^5}{(40)(50)}} = 11.5\,\text{mm} \qquad \text{or} \qquad 0.0115\,\text{m}$$

The axial stress for such an axially loaded bar was found in Eq. (10.18) to be

$$\sigma_{cr} = \frac{\pi^2 E}{(L/r)^2}$$

The minimum length for which Euler's equation may be applied is found by placing the critical stress in the above formula equal to 230 MPa. Doing this, we obtain

$$230 \times 10^6 = \frac{\pi^2 (200 \times 10^9)}{(L/0.0115)^2} \quad \text{or} \quad L = 1.135 \text{ m}$$

10.4. Consider again a rectangular steel bar 40 mm × 50 mm in cross section, pinned at each end and subject to axial compression. The bar is 2 m long and $E = 200$ GPa. Determine the buckling load using Euler's formula.

SOLUTION: The minimum second moment of this cross section was found in Problem 10.3 to be 2.67×10^5 mm^4. Applying the expression for buckling load given in Eq. (10.10), we find

$$P_{cr} = \frac{\pi^2 EI}{L^2} = \frac{\pi^2 (200 \times 10^9)(2.67 \times 10^5) \times 10^{-12}}{2^2} = 132 \times 10^3 \text{ N} \quad \text{or} \quad 132 \text{ kN}$$

The axial stress corresponding to this load is

$$\sigma_{cr} = \frac{P_{cr}}{A} = \frac{132 \times 10^3}{0.04 \times 0.05} = 66 \times 10^6 \text{ Pa} \quad \text{or} \quad 66 \text{ MPa}$$

10.5. Determine the critical load for a W203 × 28 section acting as a pinned end column. The bar is 4 m long and $E = 200$ GPa. Use Euler's theory.

SOLUTION: From Table 7-1 we find the minimum moment of inertia to be 3.28×10^6 mm^4. Thus,

$$P_{cr} = \frac{\pi^2 EI}{L^2}$$

$$= \frac{\pi^2 (200 \times 10^9)(3.28 \times 10^6) \times 10^{-12}}{4^2} = 405 \times 10^3 \text{ N} \quad \text{or} \quad 405 \text{ kN}$$

10.6. A long thin bar of length L and rigidity EI is pinned at end A, and at end B rotation is resisted by a restoring moment of magnitude λ per radian of rotation at that end. Derive the equation for the axial buckling load P. Neither A nor B can displace in the y-direction, but A is free to approach B.

SOLUTION: The buckled bar is shown in Fig. 10-8, where M_L represents the restoring moment. The differential equation of the buckled bar is

$$EI \frac{d^2 y}{dx^2} = Vx - Py$$

or

$$\frac{d^2 y}{dx^2} + \frac{P}{EI} y = \frac{V}{EI} x$$

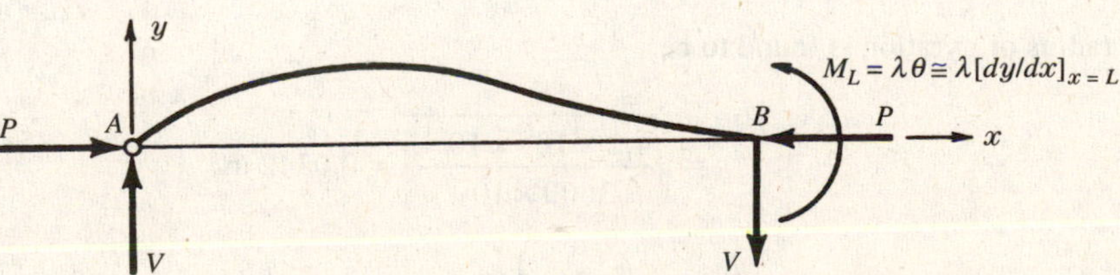

Fig. 10-8

Let $k^2 = P/EI$. Then

$$\frac{d^2y}{dx^2} + k^2 y = \frac{V}{EI} x$$

The general solution of this equation is the homogeneous solution plus the particular solution:

$$y = C \sin Lx + D \cos kx + \frac{V}{P} x \tag{1}$$

As the first boundary condition, when $x = 0$, $y = 0$; hence $D = 0$. As the second boundary condition, when $x = L$, $y = 0$; hence from Eq. (1) we obtain

$$0 = C \sin kL + \frac{VL}{P} \qquad \text{or} \qquad \frac{V}{P} = -\frac{C}{L} \sin kL$$

Thus

$$y = C\left[\sin kx - \frac{x}{L} \sin kL\right] \tag{2}$$

From Eq. (2) the slope at $x = L$ is found to be

$$\left[\frac{dy}{dx}\right]_{x=L} = C\left[k \cos kL - \frac{1}{L} \sin kL\right] \tag{3}$$

The restoring moment at end B is thus

$$M_L = C\lambda\left[k \cos kL - \frac{1}{L} \sin kL\right] \tag{4}$$

Also, since in general $M = EI(d^2y/dx^2)$, from Eq. (2) we have

$$M_L = -Ck^2 EI \sin kL \tag{5}$$

Equating expressions in Eqs. (4) and (5) after carefully noting that as M_L increases, dy/dx at that point decreases (necessitating the insertion of a negative sign), we have

$$-Ck^2 EI \sin kL = -\left[C\lambda k \cos kL - \frac{C\lambda}{L} \sin kL\right] \tag{6}$$

Simplifying, the equation for determination of the buckling load P becomes

$$P = \lambda k \cot kL - \frac{\lambda}{L} \tag{7}$$

This equation would have to be solved numerically for specific values of EI, L, and λ.

10.7. Use the AISC design recommendation to determine the allowable axial load on a W203 × 28 section 4 m long. The ends are pinned, the yield point is 240 MPa, and $E = 200$ GPa.

SOLUTION: From Table 7-1 we have the properties of the cross section as

$$I_{min} = 4.2 \times 10^6 \text{mm}^4 \qquad A = 3600 \text{ mm}^2$$

The radius of gyration is found to be

$$r = \sqrt{\frac{4.2 \times 10^6 \times 10^{-12}}{3600 \times 10^{-6}}} = 0.0342 \text{ m}$$

Thus,

$$\frac{L}{r} = \frac{4}{0.0342} = 117.0$$

We have from the definition of C_c

$$C_c = \sqrt{\frac{2\pi^2 E}{\sigma_{yp}}} = \sqrt{\frac{2\pi^2(200\times 10^9)}{240\times 10^6}} = 128.3$$

For both ends pinned, $K = 1$ and thus $K(L/r) < C_c$ so that the allowable axial stress is given by Eq. (10.28) to be

$$\sigma_a = \frac{\left[1-\frac{(KL/r)^2}{2C_c^2}\right]\sigma_{yp}}{\frac{5}{3}+\frac{3(KL/r)}{8C_c}-\frac{(KL/r)^3}{8C_c^3}} = \frac{\left[1-\frac{117^2}{2\times 128.3^2}\right](240\times 10^6)}{\frac{5}{3}+\frac{3\times 117}{8\times 128.3}-\frac{117^3}{8\times 128.3^3}} = 73.2\times 10^6 \text{ Pa}$$

The allowable axial load is

$$P_a = (3600\times 10^{-6})(73.2\times 10^6) = 264\,000 \text{ N} \qquad \text{or} \qquad 264 \text{ kN}$$

10.8. Reconsider the column of Problem 10.7 but now with a length of 6 m. Use the AISC design recommendation to determine the allowable axial load. Both ends are pinned.

SOLUTION: Now we have $L/r = 6/0.0342 = 175.4$. Thus the increased length (in comparison to that of Problem 10.7) leads to

$$K\frac{L}{r}(= 175.4) > C_c(= 128.3)$$

so that we must compute the allowable axial stress from Eq. (10.29):

$$\sigma_a = \frac{12\pi^2 E}{23(KL/r)^2} = \frac{12\pi^2(200\times 10^9)}{23\times 175.4^2} = 33.5\times 10^6 \text{ Pa} \qquad \text{or} \qquad 33.5 \text{ MPa}$$

The allowable axial load is thus

$$P_a = (3600\times 10^{-6})(33.5\times 10^6) = 121\,000 \text{ N}$$

The increased length of the column reduces the allowable axial load by more than 50%.

10.9. Use the AISC recommendation to determine the allowable axial load on a W203 × 28 section 3 m long. The ends are pinned. The material yield point is 250 MPa and $E = 200$ GPa.

SOLUTION: From Table 7-1 we have the sectional properties as

$$I_{\min} = 3.28\times 10^6 \text{ mm}^4 \qquad A = 3600 \text{ mm}^2$$

The radius of gyration is found to be

$$r = \sqrt{\frac{I}{A}} = \sqrt{\frac{3.28\times 10^6}{3600}} = 30.18 \text{ mm} \qquad \therefore \frac{L}{r} = \frac{3000}{30.18} = 99.4$$

From the definition of C_c, we have

$$C_c = \sqrt{\frac{2\pi^2 E}{\sigma_{yp}}} = \sqrt{\frac{2\pi^2(200\times 10^9)}{250\times 10^6}} = 125.7$$

For both ends pinned, $K = 1$ and thus $K(L/r) < C_c$ so that the allowable axial stress is given by Eq. (10.28) to be

$$\sigma_a = \frac{\left[1 - \dfrac{(KL/r)^2}{2C_c^2}\right]\sigma_{yp}}{\dfrac{5}{3} + \dfrac{3(KL/r)}{8C_c} - \dfrac{(KL/r)^3}{8C_c^3}} = \frac{\left[1 - \dfrac{(99.4)^2}{2(125.7)^2}\right]250 \times 10^6}{\dfrac{5}{3} + \dfrac{3(99.4)}{8(125.7)} - \dfrac{(99.4)^3}{8(125.7)^3}} = 90.35 \times 10^6 \text{ Pa}$$

The allowable axial load is

$$P = (3600 \times 10^{-6})(90.35 \times 10^6) = 325\,000 \text{ N} \qquad \text{or} \qquad 325 \text{ kN}$$

10.10. Reconsider the column of Problem 10.7 but now use the SSRC recommendation presented in Eq. (10.30) to estimate the maximum load-carrying capacity of the column.

SOLUTION: We must first compute the parameter

$$\lambda = \frac{KL}{r} \cdot \frac{1}{\pi}\sqrt{\frac{\sigma_{yp}}{E}} = \frac{1 \times 4}{0.0342\pi}\sqrt{\frac{240 \times 10^6}{200 \times 10^4}} = 1.29$$

For this value of λ, we determine the ultimate (peak) axial stress in the column from the semiempirical relation, Eq. (10.30), to be

$$\sigma_u = \sigma_{yp}\left[-0.111 + \frac{0.636}{\lambda} + \frac{0.087}{\lambda^2}\right]$$

$$= 240 \times 10^6\left[-0.111 + \frac{0.636}{1.29} + \frac{0.087}{1.29^2}\right] = 104 \times 10^6 \text{ Pa}$$

The axial load corresponding to this stress is

$$P_{max} = (3600 \times 10^{-6})(104 \times 10^6) = 374\,000 \text{ N}$$

This load represents the average of actual test values of peak loads that columns of this type were found to carry. It is to be noted that no safety factor is incorporated into these computations, so that the design load for this member is less than the 374 kN.

10.11. Select a wide-flange section from Table 7-1 to carry an axial compressive load of 750 kN. The column is 3.5 m long with a yield point of 250 MPa and a modulus of 200 GPa. Use the AISC specifications. The bar is pinned at each end.

SOLUTION: To get a first approximation, let us merely use $P = A\sigma$, from which we have

$$A = \frac{750000}{250 \times 10^6} = 0.0030 \text{ m}^2 \qquad \text{or} \qquad 3000 \text{ mm}^2$$

This tells us that any wide-flange section having an area smaller than 3000 mm^2 is unacceptable.
Next, let us try the W203 × 28 section. From Table 7-1 we find area = 3600 mm^2 and $I_{min} = 3.28 \times 10^6$ mm^4. The minimum radius of gyration is thus

$$r = \sqrt{\frac{I}{A}} = \sqrt{\frac{3.28 \times 10^6}{3600}} = 30.2 \text{ mm} \qquad \therefore \frac{L}{r} = \frac{3500}{30.2} = 116$$

From the definition of C_c, we have

$$C_c = \sqrt{\frac{2\pi^2(200 \times 10^9)}{250 \times 10^6}} = 125.6$$

Thus, since $K = 1$ for both ends pinned,

$$K\frac{L}{r}(=116) < C_c(=125.6)$$

So, using the appropriate Eq. (10.28), there results

$$\sigma_a = \frac{\left[1 - \dfrac{(116)^2}{2(125.6)^2}\right]250 \times 10^6}{\left[\dfrac{5}{3} + \dfrac{3(116)}{8(125.6)} - \dfrac{(116)^3}{8(125.6)^3}\right]} = 74.95 \times 10^6 \text{ Pa}$$

from which $P_a = (3600 \times 10^6)(74.95 \times 10^6) = 270\,000$ N or 270 kN

which indicates that this is far too light a section, since P_a is much less than 750 kN.

Next, let us try the section W254 × 72 having an area of 9280 mm^2 and $I_{min} = 38.6 \times 10^6$ mm^4. The minimum radius of gyration is found to be

$$r = \sqrt{\frac{38.6 \times 10^6}{9280}} = 64.5 \text{ mm} \qquad \therefore \frac{L}{r} = \frac{3500}{64.5} = 54.26$$

Again we have

$$K\frac{L}{R}(=54.26) < C_c(=125.6)$$

so that we again refer to Eq. (10.28) to find the allowable stress:

$$\sigma_a = \frac{\left[1 - \dfrac{(54.26)^2}{2(125.6)^2}\right]250 \times 10^6}{\left[\dfrac{5}{3} + \dfrac{3(54.26)}{8(125.6)} - \dfrac{(54.26)^3}{8(125.6)^3}\right]} = 124.6 \times 10^6 \text{ Pa}$$

for which $P_a = (9280 \times 10^{-6})(124.6 \times 10^6) = 1.15 \times 10^6$ N or 1150 kN

This section is rather heavy, so let us investigate the W254 × 54 section. Here, the area is 7010 mm^2 and $I_{min} = 17.5 \times 10^6$ mm^4. So, the minimum radius of gyration is found to be 50.0 mm and the slenderness ratio is 3500/50 = 70. Again using Eq. (10.28) we find $\sigma_a = 114$ MPa, from which the allowable load is $P_a = 799$ kN.

Investigation of the next lighter section, W254 × 43, by the above method indicates that it can carry only 478 kN.

Thus, the desired section is the W254 × 54, which can carry an axial load of 799 kN, which is in excess of the 750 kN required. A more complete table of structural shapes might well indicate a slightly lighter section than the W254 × 54.

10.12. Determine the deflection curve of a pin-ended bar subject to combined axial compression P together with a uniform normal loading as shown in Fig. 10-9.

SOLUTION: One convenient coordinate system to designate points on the deflected bar is shown in Fig. 10-9. There, the origin is situated at the point of maximum deflection Δ. The bending moment at an arbitrary point (x, y) on the deflected bar is written most easily as the sum of the moment of all forces to the *right* of (x, y).

$$M = P(\Delta - y) + \frac{wL}{2}\left(\frac{L}{2} - x\right) - \frac{w}{2}\left(\frac{L}{2} - x\right)^2 \tag{1}$$

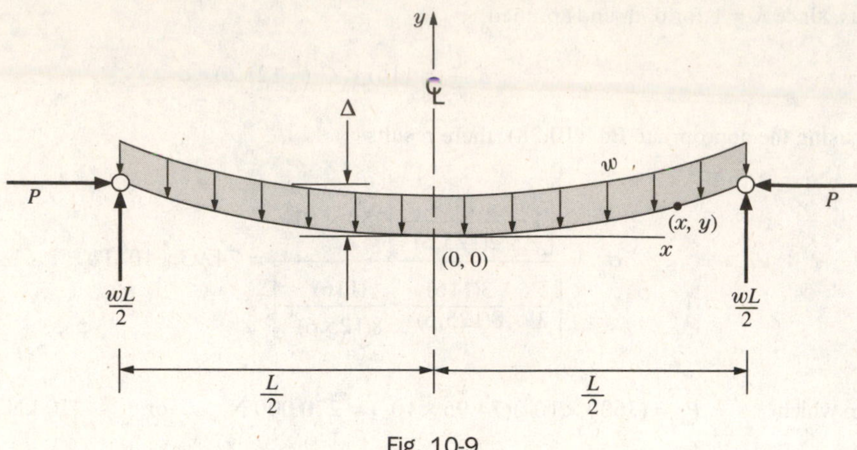

Fig. 10-9

so that the differential equation of the deflected bar is

$$EI\frac{d^2y}{dx^2} = P\Delta - Py + \frac{w}{2}\left(\frac{L^2}{4} - x^2\right) \tag{2}$$

If we introduce the notation $k^2 = P/EI$ the nonhomogeneous differential equation of the bar is

$$\frac{d^2y}{dx^2} + k^2y = \frac{w}{2EI}\left(\frac{L^2}{4} - x^2\right) + k^2\Delta$$

The solution is given as the sum of the solution of the corresponding homogeneous equation, and any particular solution of the nonhomogeneous equation:

$$y(x) = C\cos kx + D\sin kx + \frac{w}{2P}\left(\frac{L^2}{4} - x^2\right) + \frac{w}{k^2P} + \Delta$$

where C and D are constants of integration. These are easily found by realizing that, because of symmetry of the bent bar, the deflection is Δ at $x = L/2$ and also the bar has a horizontal tangent at $x = 0$. This leads to

$$y(x) = \Delta + \frac{w}{k^2P}\left[\left(\sec\frac{kL}{2}\cos kx - 1\right) + k^2\left(\frac{L^2}{8} - \frac{x^2}{2}\right)\right]$$

as the solution of Eq. (2). The peak deflection occurs at the midpoint of the bar (the origin of our coordinate system) and is given by

$$\Delta = \frac{w}{k^2P}\left[\left(\sec\frac{kL}{2} - 1\right) + \frac{k^2L^2}{8}\right]$$

SUPPLEMENTARY PROBLEMS

10.13. A 2-cm-diameter, 120-cm-long steel column with pinned ends supports a load P. Calculate the slenderness ratio and Euler's buckling load. *Ans.* 240, 10.8 kN

10.14. A wooden 2.44-m-long, 5 cm × 10 cm column is pin ended. Calculate the slenderness ratio and the Euler buckling load. Use $E = 12.7$ GPa. *Ans.* 169, 21.9 kN

10.15. The steel column of Problem 10.13 has clamped ends. What is the percentage increase in the Euler buckling load? *Ans.* 400%

10.16. Estimate the slenderness ratio and the Euler buckling load of a wooden yard stick (2.8 cm × 4 mm) if it has pinned ends. Use $E = 12$ GPa. *Ans.* 798, 21.1 N

10.17. A hollow steel bar with pinned ends has diameters of 12 cm and 10 cm. If the buckling load is 50 kN, estimate the length of the bar and calculate the slenderness ratio. Use $E = 200$ GPa. *Ans.* 14.4 m, 370

10.18. If the yield strength of the steel in Problem 10.17 is 900 MPa, calculate the smallest slenderness ratio or which Euler's equation is applicable. See Fig. 10-4. *Ans.* 46.8

10.19. A steel bar of solid circular cross section is 50 mm in diameter. The bar is pinned at each end and subject to axial compression. If the proportional limit of the material is 210 MPa and $E = 200$ GPa, determine the minimum length for which Euler's formula is valid. Also, determine the value of the Euler buckling load if the column has this minimum length. *Ans.* 1.21 m, 412 kN

10.20. The column shown in Fig. 10-10 is pinned at both ends and is free to expand into the opening at the upper end. The bar is steel, is 25 mm in diameter, and occupies the position shown at 16°C. Determine the temperature to which the column may be heated before it will buckle. Take $\alpha = 12 \times 10^{-6}$/°C and $E = 200$ GPa. Neglect the weight of the column. *Ans.* 29.3 °C

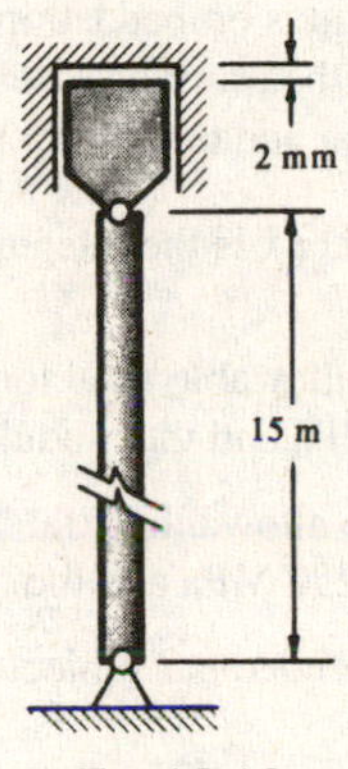

Fig. 10-10

10.21. A long slender bar AB is clamped at A and supported at B in such a way that transverse displacement is impossible as in Fig. 10-11, but the end of the bar at B is capable of rotating about B. Determine the differential equation governing the buckled shape of the bar. *Ans.* $\tan kL = kL$ where $k^2 = P/EI$

Fig. 10-11

10.22. A bar of length L is clamped at its lower end and subject to both vertical and horizontal forces at the upper end, as shown in Fig. 10-12. The vertical force P is equal to one-fourth of the Euler load for this bar. Determine the lateral displacement of the upper end of the bar. *Ans.* $16(4 - \pi)RL^3/\pi^3 EI$

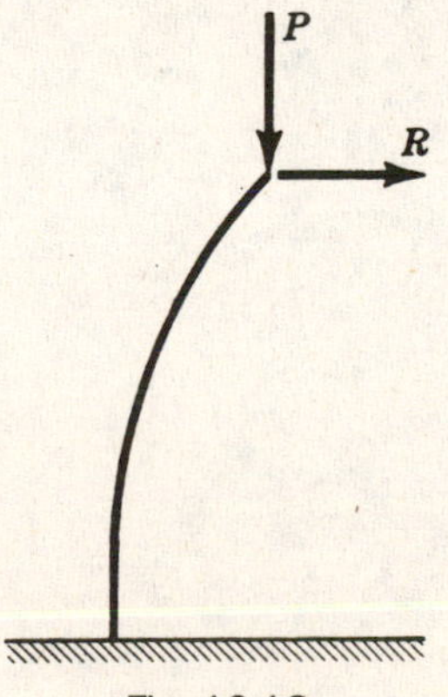

Fig. 10-12

10.23. A bar of length L and flexural rigidity EI has pinned ends. An axial compressive force of $P = \pi EI/4L^2$ is applied to the beam and a bending moment M is applied at one end. Determine the rotational stiffness, i.e., applied moment per radian of rotation at that end of the bar. Rework the problem for the case of an axial tensile force of the same numerical value.

Ans. $\dfrac{2.47EI}{L}, \dfrac{3.47EI}{L}$

10.24. An initially straight bar AC is pinned at each end and supported at the midpoint B by a spring which resists any lateral movement δ of B with a lateral force $(kEI/L^3)\ \delta$. The bar is of length $2L$ and least flexural rigidity EI. Equal and opposite thrusts P are applied at the end C as well as at the centroid of the bar at B. In any deflected form the line of action of the thrust applied at B remains parallel to the chord AC. Determine the minimum buckling load of the system.

Ans. $P_{cr} = \beta^2 \dfrac{EI}{L^2}$ where β is the smallest positive root of the equation

$$\frac{\beta}{\tan\beta} = \frac{3k + (9 + k)\beta^2 - \beta^4}{3(k - \beta^2)}$$

10.25. A long thin bar is pinned at each end and is embedded in an elastic packing which exerts a transverse force on the bar when it deflects laterally. When the transverse deflection at any point is given by y, the packing exerts a transverse force per unit length of the bar equal to ky. Determine the axial force required to buckle the bar.

Ans. $P_{cr} = \dfrac{\pi^2 EI}{L^2}\left(k^2 + \dfrac{kL^4}{k^2\pi^4 EI}\right)$ where k is the integer for which P_{cr} is minimum

10.26. Use the AISC formula to determine the allowable axial load on a W254 × 54 end-pinned column that is 4 m long. The yield point of the material is 225 MPa and the modulus is 200 GPa. *Ans.* 698 kN

10.27. Use the AISC formula to determine the allowable axial load on a W254 × 79 end-pinned column that is 14 m long. The yield point of the material is 250 MPa and the modulus is 200 GPa. *Ans.* 226 kN

10.28. Determine the deflection curve of a pin-ended bar subject to axial compression together with a central transverse force as shown in Fig. 10-13.

Ans. $y = \dfrac{Q \sin kx}{2Pn \cos \dfrac{kL}{2}} - \dfrac{Q}{2P}x$ where $k = \sqrt{\dfrac{P}{EI}}$

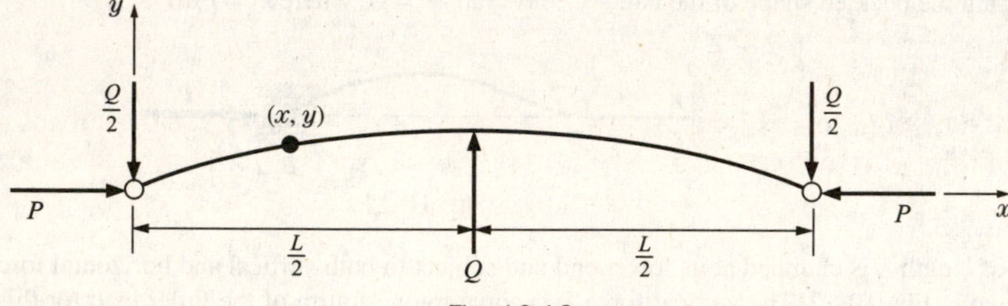

Fig. 10-13

CHAPTER 11

Fatigue

Engineers have long recognized that subjecting a metallic member to a large number of stress cycles will result in fracture of the member at stresses much lower than those required for failure under static conditions. The term *fatigue* is used to describe this failure mode due to repeated cycles of stress. Modern investigators would probably use a term such as *progressive fracture* instead of the term *fatigue*.

On the basis of this testing, two very interesting conclusions have been reached:

1. The number of cycles of stress rather than elapsed time is significant.
2. Ferrous materials stressed below a certain limiting value can withstand an indefinite number of stress cycles without failure.

The literature of fatigue studies is voluminous. These studies have proceeded along two lines: fundamental research seeking to explain the fatigue phenomenon and empirical investigations to provide information for practical analysis and design. Recent progress in *crack propagation* studies and the use of the electron microscope have enhanced fundamental research efforts, but empirical methods will continue to be widely employed for engineering purposes.

Fatigue fractures originate at points of stress concentrations, such as fillets, keyways, holes, and screw threads, or at points of internal inclusions or defects in the material. Cracks begin at these locations and then propagate through the cross section until the remaining uncracked regions are insufficient to resist the applied forces and fracture occurs suddenly.

The design of structural and machine components, where the loading is either fluctuating or repeated, must take fatigue action into consideration. For example, when traffic passes over a bridge, the structural components of the bridge experience loads that fluctuate above and below the dead loads that already exist. Also, the crankshaft of an automobile is subjected to repeated loading every time the automobile is driven.

An illustration of the type of fracture surface experienced in fatigue failures is shown in Fig. 11-1, which represents the fatigue failure of a steering shaft of a late-model automobile. The fatigue fracture originated on the outside surface of the shaft (see bottom of the figure), probably as a result of stress concentration at the root of the notch, and extended slowly almost through the entire cross section before final failure occurred (toward the top of the figure). Note the *beach marks* which are more or less concentric with the final rupture zone. These beach marks represent the locations where the propagating crack was briefly arrested and then again began to propagate.

Depending upon the specific need, fatigue testing may be performed on specimens subjected to axial, torsional bending, or combined loads. We will focus our discussion on the fatigue of members subjected to bending loads. A rotating beam fatigue testing machine is shown schematically in Fig. 11-2(*a*). The rotating specimen is subjected to a constant bending moment over the gage length, as illustrated by the moment diagram of Fig. 11-2(*b*). There are several important variables related to fatigue test specimens that include size, shape, method of fabrication, and surface finish. Thus, to minimize the effects of stress concentration, the surface of the specimen is polished and the geometry is chosen to provide a gradual change in cross-sectional dimension on either side of the *critical section* located at its center. Care is also taken to remove, by annealing, any residual stresses induced in the specimen during fabrication.

In a given test, the choice of the weight W and the spacing a of the machine bearings determines the bending moment applied to the specimen as it rotates at a constant angular speed ω. The critical cross section of

Fig. 11-1 A cross section showing fatigue failure.

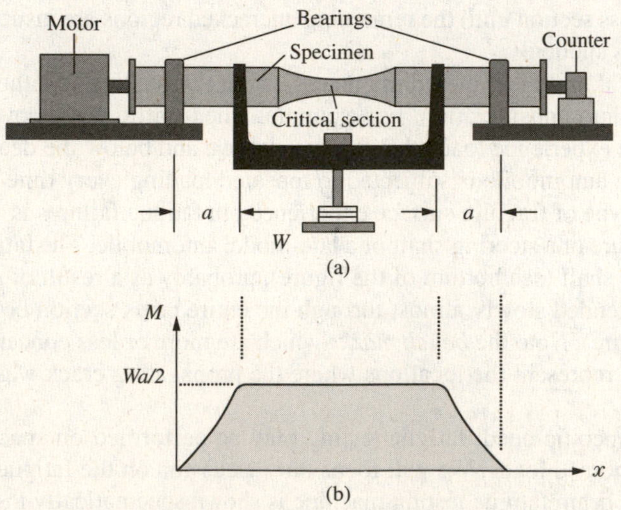

Fig. 11-2 A fatigue testing machine.

the specimen is shown in Fig. 11-3(a). At any point P on its circumference, the distance y perpendicular to the neutral axis (the z-axis), varies with time according to the relation $y = R \sin\omega t$. Therefore, as the specimen rotates with the angular speed ω, the bending stress at any point P varies according to the relation

$$\sigma = \frac{My}{I} = \frac{(Wa/2)R\sin\omega t}{\pi R^4/4} = \frac{2Wa}{\pi R^3}\sin\omega t \qquad \sigma = \sigma_R \sin\omega t \tag{11.1}$$

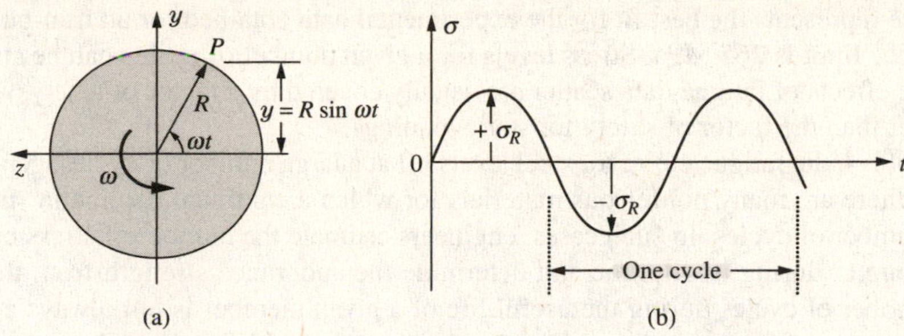

Fig. 11-3 (*a*) The cross section with (*b*) sinusoidal loading.

where $\sigma_R = 2Wa/\pi R^3$ is the amplitude of the stress which varies sinusoidally with time according the relation expressed in Eq. (11.1). A sketch of this equation is given in Fig. 11-3(*b*). This variation between maximum tension ($+\sigma_R$) and maximum compression ($-\sigma_R$) is known as *complete stress reversal* and is the most severe form of stress variation. Provided that the stress σ_R is not set too low, the specimen will experience fatigue failure after a certain number of cycles N that depends upon the applied stress level of σ_R. The testing machine stops automatically when the specimen fractures after a number N of cycles. This type of test is repeated on a large number of specimens with different levels of applied stress and the results are plotted on a σ_R versus N diagram. Actually, because of the large numbers involved, the test results are plotted on log-log paper to create a $\log\sigma_R$ versus $\log N$ diagram, as shown for a cast-alloy steel material in Fig. 11-4.

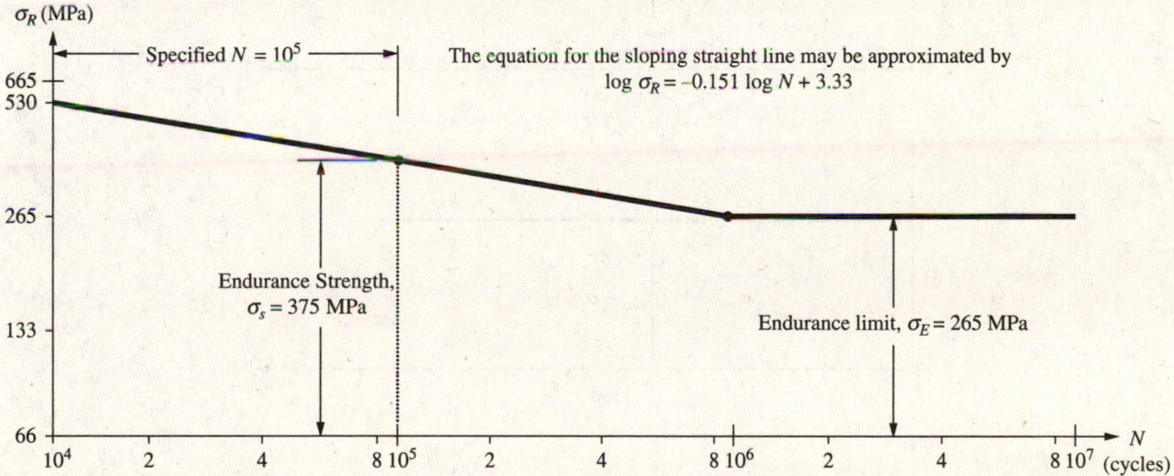

Fig. 11-4 The diagram $\log \sigma_R$ versus $\log N$ for a cast alloy steel under completely reversed bending.

Information obtained from diagrams such as that shown in Fig. 11-4, forms the basis for the analysis and design of fatigue loadings. Fatigue testing machines are also available for axial, torsional, and combined loadings; appropriate fatigue diagrams may be constructed for these other loadings. Let us refer now to the diagram in Fig. 11-4 in order to introduce two definitions related to fatigue phenomenon.

Endurance Limit, σ_E: The *endurance limit* is the maximum stress that can be completely reversed indefinitely number of times without producing fatigue failure. For the material of Fig. 11-4, the endurance limit is 265 MPa.

Endurance Strength, σ_S: The *endurance strength* is the fatigue strength corresponding to a specified number of cycles. For example, for the material in Fig. 11-4, if the specified number of cycles is 10^5, the fatigue strength is about 375 MPa.

Experimental results from fatigue tests do not lay perfectly on a straight line as may be erroneously surmised from Fig. 11-4, but rather they scatter within a relatively narrow band above and below the solid line. This solid line represents the best fit for the experimental data obtained for an iron-base super alloy for which the endurance limit is 265 MPa. Stress levels for a given number of cycles can be stated with reasonable accuracy. The effects of fatigue data scatter are usually covered by a factor of safety, which is generally considerably larger than the factor of safety for static loadings.

As seen in Fig. 11-4, the fatigue curve for steel levels off at a large number of cycles, generally around 10^6 cycles. However, there are many nonferrous materials for which a continual decline in stress levels occurs with increasing number of cycles. In such cases, engineers estimate the number of stress cycles that a given member will encounter during its lifetime and determine the endurance strength from the fatigue curves. Estimating the number of cycles during the useful life of a given member is not always a simple task; it is further complicated when a member is subjected to high-frequency vibrations during which the number of cycles can increase significantly in a very short time. Fatigue failures under such a condition have been blamed for a number of jet and helicopter crashes.

In certain applications, the stress level varies sinusoidally with time between a maximum tensile value σ_{max} and a minimum tensile value σ_{min} as shown in Fig. 11-5*a*. This stress variation may be decomposed into a static component σ_{ave}, shown in Fig. 11-5*b*, and a completely reversed sinusoidal variation σ_R, shown in Fig. 11-5*c*. This type of decomposition is desirable because a large portion of the experimental fatigue data has been obtained for completely reversed cycling.

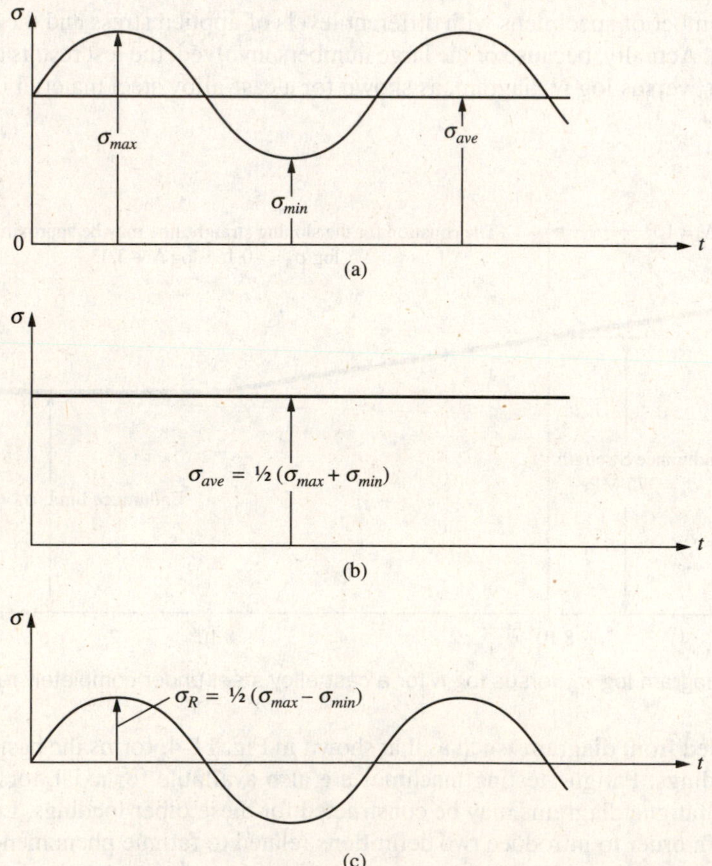

Fig. 11-5 Decomposition of (*a*) a time varying stress into (*b*) an average stress and (*c*) a sinusoidal stress.

The relationship between the static component and the completely reversed component of Fig. 11-5 forms the basis for the three theories of fatigue failure shown in Fig. 11-6, the *Gerber Parabola*, the *Goodman Straight Line*, and the *Soderberg Straight Line*, are drawn in an attempt to describe the experimental fatigue behavior under combined static and completely reversed stresses. Note that when the static component is absent, all three theories converge at one point. The mathematical expressions for the three fatigue theories just stated are

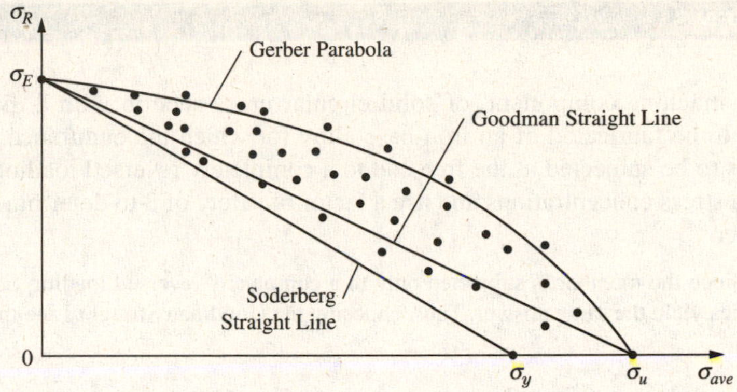

Fig. 11-6 Three theories of fatigue failure.

Gerber Parabola: The Gerber Parabola may be stated in the following dimensionless mathematical form:

$$\frac{\sigma_R}{\sigma_E}=1-\left(\frac{\sigma_{ave}}{\sigma_u}\right)^2 \tag{11.2}$$

In practice, for design purposes, a factor of safety (*FS*) is added to Eq. (11.2). This leads to

$$\frac{\sigma_R}{\sigma_E/FS}=1-\left(\frac{\sigma_{ave}}{\sigma_u/FS}\right)^2 \tag{11.3}$$

It should be noted here that the Gerber Parabola has not been widely used in American engineering practice because it is not conservative with respect to a great deal of the experimental data.

Goodman Straight Line: A dimensionless mathematical expression for the Goodman Straight Line may be stated as follows:

$$\frac{\sigma_{ave}}{\sigma_u}+\frac{\sigma_R}{\sigma_E}=1 \tag{11.4}$$

As in the case of the Gerber Parabola, a factor of safety may be included for purposes of design. This yields,

$$\frac{\sigma_{ave}}{\sigma_u/FS}+\frac{\sigma_R}{\sigma_E/FS}=1 \tag{11.5}$$

The Goodman Straight Line theory has been widely used in American engineering practice.

Soderberg Straight Line: The mathematical expression for the Soderberg Straight Line is similar to that of the Goodman Straight Line. It is

$$\frac{\sigma_{ave}}{\sigma_y}+\frac{\sigma_R}{\sigma_E}=1 \tag{11.6}$$

When a factor of safety is added, Eq. (11.6) becomes,

$$\frac{\sigma_{ave}}{\sigma_y/FS}+\frac{\sigma_R}{\sigma_E/FS}=1 \tag{11.7}$$

This fatigue theory provides a very conservative design approach because, unlike the Goodman Straight Line theory, it is based on the yield strength σ_y, instead of the ultimate strength σ_u for the static component of stress.

SOLVED PROBLEMS

11.1. A cantilever machine component, of solid circular cross section with a diameter d and a length of 60 cm, is to be fabricated of an iron-base alloy for which the endurance limit is 360 MPa. The component is to be subjected at the free end to a completely reversed loading of 8800 N maximum value. Ignore stress concentrations and use a factor of safety of 3 to determine the required diameter of the member.

SOLUTION: Since the member is subjected only to a completely reversed loading (i.e., no static component), all three theories yield the same answer. Thus, choosing the Goodman Straight Line theory, Eq. (11.3) we have, with $\sigma_{ave} = 0$,

$$\frac{\sigma_R}{\sigma_E/FS} = 1 \qquad \therefore \sigma_R = \frac{\sigma_E}{FS} = \frac{360}{3} = 120 \text{ MPa}$$

Now,

$$\sigma_R = \frac{My}{I} \qquad 120\times10^6 (120\,000\,000) = \frac{8800(0.6)d/2}{\pi/d^{4/64}} \qquad \therefore d = 0.0765 \text{ m} \qquad \text{or} \qquad \cong 76 \text{ mm}$$

11.2. The beam shown is fabricated of a nickel alloy for which the endurance limit in bending is 289 MPa, the yield strength is 300 MPa, and the ultimate strength is 570 MPa. The thickness of the beam is uniform and equal to 50 mm. The stress concentration factor due to the fillet at B is 1.3. The applied load ranges from a downward load P to an upward load ½ P as shown. Use factors of safety of 4 with respect to fatigue, 3 with respect to the ultimate strength, and 2 with respect to yielding and determine the maximum permissible load P using (a) the Gerber Parabola, (b) the Goodman Straight Line, and (c) the Soderberg Straight Line.

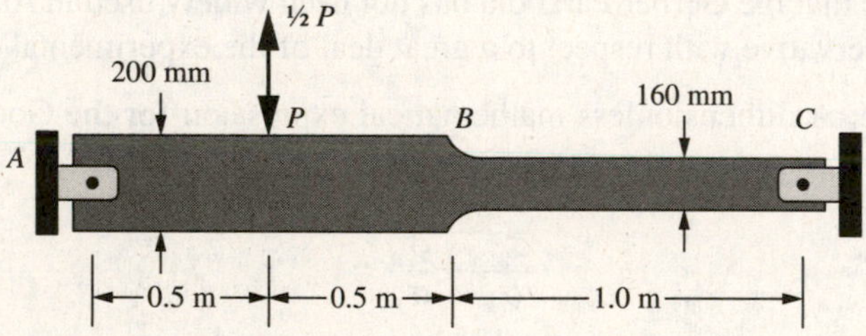

SOLUTION: The critical section is at B where the bending moment is $0.25P$. The maximum stress that occurs during the downward part of the cycle at this location is

$$\sigma_{max} = k\left(\frac{My}{I}\right) = 1.3\left[\frac{0.25P\times0.08}{0.05\times0.16^3/12}\right] = 1520\ P$$

Therefore, since the upward stroke is ½ P, it follows that

$$\sigma_{min} = -760P$$

The static component and the completely reversed component of the stress become

$$\sigma_{ave} = \frac{1}{2}(\sigma_{max} + \sigma_{min}) = 380P \qquad \sigma_R = \frac{1}{2}(\sigma_{max} - \sigma_{min}) = 1140\ P$$

(a) The Gerber Parabola containing a factor of safety is expressed by Eq. (11.3). Thus,

$$\frac{\sigma_R}{\sigma_E/FS} = 1 - \left(\frac{\sigma_{ave}}{\sigma_u/FS}\right)^2 \qquad \frac{1140P}{289\times10^6/4} = 1 - \left(\frac{380P}{570\times10^6/3}\right)^2$$

Solving this quadratic equation yields

$$P = 62.4 \text{ kN}$$

(b) The Goodman Straight Line containing a factor of safety is given by Eq. (11.5):

$$\frac{\sigma_{ave}}{\sigma_u/FS}+\frac{\sigma_R}{\sigma_E/FS}=1 \qquad \frac{380P}{570\times10^6/3}+\frac{1140P}{289\times10^6/4}=1$$

The solution of the above equation leads to

$$P=56.2\text{ kN}$$

(c) The Soderberg Straight Line with a factor of safety is expressed by Eq. (11.7):

$$\frac{\sigma_{ave}}{\sigma_y/FS}+\frac{\sigma_R}{\sigma_E/FS}=1 \qquad \frac{380P}{300\times10^6/2}+\frac{1140P}{289\times10^6/4}=1$$

Solving this linear equation yields

$$P=54.6\text{ kN}$$

11.3. A shaft of solid circular cross section is to be fabricated of a material for which the torsion ultimate strength is 600 MPa and the torsion endurance limit is 160 MPa. It is to carry an alternating torque varying between a maximum of 6000 N · m and a minimum of 1500 N · m. A small keyway in the shaft provides for a stress concentration factor of 1.3 but is sufficiently small to permit the use of the gross cross-sectional properties in the calculations. Use factors of safety of 4 relative to the endurance limit and 3 relative to the ultimate strength. Compute the least acceptable diameter of the shaft by the Goodman Straight Line theory.

SOLUTION: Since we are dealing with a torsion problem, the equations developed earlier are modified by replacing σ with τ. Thus, the maximum and minimum shearing stresses are computed as follows:

$$\tau_{max}=k\frac{T_{max}R}{J}=1.3\frac{6000(d/2)}{\pi d^4/32}=\frac{3.06\times10^4}{d^3}$$

$$\tau_{min}=\frac{\tau_{max}}{4}=\frac{7640}{d^3}$$

Therefore,

$$\tau_{ave}=\frac{1}{2}(\tau_{max}+\tau_{min})=\frac{1.912\times10^4}{d^3} \qquad \tau_R=\frac{1}{2}(\tau_{max}-\tau_{min})=\frac{1.148\times10^4}{d^3}$$

We now modify the Goodman Straight Line theory given by Eq. (11.5). Thus,

$$\frac{\tau_{ave}}{\tau_u/FS}+\frac{\tau_R}{\tau_E/FS}=1 \qquad \frac{1.912\times10^4}{d^3(600\times10^6/3)}+\frac{1.148\times10^4}{d^3(150\times10^6/4)}=1$$

The solution of this equation for the diameter d yields

$$d=0.0738\text{ m}\cong 74\text{ mm}$$

11.4. A rod of circular cross section, 65 mm in diameter, is to be subjected to alternating tensile forces that vary from a minimum of 160 kN to P_{max}. It is to be fabricated of a material with an ultimate tensile strength of 700 MPa and an endurance limit for complete stress reversal of 560 MPa. Use the Goodman Straight Line theory to find P_{max} if the factor of safety with respect to the ultimate strength is 3.0 and with respect to the endurance limit is 3.5. A small hole in the rod provides for a stress concentration factor of 1.5 but is sufficiently small so that the gross cross-sectional area may be used in the calculations.

SOLUTION: The maximum and minimum tensile stresses are found as follows:

$$\sigma_{max}=k\frac{P_{max}}{A}=1.5\times\frac{P_{max}}{\pi\times0.065^2/4}=452\ P_{max}$$

$$\sigma_{min}=k\frac{P_{min}}{A}=1.5\times\frac{160\times10^3}{\pi\times0.065^2/4}=72.3\times10^6\text{ Pa}$$

Therefore,

$$\sigma_{ave} = \frac{1}{2}(\sigma_{max} + \sigma_{min}) = 226P_{max} + 36.2\times10^6$$

$$\sigma_R = \frac{1}{2}(\sigma_{max} - \sigma_{min}) = 226P_{max} - 36.2\times10^6$$

We now apply the Goodman Straight Line theory given by Eq. (11.5). Thus,

$$\frac{\sigma_{ave}}{\sigma_u/F.S.} + \frac{\sigma_R}{\sigma_E/F.S.} = 1 \qquad \frac{226P_{max} + 36.2\times10^6}{700\times10^6/3.0} + \frac{226P_{max} - 36.2\times10^6}{560\times10^6/3.5} = 1$$

Solving, we obtain P_{max} as 450 000 N or

$$P_{max} = 450 \text{ kN}$$

SUPPLEMENTARY PROBLEMS

11.5. A machine component, with a square cross section ($b \times b$), is to be subjected to alternating axial loading varying from a minimum of 350 kN to a maximum of 1700 kN. Using the Goodman Straight Line theory with factors of safety of 4 with respect to fatigue and 3 with respect to the ultimate strength, determine the least acceptable value of b. The endurance limit for completely reversed axial fatigue is 400 MPa and the static ultimate strength is 860 MPa. Construct a scaled plot of the Goodman Straight Line showing the point that represents your solution. Ignore effects of stress concentration. *Ans.* 99 mm

11.6. A simply supported 4-m-long beam is to carry a downward alternating load, at 1.4 m from one support, which varies from 18 kN to 62 kN. It is to have a rectangular cross section with a depth d four times its width w. The following material properties are provided: endurance limit = 130 MPa and yield strength = 260 MPa. Using the Soderberg Straight Line theory with a factor of safety of 3 with respect to fatigue and 2 with respect to yielding, determine the least acceptable cross-sectional depth d. Ignore effects of stress concentration. *Ans.* 26 cm

11.7. A hollow shaft of circular cross section (inside diameter is 0.8 of outside diameter) is to be fabricated of a material for which the endurance limit is 165 MPa and the static ultimate strength is 620 MPa. It is to be subjected to an alternating torque that varies from a minimum of 1000 N · m to a maximum of 5000 N · m. Using the Goodman Straight Line theory with a factor of safety of 4 with respect to fatigue and 3 with respect to the ultimate strength, determine the least acceptable outside diameter of the hollow section. A small keyway provides for a stress-concentration factor of 1.25 but is sufficiently small to allow use of the gross sectional properties. *Ans.* 82 mm

11.8. A hollow shaft of circular cross section (inside diameter is 0.7 of outside diameter) is to be designed for a lifespan of 10^5 for which the material has an endurance strength of 230 MPa and a static torsional ultimate strength of 580 MPa. It is to be subjected to an alternating torque that varies from a maximum of 4400 N · m to a minimum of −1500 N · m. Ignore effects of stress concentration. Assume the Goodman Straight Line theory is valid and use it with a factor of safety of 3.5 with respect to fatigue and 2.8 with respect to the static ultimate strength to compute the least acceptable outside diameter of the hollow section. *Ans.* 70 mm

11.9. A 2-m-long cantilever beam is to be subjected at its free end to a downward alternating load that alternates from 1 kN to 12 kN. Use the Goodman Straight Line theory with a factor of safety of 4 relative to fatigue (endurance limit = 160 MPa) and 3 relative to the ultimate strength (ultimate strength = 440 MPa) to find the least acceptable depth of a rectangular cross section whose depth d is four times its width w. Ignore effects of stress concentration. *Ans.* 30 cm

11.10. An 3-m-long simply supported beam, of hollow circular cross section (inside diameter is 0.7 of outside diameter) is to be designed for a lifespan of 10^5 cycles and is to be made of the material described in Fig. 11.4 for which the static bending ultimate strength is 590 MPa. It is to be subjected at midspan to a downward alternating force that varies from a maximum of 44 kN to a minimum of 9 kN. Assume the Gerber Parabolic theory is valid and use it with a factor of safety of 3.0 with respect to fatigue and 2.0 with respect to the static ultimate strength to compute the least acceptable outside diameter of the hollow section. Ignore effects of stress concentration. *Ans.* 12.3 cm

11.12. A 4-m-long cantilever beam with a rectangular cross section (the depth d is three times the width w) is to be designed for a lifespan of 10^5 cycles and is to be made of the material described in Fig. 11.4 for which the static bending ultimate strength is 600 MPa. It is to be subjected at the free end to a downward alternating force that varies from a maximum of 20 kN to a minimum of 8 kN. A small hole in the beam provides for a stress-concentration factor of 1.3 but is small enough to allow the use of gross dimensions in the calculations. Using the Goodman Straight Line theory, with a factor of safety of 4 with respect to fatigue and 3 with respect to the static ultimate strength, compute the least acceptable depth d of the section. *Ans.* 23.2 cm

11.13. A 49-mm-diameter steel rod is subjected to an alternating axial load that varies from a maximum of P_{max} to a minimum of 90 kN. The steel of which the rod is fabricated has a static ultimate strength of 560 MPa and an endurance limit for complete reversal of tensile loading of 300 MPa. Use the Goodman Straight Line theory with factors of safety of 3 relative to fatigue and 2 relative to the static ultimate strength and determine P_{max}. Neglect effects of stress concentration. *Ans.* 320 kN

Tables from Various Chapters for Quick Reference

Table A.1. Conversion Factors

Quantity	Symbol	SI Units	English Units	To Convert from English to SI units Multiply by
Length	L	m	ft	0.3048
Mass	m	kg	lbm	0.4536
Time	t	s	sec	1
Area	A	m^2	ft^2	0.09290
Volume	V	m^3	ft^3	0.02832
Velocity	V	m/s	ft/sec	0.3048
Acceleration	a	m/s^2	ft/sec^2	0.3048
Angular velocity	ω	rad/s	rad/sec	1
		rad/s	rpm	9.55
Force, Weight	F, W	N	lbf	4.448
Density	ρ	kg/m^3	lbm/ft^3	16.02
Specific weight	γ	N/m^3	lbf/ft^3	157.1
Pressure, stress	P	kPa	psi	6.895
Work, Energy	W, E, U	J	ft-lbf	1.356
Power	W	W	ft-lbf/sec	1.356
		W	hp	746

Table A.2. Prefixes for SI Units

Multiplication Factor	Prefix	Symbol
10^{12}	tera	T
10^{9}	giga	G
10^{6}	mega	M
10^{3}	kilo	k
10^{-2}	centi*	c
10^{-3}	mili	m
10^{-6}	micro	μ
10^{-9}	nano	n
10^{-12}	pico	p

*Discouraged except in cm, cm^2, cm^3, or cm^4.

Table A.3. Properties of Common Engineering Materials at 20°C (68°F)

Material	Specific weight		Young's modulus		Ultimate stress		Coefficient of linear thermal expansion		Poisson's ratio
	lb/in^3	kN/m^3	lb/in^2	GPa	lb/in^2	kPa	10*e*–6/°F	10*e*–6/°C	
I. Metals in slab, bar, or block form									
Aluminum alloy	0.0984	27	10–12*e*6	70–79	45–80*e*3	310–550	13	23	0.33
Brass	0.307	84	14–16*e*6	96–110	43–85*e*3	300–590	11	20	0.34
Copper	0.322	87	16–18*e*6	112–120	33–55*e*3	230–380	9.5	17	0.33
Nickel	0.318	87	30*e*6	210	45–110*e*3	310–760	7.2	13	0.31
Steel	0.283	77	28–30*e*6	195–210	80–200*e*3	550–1400	6.5	12	0.30
Titanium alloy	0.162	44	15–17*e*6	105–120	130–140*e*3	900–970	4.5–5.5	8–10	0.33
II. Nonmetallics in slab, bar, or block form									
Concrete (composite)	0.0868	24	3.6*e*6	25	4000–6000	28–41	6	11	
Glass	0.0955	26	7–12*e*6	48–83	10000	70	3–6	5–11	0.23
III. Materials in filamentary (whisker) form: [dia. < 0.001 in (0.025 mm)]									
Aluminum oxide	0.141	38	100–350*e*6	690–2410	2–4*e*6	13800–27600			
Barium carbide	0.090	25	65*e*6	450	1*e*6	6900			
Glass			50*e*6	345	1–3*e*6	7000–20000			
Graphite	0.081	22	142*e*6	980	3*e*6	20000			
IV. Composite materials (unidirectionally reinforced in direction of loading)									
Boron epoxy	0.071	19	31*e*6	210	198000	1365	2.5	4.5	
S-glass-reinforced epoxy	0.0766	21	9.6*e*6	66.2	275000	1900			
V. Others									
Graphite-reinforced epoxy	0.054	15	15*e*6	104	190000	1310			
Kevlar-49 epoxy*	0.050	13.7	12.5*e*6	86	220000	1520			

*Tradename of E. I. duPont Co.

Note: 12*e*–6°/C = 12×10^{-6}/°C

Table A.4. Properties of Selected Areas

Shape	Centroid	Moment of Inertia
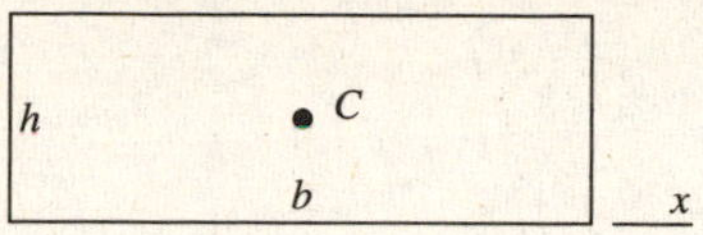 h, C, b, x	$x_C = b/2$	$I_C = bh^3/12$ $I_x = bh^3/3$
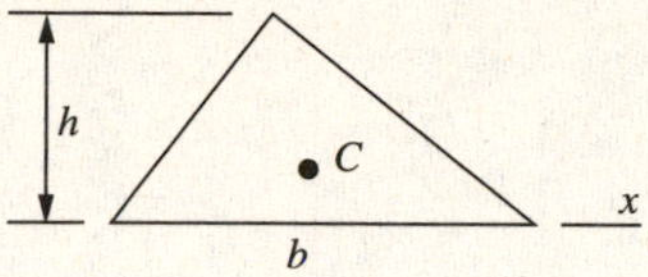 h, C, b, x	$y_C = h/3$	$I_C = bh^3/36$ $I_x = bh^3/12$
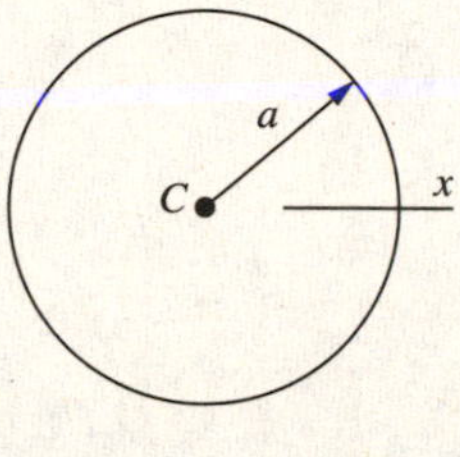 a, C, x	$x_C = 0$	$I_C = \pi a^4/4$ $J = \pi a^4/2$
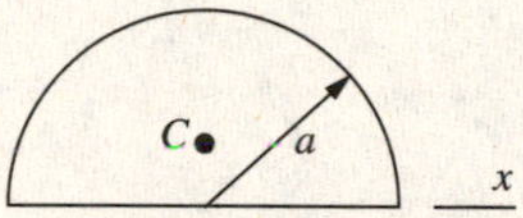 C, a, x	$y_C = 4a/3\pi$	$I_x = \pi a^4/8$

Table A.5. Properties of Selected Wide-Flange Sections, USCS Units

Designation*	Weight per foot, lb/ft	Area, in^2	I (about x-x axis), in^4	S, in^3	I (about y-y axis), in^4
W 18 × 70	70.0	20.56	1153.9	128.2	78.5
W 18 × 55	55.0	16.19	889.9	98.2	42.0
W 12 × 72	72.0	21.16	597.4	97.5	195.3
W 12 × 58	58.0	17.06	476.1	78.1	107.4
W 12 × 50	50.0	14.71	394.5	64.7	56.4
W 12 × 45	45.0	13.24	350.8	58.2	50.0
W 12 × 40	40.0	11.77	310.1	51.9	44.1
W 12 × 36	36.0	10.59	280.8	45.9	23.7
W 12 × 32	32.0	9.41	246.8	40.7	20.6
W 12 × 25	25.0	7.39	183.4	30.9	14.5
W 10 × 89	89.0	26.19	542.4	99.7	180.6
W 10 × 54	54.0	15.88	305.7	60.4	103.9
W 10 × 49	49.0	14.40	272.9	54.6	93.0
W 10 × 45	45.0	13.24	248.6	49.1	53.2
W 10 × 37	37.0	10.88	196.9	39.9	42.2
W 10 × 29	29.0	8.53	157.3	30.8	15.2
W 10 × 23	23.0	6.77	120.6	24.1	11.3
W 10 × 21	21.0	6.19	106.3	21.5	9.7
W 8 × 40	40.0	11.76	146.3	35.5	49.0
W 8 × 35	35.0	10.30	126.5	31.1	42.5
W 8 × 31	31.0	9.12	109.7	27.4	37.0
W 8 × 28	28.0	8.23	97.8	24.3	21.6
W 8 × 27	27.0	7.93	94.1	23.4	20.8
W 8 × 24	24.0	7.06	82.5	20.8	18.2
W 8 × 19	19.0	5.59	64.7	16.0	7.9
W 6 × $15\frac{1}{2}$	15.5	4.62	28.1	9.7	9.7

*The first number after the W is the nominal depth of the section in inches. The second number is the weight in pounds per foot of length.

Index

About the Authors

WILLIAM A. NASH (deceased) was Professor of Civil Engineering at the University of Massachusetts, Amherst. He received his B.S. and M.S. from the Illinois Institute of Technology and his Ph.D. from the University of Michigan. He served as Structural Research Engineer at the David Taylor Research Center of the Navy Department in Washington, D.C., and was a faculty member at the University of Florida for 13 years prior to his last affiliation. He had extensive consulting experience with the U.S. Air Force, the U.S. Navy Department, Lockheed Aerospace Corp., and the General Electric Co. His special areas of interest were structural dynamics and structural stability.

MERLE C. POTTER has engineering degrees from Michigan Technological University and The University of Michigan. He has coauthored *Fluid Mechanics, The Mechanics of Fluids, Thermodynamics for Engineers, Thermal Sciences, Differential Equations, Advanced Engineering Mathematics,* and *Jump Start the HP-48G* in addition to numerous exam review books. His research involved fluid flow stability and energy-related topics. He has received numerous awards, including the ASME's 2008 James Harry Potter Gold Medal. He is Professor Emeritus of Mechanical Engineering at Michigan State University.